THE ALASKAN BEAUFORT SEA

Ecosystems and Environments

Academic Press Rapid Manuscript Reproduction

THE ALASKAN BEAUFORT SEA

Ecosystems and Environments

Edited by

Peter W. Barnes

U.S. Geological Survey
Branch of Pacific Marine Geology
Menlo Park, California

Donald M. Schell

Institute of Water Resources
University of Alaska
Fairbanks, Alaska

Erk Reimnitz

U.S. Geological Survey
Branch of Pacific Marine Geology
Menlo Park, California

1984

ACADEMIC PRESS, INC.
(Harcourt Brace Jovanovich, Publishers)
Orlando   San Diego   New York   London
Toronto   Montreal   Sydney   Tokyo

ACADEMIC PRESS, INC.
Orlando, Florida 32887

United Kingdom Edition published by
ACADEMIC PRESS, INC. (LONDON) LTD.
24/28 Oval Road, London NW1 7DX

Library of Congress Cataloging in Publication Data

Main entry under title:

The Alaskan Beaufort Sea.

Includes index.
1. Oceanography--Beaufort Sea--Addresses, essays,
lectures. 2. Sea ice--Beaufort Sea--Addresses, essays,
lectures. 3. Marine ecology--Beaufort Sea--Addresses,
essays, lectures. 4. Petroleum in submerged lands--
Beaufort Sea--Addresses, essays, lectures. I. Barnes,
Peter W. II. Schell, Donald. III. Reimnitz, Erk.
GC413.A4 1984   551.46'87   84-48447
ISBN 0-12-079030-0 (alk. paper)

PRINTED IN THE UNITED STATES OF AMERICA

84 85 86 87   9 8 7 6 5 4 3 2 1

Dedication

The genesis of this book was in the Outer Continental Shelf Environmental Assessment Program (OCSEAP), a joint venture of the National Oceanic and Atmospheric Administration and the Bureau of Land Management. The specific nature of the work in the Beaufort Sea and the camaraderie among the investigators who worked in arctic Alaska, both necessary preconditions for this volume, were products of inspired leadership in the OCSEAP Arctic Project Office. Gunter Weller and David Norton, through 7 years of effort, encouraged and inspired objectivity in science, cross-fertilization by interdisciplinary discussions and research, and excellent personal relations between investigators and administrators. They further forged a scientific team that worked as a unit in research and in the writing of the present volume.

Both Gunter, as an outstanding program manager, and David had a clear vision of what needed to be done and of the vital and essential role of basic scientific studies in the mission. Without fuss or cant, they formed an indispensable buffer or shield, absorbing most of the pressure for quick practical results, and had the courage to cut through bureaucratic barriers whenever necessary.

Finally, the synthesis sessions (discussed in the first chapter) were invaluable in their contribution to cross-disciplinary and researcher–developer communication. They resulted in a remarkable level of interdisciplinary exchanges of information and ideas, and served to give us a sense of importance and a genuine feeling for the value of our work to society.

With respect and affection, the authors, colleagues, and editors dedicate this work to Gunter and David.

Contents

Biological Interactions

Man's Interaction

Contributors

Numbers in parentheses indicate the pages on which the authors' contributions begin.

Knut Aagaard (47), School of Oceanography WB-10, University of Washington, Seattle, Washington 98195

Ronald M. Atlas (327), Department of Biology, University of Louisville, Louisville, Kentucky 40292

Peter W. Barnes (185, 213), U.S. Geological Survey, Branch of Pacific Marine Geology, Menlo Park, California 94025

Peter G. Connors (403), Bodega Marine Laboratory, University of California, Bodega Bay, California 94923

Peter C. Craig (347), LGL Ecological Research Associates, Juneau, Alaska 99801

George J. Divoky[1] (417), College of the Atlantic, Bar Harbor, Maine 94609

Kenneth H. Dunton (311), Institute of Water Resources, University of Alaska, Fairbanks, Alaska 99701

Kathryn J. Frost (381), Alaska Department of Fish and Game, Fairbanks, Alaska 99701

Joan P. Gosink (73), Geophysical Institute, University of Alaska, Fairbanks, Alaska 99701

John W. Govoni (115), United States Army Cold Regions Research and Engineering Laboratory, Hanover, New Hampshire 03755

Robert P. Griffiths (327), Department of Microbiology, Oregon State University, Corvallis, Oregon 97331

William B. Griffiths (347), LGL Ecological Research Associates, Sidney, British Columbia V8L 1X7, Canada

[1]Present address: Institute of Arctic Biology, University of Alaska, Fairbanks, Alaska 99701.

Arnold Hanson (137), Department of Atmospheric Sciences AK40, University of Washington, Seattle, Washington 98195

Rita Horner (295), 4211 NE 88th Street, Seattle, Washington 98115

Jerome B. Johnson[2] (137), Geophysical Institute, University of Alaska, Fairbanks, Alaska 99701

Stephen R. Johnson[3] (347), LGL Ecological Research Associates, Edmonton, Alberta, Canada

Edward W. Kempema (159), U.S. Geological Survey, Branch of Pacific Marine Geology, Menlo Park, California 94025

Thomas L. Kozo (23), Vantuna Research Group, Occidental College, Los Angeles, California 90041

Lloyd F. Lowry (381), Alaska Department of Fish and Game, Fairbanks, Alaska 99701

Ronald C. Metzner (137), Geophysical Institute, University of Alaska, Fairbanks, Alaska 99701

John L. Morack (259), Physics Department and Geophysical Institute, University of Alaska, Fairbanks, Alaska 99701

Thomas C. Mowatt (275), Institute of Marine Science, University of Alaska, Fairbanks, Alaska 99701

A. Sathy Naidu (275), Institute of Marine Science, University of Alaska, Fairbanks, Alaska 99701

K. Gerard Neave (237), Northern Seismic Analysis, R.R. #1, Echo Bay, Ontario P0S 1C0, Canada

David Norton[4] (3), Geophysical Institute, University of Alaska, Fairbanks, Alaska 99701

Thomas E. Osterkamp (73), Geophysical Institute, University of Alaska, Fairbanks, Alaska 99701

Robert S. Pritchard[5] (95), Research and Technology Division, Flow Industries, Inc., Kent, Washington 98031

Stuart E. Rawlinson (275), Division of Geological and Geophysical Surveys, Alaska Department of Natural Resources, Fairbanks, Alaska 99701

[2]Present address: United States Army Cold Regions Research and Engineering Laboratory, Hanover, New Hampshire 03755.
[3]Present address: 10943 University Avenue, Edmonton, Alberta T6G 1Y1, Canada.
[4]Present address: Institute of Arctic Biology, University of Alaska, Fairbanks, Alaska 99701.
[5]Present address: IceCasting, Inc., Seattle, Washington 98103.

Douglas M. Rearic (185, 213), U.S. Geological Survey, Branch of Pacific Marine Geology, Menlo Park, California 94025

Erk Reimnitz (159, 185, 213), U.S. Geological Survey, Branch of Pacific Marine Geology, Menlo Park, California 94025

James C. Rogers (259), Department of Electrical Engineering, Michigan Technological University, Houghton, Michigan 49931

Donald M. Schell (347), Institute of Marine Science, Institute of Water Resources, University of Alaska, Fairbanks, Alaska 99701

Paul V. Sellmann (237), United States Army Cold Regions Research and Engineering Laboratory, Hanover, New Hampshire 03755

Lewis H. Shapiro (137), Geophysical Institute, University of Alaska, Fairbanks, Alaska 99701

Devinder S. Sodhi (115), United States Army Cold Regions Research and Engineering Laboratory, Hanover, New Hampshire 03755

Donald R. Thomas (441), Research and Technology Division, Flow Industries, Inc., Kent, Washington 98031

Walter B. Tucker III (115), United States Army Cold Regions Research and Engineering Laboratory, Hanover, New Hampshire 03755

W. F. Weeks (213), United States Army Cold Regions Research and Engineering Laboratory, Hanover, New Hampshire 03755

Herbert V. Weiss (275), Department of Chemistry, San Diego State University, San Diego, California 92182

Gunter Weller (3), Geophysical Institute, University of Alaska, Fairbanks, Alaska 99701

Preface

In 1968 oil was discovered at Prudhoe Bay on the north coast of Alaska. Shortly thereafter, the delineation of the discovery as North America's largest petroleum reservoir set the stage for rapid industrial development of this wilderness region. Since then, an additional ten smaller fields have been found, including some that extend offshore into the Beaufort Sea. As the oil industry sought to extend development offshore, the Bureau of Land Management (BLM) and later the Minerals Management Service (MMS) funded large biological and physical studies of the marine ecosystem to provide background material for environmental impact statements. These studies were managed under an interagency agreement by the National Oceanic and Atmospheric Administration (NOAA) and its Outer Continental Shelf Environmental Assessment Program (OCSEAP). Additional studies supported by industry and local and state governments also contributed to a burgeoning collection of facts and ideas from this little-known and harsh environment.

This volume represents a collection of papers written after the culmination of NOAA's OCSEAP, which brought together about 100 scientists from many different disciplines. Since much of the pioneering work was done under severe time constraints imposed by leasing schedules, the results often remained in the gray literature, known only to government agencies involved in the leasing process or to industry. In this volume, we present to the scientific community some of the findings from this unique environment. The broad scope of the material will also be of interest to managers of public and private lands, conservationists, and developers.

The volume is organized into four sections. The first section, an introduction to the Beaufort Sea, is followed by 13 chapters describing the physical environment of the area: meteorology, oceanography, offshore permafrost, sea ice, and geology. Built on this physical framework, the chapters in the biological section cover topics ranging from microbiology up through the food chain to birds and mammals. The final section deals with the potential impact of man as he moves to extract oil from offshore areas.

In a single volume, this work provides the reader with an interdisciplinary glimpse into almost all aspects of the environment, with a thorough survey of the background literature. Although the volume focuses on the Alaska Beaufort Shelf environment, much of the knowledge is applicable to other seas surrounding the Arctic Ocean, such as the Canadian Beaufort Sea to the east and the Chukchi Sea to the west. In terms of physical environment and geographic setting, perhaps the most similar example exists along vast stretches of the arctic coast of the Soviet Union.

The funding for most of the work presented in this volume, including partial production of this book, came from BLM/MMS through NOAA. All of us are grateful for the funding and the logistics support we received from these agencies, as well as for the encouragement to pursue our research.

We sincerely hope that this volume will contribute toward a better understanding of high latitude marine environments, in general, and the Alaskan Beaufort Sea in particular, and will lead to an equitable and effective use of these environments.

To those colleagues who have participated in OCSEAP and contributed papers to this volume, to Andrew Alden for his meticulous copy editing, and especially to those numerous associates who critically read manuscripts for us, we say, "thank you."

PETER W. BARNES
DONALD M. SCHELL
ERK REIMNITZ

Introduction

THE BEAUFORT SEA: BACKGROUND, HISTORY, AND PERSPECTIVE

David Norton[1]
Gunter Weller

Geophysical Institute
University of Alaska
Fairbanks, Alaska

I. INTRODUCTION

In the summer of 1968 British Petroleum and Atlantic Richfield Companies announced their discovery of oil beneath the arctic coast of Alaska. The Prudhoe Bay discovery (10 billion barrels of oil and 25 trillion cubic feet of natural gas) was gigantic by North American standards and big even by Middle Eastern standards. Within ten years, the remote and rarely visited region around Prudhoe Bay became a complex of technological sophistication, the trans-Alaska pipeline was built, and extensive seismic exploration occurred both on land and offshore on the continental shelf.

The pace of industrial activities has recently increased even further as a result of offshore lease sales and the successful drilling for petroleum from natural and artificial islands. Concurrently, a group of scientists has conducted environmental and hazards research, designed to pinpoint problem areas and help in the protection of the environment as well as the safe exploitation of petroleum resources. This research began in the 1960's and even earlier but gathered momentum in 1975 with the studies of the Outer Continental Shelf Environmental Assessment Program (OCSEAP). The pre-OCSEAP status of knowledge was summarized in the proceedings of a symposium on the coast and shelf of the Beaufort Sea (Reed and Sater, 1974), and the OCSEAP-generated information is contained in numerous publications, including this volume and several synthesis volumes (edited by Weller et al., 1977, 1978, 1979, and Norton and Sackinger, 1981).

This book presents some highlights of recent studies. This introductory chapter sets the stage by describing the characteristics and history of the

---

[1]*Present address:* *Institute of Arctic Biology, University of Alaska, Fairbanks.*

THE ALASKAN BEAUFORT SEA:
ECOSYSTEMS AND ENVIRONMENTS

3

region in which the research took place and the objectives of the studies program.

II. REGIONAL SETTING

The area of interest in this book is the coast and continental shelf of the Beaufort Sea between Point Barrow, the northernmost point in the United States, and the Canadian border, a linear distance of about 600 km. The low tundra coastline is heavily indented with shallow bays and lagoons, and the continental shelf is relatively narrow, extending 50-100 km off the coast (Fig. 1). The adjoining Canadian Abyssal Plain is more than 3,000 m deep.

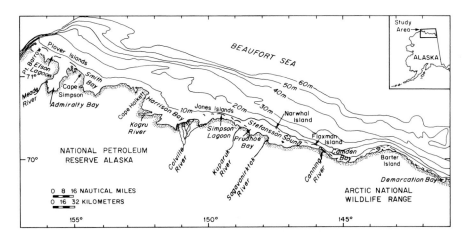

FIGURE 1. The coast and shelf of the Alaskan Beaufort Sea.

Governance and ownership patterns are complex. Politically, the entire coastline and its hinterland are now part of the municipality of the North Slope Borough. Ownership of submerged lands is claimed by the State of Alaska from the coast out to 3 miles (4.84 km) offshore. Onshore, ownership is divided between the United States government, the State of Alaska, and minuscule private holdings. By far the largest acreage belongs to the United States government, which owns most of the land between the Colville River and Chukchi Sea coast to the west, the National Petroleum Reserve Alaska (NPRA, formerly Naval Petroleum Reserve No. 4), and the Arctic National Wildlife Range, located between the Canning River and the Canadian border. The small central section of the coast, including the area around Prudhoe Bay, is largely the property of the State of Alaska, acquired at the time of statehood in 1959. It is on this wedge of state-owned land that most of the significant petroleum discoveries, both onshore and offshore, had been made by the end of 1982.

The Beaufort Sea coast is predominantly low-lying wetland tundra, dotted by numerous thaw lakes. Coastal bluffs are mostly less than 5 m high, and extensive, shallow deltas are formed at the mouths of major rivers. Offshore islands and shoals determine the nature of much of the nearshore physical and biological environment along the Beaufort coast. These islands effectively moderate the influence of polar pack ice where they occur, and in the few weeks of summer partially separate the cold saline waters of the open ocean from the warmer, brackish waters that generally flow from east to west in a narrow band along the mainland coast. Most of these islands are sand and gravel barrier islands, bounding shallow lagoons (*e.g.*, Jones Islands and Simpson Lagoon), while others are relicts of earlier coastal retreat processes and lie farther offshore, with deeper water between them and the mainland (*e.g.*, Narwhal Island and Stefansson Sound). The islands themselves, and the mainland coast where unprotected by islands, are subject to considerable erosion by wave action. Aided by thermal erosion of the tundra, erosion rates average 1-3 m per year along the mainland coastline but may reach 18 m in single, severe storms in some locations. Subsea permafrost is extensive in the nearshore areas and extends in an irregular pattern a few tens of kilometers offshore (Morack and Rogers, Neave and Sellmann, this volume).

Sea ice dominates the entire Beaufort Sea area (Fig. 2). Ice cover is almost 100% for nine to ten months each year and freezes up to 2.4 m in thickness in one season. Multiyear ice in the Beaufort Sea averages 4 m in thickness, and ice islands (icebergs) and ice ridges probably occur with drafts of as much as 50 m. Landfast ice forms gradually in the fall and by the end of winter may range in extent from less than 1 km to as much as 50 km offshore. The ice pack, moved westward past the Alaskan coast by the clockwise Beaufort Gyre, shears against the stationary landfast ice, forming an extensive pressure-ridge system that is commonly aground along the inshore edge. Pressure-ice ridges and hummocks may exceed 10 m in height and be matched on the underside by ice keels several tens of meters deep (see Tucker *et al.*, this volume). Figure 3 shows the various ice types at the end of winter and indicates the terms used in identifying the ice zonation. The seafloor of the continental shelf is scoured by dragging ice keels which form deep gouges; some gouges are more than 4 m deep. Ice gouges of indeterminate age have been found as far out as a depth of 62 m, although they are more numerous and frequent in shallower waters, especially along the stamukhi zone (see Barnes *et al.*, Weeks *et al.*, this volume). Ice is also known to ride up onto the coast (Shapiro *et al.*, this volume).

Oceanographic processes on the Beaufort Sea shelf are greatly influenced by the circulation patterns of the Arctic Ocean. At the shelf margin, currents flow westward between Mackenzie Bay and Point Barrow under the influence of the clockwise Beaufort Gyre driven by the dominant northeasterly winds, but a deeper intrusion of Bering Sea water flows eastward on the outer continental shelf (see Aagaard, this volume). Inside the mid-shelf area a net westward wind-driven coastal current dominates (see Kozo, this volume). The extent of ice-free open water in summer

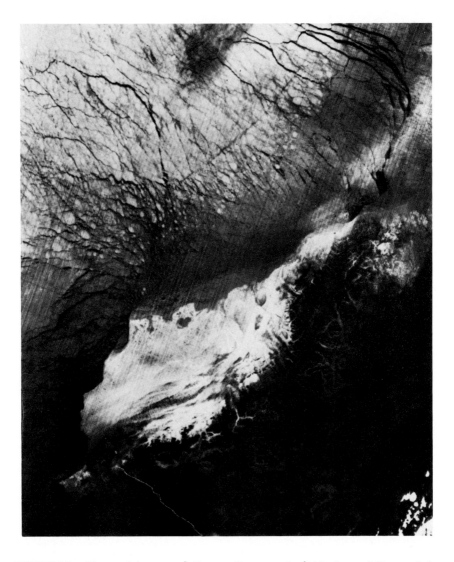

FIGURE 2. *Thermal image of the northern part of Alaska and the pack ice of the Beaufort Sea taken on March 8, 1979 by the Defense Meteorological Satellite Program (DMSP). The nearshore ice is stable but the large crescent-shaped leads offshore indicate large-scale displacement of the ice to the west.*

affects the degree of wave action on the permafrost of the coast, thus influencing the coastal erosion rate. Storms, frequent during late summer and fall, occasionally generate storm surges. A storm surge 5 m high once destroyed 19 buildings at Barrow, eroded coastal bluffs by up to 3 m, and

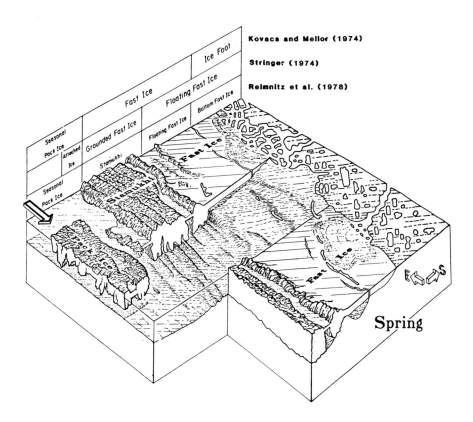

FIGURE 3. Ice zonation in spring in relation to the coast and bottom morphology of the Beaufort Sea (from Reimnitz et al., 1978. Drawing by Tau Rho Alpha).

caused a local shoreline retreat of about 18 m (Hume and Schalk, 1967). Lunar tides, by comparison, reach a maximum of 30 cm.

The variety and distribution of biota along the arctic coast reflect habitat conditions: the combined effects of water and ice movements, water mass and bottom characteristics, and the availability of suitable food. The presence or absence of ice profoundly affects fish, bird, and marine mammal movement and behavior in this region. Many species congregate near the edge of the pack ice and move with the ice. In the shallow and relatively warm coastal environment along and within the barrier islands, relatively simple and direct food webs reach summer peaks of secondary biological productivity greater than those of the open arctic seas (see Craig et al., this volume), although the primary productivity of open oceanographic systems has recently been shown to be a more important source of carbon to the coastal shallows than terrestrial production (Fig. 4). In Beaufort Sea trophic systems generally, species are few and food webs are simpler than in temperate seas (Fig. 5).

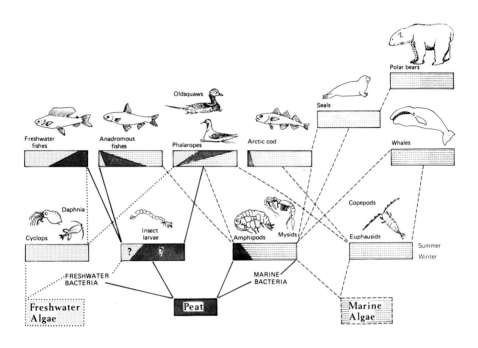

FIGURE 4. Generalized foodweb structure and major seasonal energy dependencies in Alaskan Beaufort Sea coastal ecosystems (from Schell et al., 1982).

In common with coastal areas at lower latitudes, the macrofauna of Beaufort Sea inshore waters includes crustaceans, mollusks, and polychaetes. The limited number of fish species includes arctic char, whitefishes, ciscoes, cod, smelt, flatfishes, and sculpins. The nearshore waters are also important to arctic shorebirds and waterfowl; a large portion of the bird population of the Canadian Arctic islands passes through this region, including large numbers of four species of eider ducks, black brant, oldsquaw ducks, several species of gulls and terns, and many species of shorebirds. Eiders, gulls, and terns nest densely on some of the barrier islands. Marine mammals of the area include seals, mainly bearded, ringed and harbor; whales, mainly bowhead and beluga; and walruses. Walruses, however, are not as abundant in the Beaufort Sea as they are to the south, in the Chukchi and northern Bering Seas. Bowhead and belukha whales follow the ice leads in spring. Polar bears frequent the pack ice, and pregnant bears den in heavily snowdrifted areas along the coast as far as 40 km inland. Arctic foxes also inhabit the sea ice in winter, where some appear to survive by following bears to pick over remains of their kills.

Weather and logistics constrain scientific studies in the area. Access is primarily via major airstrips at Barrow at the western end of the area, Prudhoe Bay in the center, and Barter Island (Kaktovik) near the eastern

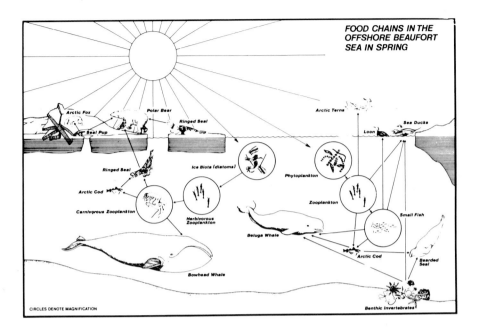

FIGURE 5. Generalized food chains in the offshore regions of the Beaufort Sea in spring (from Blood, 1977; illustration by Mark Blaney).

end. Prudhoe Bay, the northern terminus of the trans-Alaska oil pipeline, can also be reached year-round by road from Fairbanks. Finally, heavy barge transport by sea in most years can reach all of these destinations during the short ice-free season in August and September. With the closing of the Naval Arctic Research Laboratory at Barrow in 1981, there are now essentially no government scientific support facilities along the entire coast, although helicopters, fixed-wing aircraft, and boats can be chartered (Fig. 6), and commercial accommodation is available at Barrow and Prudhoe Bay. The only major ships available are Coast Guard icebreakers (Fig. 7), which have limitations in the equipment and scientific personnel they can carry. There is presently no other U.S. polar research vessel available for work in the Arctic. Logistical constraints are thus of serious concern for continued scientific studies in the region.

III. HISTORY OF THE REGION

Tareumiut people (Eskimo of the Ipiutak culture) have occupied the northern Alaska coast for at least 2,000 years and were identified by their proximity to the major rivers crossing the coastal tundra plain. These rivers, including the Meade, Colville, Kuparuk, Sagavanirktok, and Canning Rivers (Fig. 1), were used for hunting, trading, and transportation. The

FIGURE 6.   Helicopter deployment of a building used to support sea ice telemetry and biological investigations from Narwhal Island.

Tareumiut numbered between 1,500 and 2,000 in the 1830's. Another major group, the Nunamiut, occupied the huge interior region between the coastal zone and the Brooks Range.   Estimates of their population in range from 1,000 to about 3,000 in 1800 (Nielson, 1977).

The western cultures had a profound influence on northern Eskimo societies after contact.   Russians only once (in 1838) ventured as far north as Simpson Lagoon, but their trade goods--copper kettles, tobacco, knives, and tinware--preceded them as far eastward as the Mackenzie River, where they were in use prior to British explorations of the late 1700's.   But there was no real contact with Western civilization until Captain W. F. Beechey of Britain rounded Point Barrow in 1826 and named it in honor of Admiral Sir John Barrow.   For the next 86 years the exploration impulse brought at least 50 separate expeditions to the arctic coast beyond Point Barrow as Britons, Danes, Norwegians, and Americans searched for the Northwest Passage.

During the period 1848-1914 the American whaling industry sent ships in increasing numbers into the Arctic Ocean.   By 1880 more than 300 vessels were likely to be operating off Point Barrow and to the east annually, with a few even wintering in the ice because they failed to get south in time.   The introduction of whiskey and other alcoholic beverages by whalers into northern Eskimo society during this period was perhaps the most obvious feature of white culture so destructive to the Eskimo.   A whaler's life also had its dangers.   In the years prior to the industry's zenith

FIGURE 7. U.S. Coast Guard Icebreaker Glacier, operating off Prudhoe Bay in 1977, supporting OCSEAP studies.

in the Arctic, fleets suffered three major disasters, and almost yearly ships were lost to ice floes and storms. In 1871, 31 ships were lost due to ice (Nielson, 1977).

Traders and missionaries followed next, the former bringing guns and whiskey and the latter Christian religion, its value systems, and teaching. Both had a profound influence on the way of life of Eskimos, who quickly adopted many of the newly available methods and materials (Nielson, 1977). Also introduced by these contacts were diseases to which the Eskimos had no immunity, and before modern health care could catch up the native population had decreased drastically.

Scientific explorations started in earnest in the late 1800's in the arctic regions of the North Slope and the Beaufort Sea. They included (Nielson, 1977): (1) the International Polar Expedition to Barrow, 1881-1885; (2) the Alaska Boundary Survey, U.S. Coast and Geodetic Survey, J. H. Turner, 1880; (3) Frederick Funston's expedition to the eastern boundary of Alaska in 1894; (4) the reconnaissance of F. C. Schrader and W. J. Peters to the arctic coast via the Anaktuvuk and Colville Rivers, 1901; (5) the explorations of S. J. Marsh, F. G. Carter, and H. T. Arey in the interior from Barrow to the Canning River, 1901-1903; (6) the Anglo-American Polar Expedition to the Flaxman Island region, Ejnar Mikkelson and Ernest de K. Leffingwell, 1906-1907, and Leffingwell's explorations of 1909-1911 and 1913-1914; (7) the Arctic expedition of Hudson Stuck, archdeacon of the Episcopal missions of Alaska, 1901-1920; (8) the Canadian Alaska Boundary Survey of 1912; and (9) the Canadian Arctic Expedition of 1913.

Leffingwell's work in particular produced a great deal of information on the geology of the Arctic coast and the barrier islands, providing the base of our present-day studies (Leffingwell, 1919). He was also responsible for making the first accurate surveys and maps, publishing correct place names and standardizing English-Eskimo usage in many cases.

The 1930's were notable because of the beginning of aircraft and airship exploration of the Arctic and the Brooks Range, including extensive flights and searches for the missing Soviet aviator Levanevsky. Not until after World War II did such flights become a regular feature of "bush" living. The only other major intrusion of Western technology in the pre-war era was the executive order in 1923 which created the 23-million-acre NPRA. Exploration of this vast reserve began almost immediately and has continued to the present, after being interrupted during the Second World War.

World War II signaled a massive military buildup throughout Alaska, which culminated in the Arctic with the construction in the early 1950's of the Distant Early Warning (DEW) Line network of radar stations. The Arctic Research Laboratory at Barrow (later called the Naval Arctic Research Laboratory, NARL) was established in 1947. This scientific facility provided the logistics and facilities to allow a wide range of scientific investigations to be carried out along the coast and in the center of the Arctic Ocean, where numerous floating stations were established on the pack ice. The demise of the laboratory began with increased costs to users during the mid-1970's and ended with the closing of the laboratory as a scientific support facility in 1981.

The discovery of oil at Prudhoe Bay in 1968 brought radical changes to arctic Alaska, by propelling money, men, and a sophisticated technology into the area.

IV. OIL IN THE BEAUFORT SEA

When the Prudhoe Bay oil discovery was announced in 1968, and was confirmed to be a giant field, virtually every facet of life in Alaska—including science—was destined to be affected. Signs of petroleum on the North Slope were reported as early as 1837, and gas fields were documented by the Navy's exploration of the NPRA between 1944 and 1953. Except for the South Barrow Gas Field, however, no production and marketing of petroleum products was feasible because the costs of exploitation and transportation from the small proven reserves on the North Slope were prohibitive (Naske and Slotnick, 1979). Federal and State leasing had previously resulted in moderate industry interest. When confirmation drilling established that the Prudhoe Bay field contained 9.6×10^9 barrels of crude oil, and that single wells would produce in excess of 2,000 barrels per day, industry interest soared. The first state competitive lease sale on Prudhoe Bay land following the discovery netted $900 million in bonus bids. This was equivalent to about $2,200 per acre for the rights to explore, in contrast to $13 per acre for similar rights paid one year prior to the discovery (Alaska Dept. of Natural Resources, 1977).

Development of the Prudhoe Bay field, however, ran into two procedural obstacles. One was the enactment of the National Environmental Policy Act in 1970, the other the unsettled issue of Alaska native land claims. In these obstacles, both the oil industry and the Department of the Interior were saddled with problems generated earlier in Alaska and elsewhere when large development projects came into conflicts with environmental quality or with rights of indigenous peoples, respectively. By the time that construction of the Trans-Alaska Pipeline finally began in 1974, the National Environmental Policy Act had been sidestepped by Congress acting during the Arab oil embargo in 1973-74. The Alaska Native Land Claims Settlement Act had become law in 1971. Leaseholding companies in the Prudhoe Bay field had lobbied Congress in support of this legislation, realizing that their sizable investments were hostage to the land claims issue insofar as the issue held up permission to build the pipeline (Naske and Slotnick, 1979).

Effects of the Prudhoe Bay discovery radiated outward during the same period in which the pipeline construction project was stalled. The International Biological Program took form in Alaska in 1969 as the U.S. Tundra Biome studies funded by the National Science Foundation. In response to Prudhoe Bay, this large interdisciplinary scientific undertaking expanded to consider some specific applied problems inherent in oil development and transportation on the North Slope (Brown et al., 1980). The Arctic Institute of North America sponsored a symposium on Beaufort Sea Coast and Shelf Research in 1974 (Reed and Sater, 1974) that summarized, distilled, and organized the environmental knowledge of the U.S. and Canadian Beaufort Sea up to that time. The eventual existence of a petroleum transportation system originating at Prudhoe Bay made it possible to begin thinking realistically of the nearshore Beaufort Sea as a viable source of additional petroleum.

By 1974, the State of Alaska feared revenue shortfalls stemming in part from delays in the pipeline project, and began planning for a further lease sale of submerged state lands in the Beaufort Sea adjacent to Prudhoe Bay as early as 1975. In this planning, the state was encouraged by the Alaska Oil and Gas Association, which professed technological preparedness to adapt arctic land-based technology to shallow Beaufort Sea conditions.

In this same year, the Outer Continental Shelf Environmental Assessment Program (OSCEAP) was being organized by the National Oceanic and Atmospheric Administration (NOAA) for the Department of the Interior, as an Alaskan variant of the national Outer Continental Shelf Environmental Studies Program (Englemann, 1976). Thereafter, the history of oil development and the scientific evaluation of the Beaufort Sea and North Slope became closely interactive: leasing decisions and environmental stipulations advanced by the Department of the Interior, the State of Alaska, and the local government of the North Slope Borough, all drew upon the knowledge gained by OCSEAP-contracted scientists (see OCSEAP history below). In turn, the scope of scientific research in any given year was increasingly shaped by specific questions raised by oil and gas development options.

In 1975, at the outset of an infusion of federal money to support studies in the Beaufort Sea, the State of Alaska and the Department of the

Interior were on separate, parallel courses of action to hold early lease sales of their respective submerged lands in the Beaufort Sea. The state was looking for a solution to its cash-flow problems, while the federal motivation was to increase national energy self-reliance. Both would-be lessors began to encounter new unforeseen obstacles, and their scheduled lease sale dates slipped correspondingly. The obstacles included uncertainty over technology for coping with sea ice hazards, articulations of concern by subsistence whalers and fishermen, conflicts between oil exploration and Coastal Zone Management Programs, and disputes between the two governments over ownership of key acreage along the convoluted coastline. Obstacles and ownership disputes had been set aside by the time the Joint State/Federal Sale was eventually held in December 1979. This offering of half a million acres (about 200,000 ha) resulted in just over $1000 million in bonus bids.

By the end of 1981 some 15 exploratory wells had been drilled by the lessees, or were in the planning stages on Joint Sale acreage. The Sagavanirktok Delta area provided the first strike of oil within the Joint Sale Lease area, and this is the offshore area judged likely to go into production first. Other promising hydrocarbon discoveries have been made in and adjacent to the Joint Sale area's eastern end, around Challenge Island and Flaxman Island (Oil and Gas Journal, 1981).

In addition to offshore leasing and exploration activities, onshore events are proceeding rapidly. The Kuparuk field just west of Prudhoe Bay has been brought into production. The first NPRA lease sale was held in January 1982, and a second sale took place in May 1982.

Offshore leasing in the Beaufort Sea is expected to accelerate. A State of Alaska sale coincided with the NPRA sale in May 1982, reoffering tracts that were not sold in the 1979 Joint Sale, and included the Flaxman Island tracts originally to have been sold in a previously postponed lease sale. In October 1982, the Department of the Interior's Sale 71 offered about 2 million acres (800,000 ha) of offshore tracts adjoining the Joint Sale leases offshore and westward to beyond Cape Halkett. Bidding was brisk, and the successful bidders offered a total of $2,100 million. Current Interior Department planning calls for additional sales every two years. This pace of petroleum leasing and development activities is in remarkable contrast to the cautious approach of the previous decade. It had taken 10 years from the last previous Prudhoe Bay leasing in 1969 (State of Alaska, 1975) to arrive at the point of readiness for the offshore Joint Sale in 1979. Now, far larger tracts are to be opened at intervals of two years or less while they last.

V. OCSEAP IN THE ALASKAN BEAUFORT SEA

OCSEAP-sponsored field investigations in the Beaufort Sea began in 1975. NOAA contracted with scientists for the work, and managed the program under an interagency agreement with the Bureau of Land Management (BLM) of the Department of the Interior. OCSEAP's role was to furnish information to the BLM for use in compiling Environmental

Impact Statements on offshore lease areas contemplated for sales and petroleum development. This type of analysis is mandated by the National Environmental Policy Act of 1970. At the outset, offshore leasing nationwide was regarded by the Nixon administration as a quick solution to the country's reliance on foreign petroleum sources. The administration's stated goal (however unrealistic in retrospect) was to achieve energy self-sufficiency by 1980. Since Alaskan continental shelf lands comprise 74 percent of the total acreage of the U.S. continental shelf, OCSEAP was a formidable and urgent undertaking, within which the Beaufort Sea was but one of nine Alaskan lease regions selected for study.

In 1975 and 1976, it seemed that more money was available to OCSEAP in Alaska than either time or forethought in planning a coherent program (Weller and Norton, 1977). In the Beaufort Sea, however, OCSEAP inherited rather than created a community of capable arctic scientists. The trauma and chaos that might have attended the start of such a crash program were minimal in the Beaufort Sea because these scientists (including several authors of papers in this book) were experienced with one or another of the recent antecedents of OCSEAP, in arctic multidisciplinary science. These included the Tundra Biome of the International Biological Program (IBP) and the Arctic Ice Dynamics Joint Experiment (AIDJEX), both funded by the National Science Foundation. During the initial phase of OCSEAP studies in the Beaufort Sea, the Canadians were in the midst of their Beaufort Sea Project, an environmental assessment effort that partially paralleled OCSEAP's studies in Alaska. Moreover, thanks to the Beaufort Sea Symposium sponsored by the Arctic Institute of North America (Reed and Sater, 1974) and to a National Science Foundation symposium on Beaufort Sea research needs in 1974 (the so-called Arctic Offshore Program, unpublished) there was a minimum of uncertainty among OCSEAP investigators as to what needed to be done. These antecedents had the welcome effect of providing arctic scientists with a sense of community and ease of communication transcending disciplinary, institutional, and national boundaries.

Survey and inventory of the Beaufort Sea resources and hazards to development were the dominant activities of OCSEAP investigators through 1976 in most disciplines. Concurrently with the second OCSEAP field season in the Beaufort Sea, the need became apparent to focus and integrate the results of OCSEAP's substantial investments in Beaufort Sea science (Norton, 1977; Weller and Norton, 1977). The catalysts that allowed OCSEAP investigations to fashion an integrated research program that went beyond the pedestrian and uninspired stages of surveys, seem to have been three activities conceived or encouraged in 1976-77:

(1) A Beaufort Sea technology scenario (Clarke, 1976) was the first public analysis of the petroleum industry's most likely approaches to exploration, development, and production of offshore oil and gas in the Beaufort lease region. The predictions of this modest study have proved remarkably accurate since then. In essence, the report encouraged OCSEAP, the State of Alaska, and the federal leasing agencies to focus on coastal and nearshore dynamics because the technological capability to exploit petroleum resources safely in deeper waters and offshore areas of more severe sea ice stresses would take years to develop. The report also

correctly emphasized the reliance that would be placed on local gravel resources to build artificial islands, causeways, and roads in the nearshore zones most likely to be developed early.

(2)   Clarke's (1976) report showed that it was clearly important to investigate the sensitivities of "typical" nearshore ecological units or systems. Accordingly, a large interdisciplinary project was begun in 1976-77, to analyze physical and biological processes in a representative barrier island-lagoon system. The resulting "Simpson Lagoon Study" coordinated university scientists and consultants in many innovative approaches to OCSEAP, both in field investigations emphasizing processes, and in subsequent analysis of results in an interdisciplinary mode (Truett, 1979, 1981).

(3)   The final integrating mechanism developed in 1976-77 was the synthesis exercise to draw together research results from participating scientists. At the first such synthesis exercise the group successfully developed an approach involving disciplinary sessions followed by interdisciplinary sessions grouped around discrete development "scenarios." The first synthesis report (Weller et al., 1977) set the pattern for future investigations and integrated reports (Weller et al., 1978, 1979; Schell, 1980; Norton and Sackinger, 1981).   Industry and regulatory agencies, including the North Slope Borough, participated in these subsequent synthesis activities.   This participation effectively brought together the consumers and producers of environmental information, producing beneficial interactions. One indication of OCSEAP's success in the Beaufort Sea was the adoption by State and Federal lessors of all 13 scientific recommendations from the third synthesis exercise (Weller et al., 1979) at the time of the Joint Sale in December 1979.

The period of most direct contact between scientists and public policymakers in the Beaufort Sea, 1977-81, was also marked by pioneering approaches to arctic field studies and to analyses of the resulting information. Some of these innovative approaches are documented in the present volume, while others will be found in the general scientific literature. Many resulted from treating sea ice as a convenient platform for logistics operations, rather than a barrier preventing access to the ocean below for 9-10 months each year. For example, biologists J. J. Burns and K. H. Dunton resided and worked from semipermanent ice camps from which they made bioacoustic measurements and under-ice SCUBA explorations (Fig. 8). K. Aagaard pioneered current-meter deployment and recovery through sea ice, using helicopters, and J. B. Matthews adapted this technique to shallow water where ice poses considerable hazards to the meters. Ecological Research Associates (LGL) scientists applied Adaptive Environmental Assessment (Holling, 1978) techniques to the multidisciplinary studies of processes in Simpson Lagoon (Truett, 1979). Invertebrate biologists with LGL perfected sampling gear and techniques for quantitative sampling of epibenthic invertebrates in Simpson Lagoon. Fisheries biologists with LGL also developed a procedure to evaluate temperature preferences in arctic ciscoes by using a laboratory gradient chamber and captive fish, and subsequently applying these findings to predictive modeling of fish behavior in the wild (Feckhelm et al., 1983; Neill et al., 1983). D. M. Schell determined the functional importance of carbon from North Slope peat to the marine and coastal system, by natural abundances of carbon isotopes in various representative organisms.

FIGURE 8. *Divers prepare to descend in Stefansson Sound to investigate biological dynamics of kelp beds in the Boulder Patch (photograph by K. H. Dunton).*

P.G. Connors developed an elegant test procedure for evaluating red phalaropes' abilities to detect and avoid oil-contaminated water.

Modern tools and technologies were used and adapted for the Beaufort Sea environment, including laser profilometer flights and traverses with impulse radars to detect sea ice roughness and thickness, respectively (W. Weeks and A. Kovacs). The R/V *Karluk* (Fig. 9), equipped with range-range navigation, side-scan sonar, seismic-reflection systems, vibracorer, closed-circuit TV, and other systems to examine ice scour and other processes and properties of the ocean floor represented the state of art in marine geological research (P. Barnes and E. Reimnitz). Satellite imagery was used by W. Stringer in near-real time to assess and predict ice movement and processes. Satellite-interrogated buoys were used to assess ice motion through the arctic night by R. Pritchard, and by T. Kozo to measure atmospheric parameters. A cheap and quick way of jetting holes in the seafloor down to the subsea permafrost level was developed by T. Osterkamp and W. Harrison, and L. Shapiro developed a technique for *in situ* measurements of sea ice strength properties.

These are only some of the innovative approaches used by OCSEAP investigators in the Beaufort Sea. The following papers document some of the results of their studies in greater detail.

FIGURE 9. *Fueling operation near Pingok Island aboard the R/V Karluk, which was used extensively in studies of marine geological processes in the Beaufort Sea for OCSEAP.*

REFERENCES

Alaska Department of Natural Resources, (1977). "A Study of State Petroleum Leasing Methods and Possible Alternatives." Anchorage.
Blood, D.A. (1977). "Birds and Marine Mammals. The Beaufort Sea and the Search for Oil." Department of Fisheries and Environment, Ottawa.
Brown, J., Miller, P.C., Tieszen, L.L., and Bunnell, F.L. (eds.) (1980). "An arctic ecosystem: the coastal tundra at Barrow, Alaska," U.S. Internat. Biol. Program Synthesis Series Vol. 12. Dowden, Hutchison and Ross, Stroudsburg, PA.
Clarke, E.S. (1976). "Arctic Project Bulletin 11" (G. Weller *et al.*, eds.). NOAA, OCSEAP, Univ. Alaska, Fairbanks.
Engelmann, R.J. (1976). *In* "Proceedings of the 27th Alaska Science Conference," (D.W. Norton, ed.), Vol. 2, p. 83. AAAS Alaska Division, Fairbanks.
Feckhelm, R.G., Neill, W.J., and Gallaway, B.J. (1983). *Biol. Pap. Univ. Alaska 21,* 24.

Holling, C.S. (ed.) (1978). "Adaptive Environmental Assessment and Management." John Wiley & Sons, New York.

Hume, J.D. and M. Schalk (1967). *Arctic 20*, 86-102.

Kovacs, A., and Mellor, M. (1974). *In* "The Coast and Shelf of the Beaufort Sea," (J.C. Reed and J.E. Sater, eds.), p. 113. Arctic Institute of North America, Arlington, VA.

Leffingwell, E. de K. (1919). *Prof. Paper 109.* U.S. Geol. Survey.

Naske, C.M., and Slotnick, H.E. (1979). "Alaska - A History of the 49th State." Eerdmans, Grand Rapids, MI.

Neill, W.J., Feckhelm, R.G., Gallaway, B.J., Bryan, J.D., and Anderson, S.W. (1983). *Biol. Pap. Univ. Alaska 21*, 39.

Nielson, J.M. (1977). "Beaufort Sea Study. Historic and Subsistence Life Inventory: A Preliminary Cultural Resource Assessment." North Slope Borough, Barrow, AK.

Norton, D.W. (1977). *Interdiscipl. Sci. Rev. 2*, 207.

Norton, D.W., and Sackinger, W.M. (eds.) (1981). "Beaufort Sea Synthesis-- Sale 71. Environmental Assessment of the Alaskan Continental Shelf." Draft Preprint, NOAA-OCSEAP, Univ. Alaska, Fairbanks.

Oil and Gas Journal (1981). *79*, 21.

Reed, J.C., and Sater, Y.E. (eds.) (1974). "The Coast and Shelf of the Beaufort Sea." Arctic Inst. North America, Arlington, VA.

Reimnitz, E., Toimil, L., and Barnes, P.W. (1978). *Mar. Geol. 28*, 179.

Schell, D.J., Ziemann, P.J., Parrish, D.M., and Brown, E.J. (1982). "OCSEAP Cumulative Summary Report, Research Unit #537," NOAA, OCSEAP, Univ. Alaska, Fairbanks.

Schell, D.M. (ed.) (1980). *Arctic Project Bull. #29.* NOAA, OCSEAP, Univ. of Alaska, Fairbanks.

State of Alaska (1975). "Proposed Beaufort Sea Nearshore Petroleum Leasing, Draft Environmental Assessment." Office of the Governor, Juneau.

Stringer, W.J. (1974). *In* The Coast and Shelf of the Beaufort Sea," (J.C. Reed and J.E. Sater eds.), p. 165. Arctic Institute of North America, Arlington.

Truett, J.C. (1979). *In* "POAC-'79" Vol. 1, p. 423. Norwegian Tech. Inst., Trondheim.

Truett, J.C. (1981). *In* "Environmental Assessment of the Alaskan Continental Shelf," Final Repts., Vol. 8, p. 259. NOAA, Boulder.

Weller, G., and Norton, D.W. (1977). *Interdiscipl. Sci. Rev. 2*, 214.

Weller, G., Norton, D.W., and Johnson, T.M. (eds.) (1977). *Arctic Project Bull. 15.* NOAA-OCSEAP, Univ. Alaska, Fairbanks.

Weller, G., Norton, D.W., and Johnson, T.M. (eds.) (1978). "Environmental Assessment of the Alaskan Continental Shelf, Interim Synthesis: Beaufort/Chukchi Seas." NOAA-OCSEAP, Boulder.

Weller, G., Norton, D.W., and Johnson, T.M. (eds.) (1979). *Arctic Project Bull. 25.* NOAA-OCSEAP, Univ. Alaska, Fairbanks.

The Environment

MESOSCALE WIND PHENOMENA ALONG
THE ALASKAN BEAUFORT SEA COAST

Thomas L. Kozo

Vantuna Research Group
Occidental College
Los Angeles, California

I. INTRODUCTION

Wind is a major influence on the physical and biological environment of the Alaskan Beaufort Sea. Surface wind conditions influence the times of sea ice breakup or freeze-up, create nearshore currents, and move ice floes and oil spills. Winds affect the timing and routing of animal and plant migrations. Eskimos require favorable winds for hunting and fishing, and industry representatives require favorable wind conditions for safe offshore activities and annual resupply by sea. However, National Weather Service (NWS) synoptic-scale meteorological predictions often provide inaccurate wind velocities for critical coastal areas. Factors including thermal discontinuities at sea ice-water-tundra boundaries and topography act to generate weather phenomena small enough to remain undetected within the synoptic observational network. These local mesoscale events confront humans and most biota. This article will discuss: (1) subsynoptic meteorological networks along the Beaufort Sea coast and shelf; (2) thermally generated mesoscale effects on surface winds; and (3) orographic mesoscale effects on surface winds.

II. MESOSCALE PRESSURE AND WIND VELOCITY NETWORKS

A. Background

Results of arctic research in the 1970's have shown the inadequacies of synoptic-scale surface-pressure grids in the Alaskan Beaufort coast region. This study area had NWS stations only at Barrow and Barter Island, approximately 540 km apart. Rogers (1978) concluded that the lack of correlation between geostrophic wind direction and the distance of the ice margin from the coast (Sater *et al.*, 1974) was due to the geostrophic wind being an unusable or unreliable parameter in the study area.

Albright (1978) noted large differences between the geostrophic winds computed from the Arctic Ice Dynamics Joint Experiment (AIDJEX) polar pack network data and coarser resolution NWS analyses. Using the subsynoptic AIDJEX grid to full advantage, Albright (1980) showed that over the arctic ice pack if the geostrophic wind speed $U_g \geq 5$ m s^{-1}, then

$$U_{10} = 0.585 \ U_g, \ \alpha = 25.9^\circ \ \text{(annual average)} \qquad (1)$$

where U_{10} is the measured surface wind at 10 m height and α is the directional difference between U_g and U_{10}.

B. Techniques to Solve for Geostrophic Winds

Establishing the geometry of the surface-pressure field will allow for calculation of the geostrophic wind U_g from

$$U_g = \frac{-\nabla P \ (x,y)}{\rho f} \qquad (2)$$

where $P \ (x,y)$ is the surface-pressure field, ρ is the air density, and f is the Coriolis parameter.

A two-dimensional least squares cubic fit (Krumbein, 1959) was assumed to provide the maximum degree of variations required to resolve the pressure field for the study area size. The cubic polynomial to be solved is

$$Z = a+bx+cy+dx^2+exy+fy^2+gx^3+hx^2y+jxy^2+ky^3 \qquad (3)$$

where Z is pressure. A matrix solution determines constants a through k with x the longitude and y the latitude of each pressure input.

In the summer of 1976, a system of nine pressure stations (Kozo, 1979) including two buoys out on the sea ice provided input data for the cubic fit. These data were applied to a grid which was also supplied pressures interpolated from NWS maps at the study area boundaries. These interpolated pressures were given a 10% weight relative to the pressure-station inputs. A pressure field was generated and machine-contoured with the geostrophic wind calculated at selected sites from the derivatives of Eq. (3). In subsequent years, a three-point network comprising a plane pressure surface that needs only the first three terms of Eq. (3) was used for geostrophic wind solutions. Permanent networks were created by supplementing the Barrow and Barter Island NWS stations with two automated atmospheric pressure and temperature platforms emplaced in July 1980 at Franklin Bluffs and Narwhal Island (Kozo, 1980). With these additional platforms a year-round pressure network providing data for geostrophic wind calculation now exists. Three different pressure-station triangles with geometric solution centers at Harrison Bay (A), south of Simpson Lagoon (B), and Brownlow Point (C) are shown in Fig. 1. Analyses of pressure data from these triangles provide a real-time estimate of surface winds over the continental shelf, where in situ measurements are too costly or totally absent.

Sea ice edge movement, ice fracture, or surges caused by storms in the study area often are not detected by satellite imagery in real time due to undercast. The small-scale pressure nets can signal probable occurrence of these events and can be used to estimate the surface wind stress on the sea

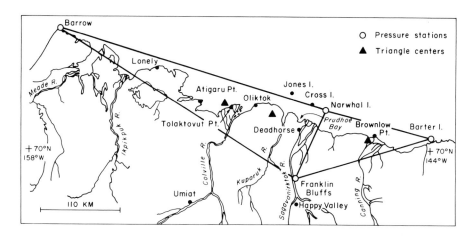

FIGURE 1.   Pressure stations furnishing year-round data to calculate geostrophic winds for three triangle combinations with centers shown.

or ice. Calculations of nearshore ice movements or oil-spill trajectories are possible during events that are otherwise obscured for days by poor atmospheric conditions.

C. Uses of Network Velocity and Pressure Data

    1. *Geostrophic Winds Versus Measured Surface Winds.* Geostrophic winds calculated for Harrison Bay (A) have been compared with measured surface winds at Tolaktovut Point for August and September 1980 (Fig. 2). Directions agree closely, and as expected, surface wind speeds are lower than calculated geostrophic winds. These particular months had an unusual preponderance of westerly winds including a three-week period starting 17 August 1980 (Fig. 2) during which moderate winds packed ice into Harrison Bay.

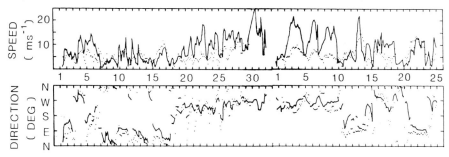

FIGURE 2.   Time series of calculated geostrophic winds (solid line) for a solution center near Tolaktovut Point compared to measured surface winds (dots) at Tolaktovut Point for 1 August to 25 September 1980.

These ice conditions were not caused by high winds associated with storm surges but merely by moderate steady westerly winds. Wind steadiness is the ratio of the magnitude of the mean wind vector $(\overline{u}^2 + \overline{v}^2)^{1/2}$ to the mean magnitude $(\overline{u^2 + v^2})^{1/2}$ expressed in percent (Halpern, 1979). Tropical trade winds, for example, can have steadiness greater than 90%. The surface wind steadiness was 72% at Tolaktovut Point during the three weeks of westerlies. Typical summers where offshore operations were free of storm surge or sea ice hazards exhibit steadiness values less than 50%.

   2. *Masking Effects.* Coastal surface wind measurements made during mesoscale events often bear little relation to large-scale wind directions and should not be used for offshore ice edge motion estimations. Figure 3 shows the positions of three pressure buoys from the 27 July to 7 August

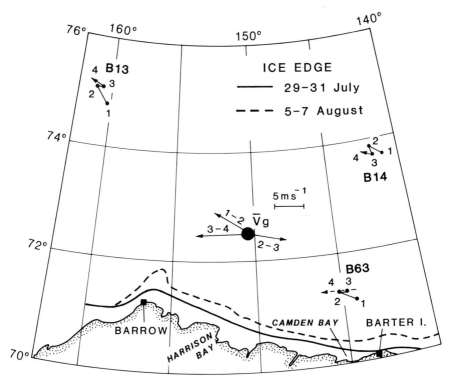

FIGURE 3. *The ice edge on 29-31 July and 5-7 August 1979 (Navy-NOAA Joint Ice Center Analyses), and corresponding movement of pressure buoys B13, B14 (National Science Foundation array) and B63 (Outer Continental Shelf Environmental Assessment Program). Position 1 refers to 27 July, position 2 to 2 August, position 3 to 5 August, and position 4 to 7 August. The dashed line for B63 indicates the position was unknown. The average geostrophic wind (V_G) velocities in the area of the large dot are indicated for the time segments 1-2, 2-3 and 3-4.*

1979. Offshore buoy data with concomitant data from Barrow and Barter Island allowed calculation of three-hourly geostrophic winds. The on-ice buoy movement corresponded closely to the calculated mean geostrophic wind direction for the designated time periods at the solution center (Fig. 3). Net summer ice movement is shoreward for westerly winds and offshore for easterly winds as mass transport in the underlying water column lies at 90° to the wind direction. The ice edge moved shoreward an average of 40 km from 2 to 5 August. In Fig. 4, three-hourly surface wind directions measured at Cross Island (Fig. 1) during this period are compared to the calculated geostrophic wind. On 3-4 August there was a total of 21 hours with a directional difference between the surface wind and the geostrophic wind of greater than 120°. The three buoy tracks and the ice edge position in Fig. 3, on the other hand, suggest that the surface wind farther offshore had a continuous westerly component during this time period, well correlated with the geostrophic wind.

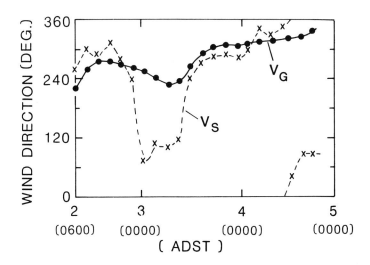

FIGURE 4. Comparison of the calculated geostrophic wind (V_G) direction with the surface wind (V_S) direction on Cross Island from 2 August to 5 August 1979. This was the time period designated 2-3 in Fig. 3. ADST = Alaska Daylight Savings Time.

III. THERMAL MESOSCALE EFFECTS

A. Large-scale Factors Affecting the Mesoscale

1. *Monsoons.* Monsoons are generally associated with warm climates, but evidence of them during the arctic summer has been documented for more than 30 years. Dzerdzeevskii (1945) described a semipermanent arctic front caused solely by the thermal contrast along the northern shores

of Alaska, Siberia, and Canada. Borisov (1959) commented that arctic seacoast winds have monsoonal character. The summer heating of the land areas causes a pressure deficit producing a tendency for air to move from sea to land. The Coriolis force imparts an easterly wind component along most arctic ocean coasts (Borizov, 1959). This tendency (not solely due to monsoons) for easterlies is reflected in summer historical wind data (Brower et al., 1977).

2. Inversions. An atmospheric layer in which the temperature increases with altitude and which exhibits strong static stability leading to diminished turbulent exchange is an inversion layer. The arctic land regions in winter are characterized by ground-based inversions due to a surface radiation deficit. Ice-covered polar seas can have such radiation inversions as well as advective inversions in all seasons. Inversions usually result in weaker surface winds, for a given atmospheric pressure gradient, than neutral or unstable conditions because little mixing occurs between the inversion layer and the faster moving layer above (Arctic Forecast Guide, 1962). Coupling between atmosphere and ocean is also reduced because of smaller drag coefficients (Roll, 1965) and large directional differences between the geostrophic and surface winds (Albright, 1980).

B. Sea Breezes

The sea breeze is air moving inland in response to differential heating across a coastline. The intensity, duration, and extent of sea breezes are mainly determined by horizontal gradients in the amount of heat supplied by the earth's surface to the atmosphere (Defant, 1951). The horizontal extent of the sea breeze is large enough to be influenced by the earth's rotation and large-scale atmospheric pressure gradients (Walsh, 1974). The sea breeze effect can be separated from larger scale wind conditions because of its characteristic diurnal pattern.

1. Experimental Evidence for Arctic Sea Breezes. Before the mid-1970's, the northernmost sea breeze was found in Finland on the shores of the Baltic Sea at 60° N (Rossi, 1957). A recent study of synoptic wind data from Barrow (Moritz, 1977) suggested that persistent unidirectional temperature gradients at Alaska's north coast add a component to the atmospheric pressure gradient that is not recorded by the NWS synoptic observation network. The sea-land temperature gradient remains constant in direction despite 15°C decreases in land temperature at night due to ocean temperatures near 0°C.

Sea-breeze data have been obtained along the Alaskan Beaufort Sea coast since 1976 (Kozo, 1979). Pressure data from buoys within the ice pack, two existing NWS stations, Distant Early Warning (DEW) sites at Oliktok and Lonely, and camps at Umiat and Happy Valley were used (Fig. 5) to approximate wind vector turning in the atmospheric boundary layer (Kozo, 1982a). Seven surface wind stations provided summer data in the primary study area centered on C in Fig. 5. Two were on the Jones Islands and one each was on Cross Island, Narwhal Island, Tolaktovut Point, Oliktok, and Deadhorse (Fig. 1).

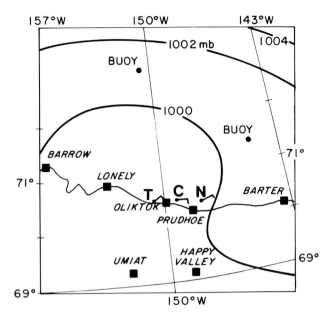

FIGURE 5. Two-dimensional least-squares fit to surface pressure network on 16 August 1976 (0000 GMT). T, C, N represent surface wind measurements from Tolaktovut Point, the Jones Islands, and Narwhal Island, respectively. Pressure contours are in millibars (mb). Wind flag = 1.5 to 3.5 m s^{-1}.

The momentum equation for synoptic-scale atmospheric flow is

$$\partial \boldsymbol{V}/\partial t + \boldsymbol{V} \cdot \nabla \boldsymbol{V} + f(\boldsymbol{k} \times \boldsymbol{V}) + \nabla P/\rho = 0 \qquad (4)$$

where $\boldsymbol{V}$ is the horizontal velocity vector, f the Coriolis parameter, $\boldsymbol{k}$ the vertical unit vector, ρ the air density, and P the pressure. The last two terms, representing the Coriolis force and pressure gradient force respectively, are generally dominant and can be used to calculate geostrophic wind velocity, which approximates the free stream flow above the planetary boundary layer parallel to the isobars. The first two terms represent the local and advective parts, respectively, of the total acceleration in a fluid.

If either of the first two terms becomes important, a correction must be applied to the geostrophic velocity. The first term becomes important during frontal passage, when large wind changes occur in less than five hours. The second term can become significant when strong curvature of flow exists (isobaric radius of curvature less than 300 km). The calculated free-stream flow can also be in error when the surface pressure gradient differs significantly from it due to large-scale horizontal quasi-steady-state temperature gradients (thermal wind).

The first approximation to the geostrophic wind field was obtained from NWS from surface pressure maps for 0000 and 1200 GMT (Greenwich

Mean Time).   To increase the resolution beyond that of the maps the
pressure network in Fig. 5 was used in conjunction with the least-squares
technique applied for a cubic surface.   Pressure data collected at these
sites provided the basic input for calculation of geostrophic velocities near
wind-measuring sites (Kozo, 1979).

Figure 5 includes pressure contours on a sea-breeze day.   There is a
weak pressure gradient in the Jones Islands area.   The land-sea temperature
difference from Deadhorse to the coast (20 km) was 7.8°C.   The calculated
geostrophic wind on the Jones Islands this day (0000 GMT) was 0.93 m s^{-1}
from 188.5° while measured surface wind was 5.28 m s^{-1} from 90°.

To compare surface winds (at C, Fig. 5) with geostrophic winds on a
three-hourly basis, the least-squares technique was used to fit a plane
surface to pressure data from Prudhoe, Lonely, Umiat, and the closest
offshore buoy.   Figure 6 shows that geostrophic winds within a 60° band
from 195° to 255° correspond to surface winds from 90° to 150°.
Planetary boundary-layer turning, usually the result of a three-way balance
between the Coriolis force, large-scale pressure gradient force, and viscous
force, should shift winds approximately 20-30° counterclockwise (CCW)
from their free-stream direction to the surface.   Thus the sea breeze had
offset weak geostrophic winds to produce an apparent average boundary-
layer turning of 120° CCW from the free-stream level to the surface.

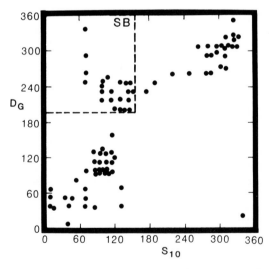

FIGURE 6.   Geostrophic wind direction (D_G) versus surface wind direction
(S_{10}) for time segments 13-23 August and 30 August - 3 September 1976.
Data points showing influence of sea breeze (SB) lie within dashed line.

Figure 7 combines three years of August data in a wind speed and
direction histogram.   Allowing for synoptic effects, Fig. 6 shows that much
of the asymmetry in the histogram must be due to sea breezes.

Time series of surface wind data from coastal stations and islands were examined with a rotary spectrum technique (Gonella, 1972). The variance for each frequency band is divided into clockwise (CW) rotating variance (negative frequency) and CCW-rotating variance (positive frequency). The sea breeze is a CW-rotating wind oscillation, and significant peaks at the 24-hour period in the CW part of the spectrum indicates presence of sea-breeze circulation (O'Brien and Pillsbury, 1974). On Fig. 8, a semilog plot of rotary spectra for August 1976 at Tolaktovut Point, the significant peak occurring near -1 cycle day $^{-1}$ is the CW-rotating contribution from the sea breeze. Similar peaks in the Deadhorse and Narwhal Island spectra (not shown) are evidence of sea-breeze influence in a band at least 40 km wide centered on the coastline.

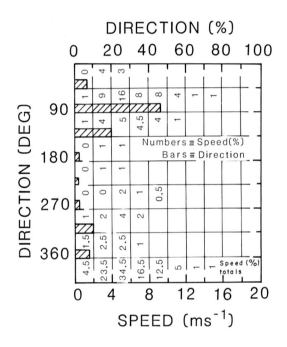

FIGURE 7. Histogram of surface wind speed and direction on the Jones Islands for August 1976, 1977, and 1978 combined. The top scale represents the percentage frequency of winds from a given direction, indicated by the length of the bars. Wind speed frequencies are read along the bottom of the graph and are indicated by numbers in blocks.

Temporal rotation of the local surface wind vector with distance from the coast is presented in Fig. 9 for 17 August 1976. All sites had CW-rotating surface wind vectors from maximum sea-breeze influence at 1500 until midnight, while from 0900 to 1500 the wind generally rotated CCW. The large-scale wind field was apparently weak, as seen from computations

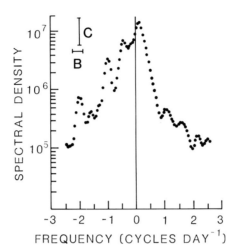

FIGURE 8. Time series rotary spectra of surface wind velocity data for August 1976 at Tolaktovut Point. The vertical axis has units of spectral density $((m^2 \ s^{-2}) \ (cycles \ (3h)^{-1})^{-1})$. C is 95% confidence limit, B is bandwidth.

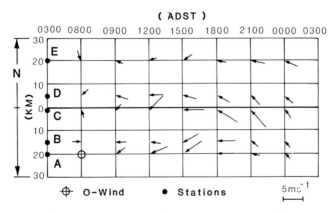

FIGURE 9. Surface wind vectors for designated stations on 17 August 1976. The calculated geostrophic wind was from 227° at 3.2 m s^{-1} for 1500 ADST and 137° and 2.2 m s^{-1} at midnight, a CCW rotation of 90°. E = Narwhal Island, C = Tolaktovut Point, D = Jones Islands, A = Deadhorse, and B = Arco Airport.

of geostrophic winds during this time interval, and had limited influence on the sea-breeze rotation.

Two theodolites were used to track standard meteorological pilot balloons during August 1977 on the Jones Islands. An example of these data converted to horizontal wind velocity versus height is presented in Fig. 10

during sea-breeze conditions. Radiosonde data indicated an inversion layer top at 200 m with a temperature of 17°C and 5.0°C surface temperature. The temperature difference from Deadhorse Airport to the coast was 12.8°C at 1500 ADST. The free-stream wind direction remained constant from approximately 220° for 4 hours during the balloon tracking. All three profiles show temporal CW rotation of the surface wind vector and also over 120° CW rotation of the wind vector from the ground to the inversion layer top. The upper level offshore winds at 1524 averaged 3 m s^{-1} and should have allowed for sea-breeze development. Schmidt (1947) found the sea breeze to be strongest during moderate offshore geostrophic winds which were evident in Fig. 10.

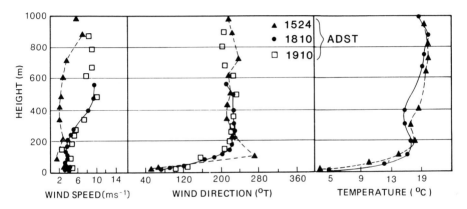

FIGURE 10. *Profiles of wind speed, wind direction, and temperature of the Jones Islands for times indicated on 15 August 1977.*

Large diurnally varying land-sea thermal contrasts, CW rotation of surface wind vectors, and surface winds in opposition to moderate offshore gradient winds characterize sea breezes in the study area. Sea breezes dominate the surface wind direction for at least 25% of the summer data collection periods.

The major implications of sea-breeze forcing along the coast are:

(1) Maintenance of westward-flowing alongshore surface currents at least 20 km from the coast which promote lagoon flushing even during weak opposing synoptic wind conditions.

(2) Production of wind-driven current shears beyond 20 km offshore where synoptic conditions again prevail.

(3) Masking of synoptic wind directions which affect water and ice movement 20 km seaward from the coast.

2. *Modeling the arctic sea breeze.* The Alaskan arctic sea breeze exhibits four important features:

(1) It is never followed by a land breeze, because the land stays warmer than the water during the summer months.

(2) The surface wind vector rotates faster than at mid-latitudes (Neumann, 1977) due to the 37% larger Coriolis force. Wind directions in

the quadrant from 25° to 115° predominate due to the orientation of the coast and the thermally induced mesoscale pressure gradient.

(3) The arctic sea breezes should be weak because initial conditions are similar to those of mid-latitude land breezes, where strong ground-based inversions produce small eddy diffusion coefficients and limit vertical circulation.

(4) The arctic sea breeze is driven by a relatively large sea-land temperature difference of 12°-14°C at the coast, which partly compensates for inversion damping.

The above four properties have been included in a time-dependent nonlinear two-dimensional numerical model (Kozo, 1982b) that allows imposition of prevailing large-scale wind conditions. The model, using typical arctic conditions as input, gives results that reproduce measurements of atmospheric boundary layer wind turning, temporal inversion height variation, and increased sea-breeze circulation due to

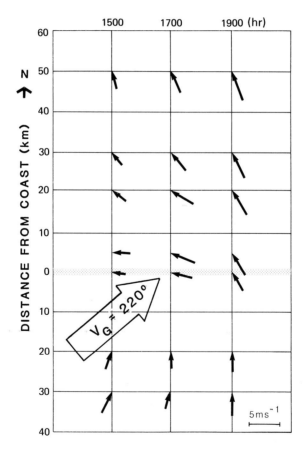

FIGURE 11. Model output of surface wind vectors for 1500, 1700, and 1900 ADST with geostrophic wind (V_G) from 220° at 5 m s^{-1}.

weak opposing synoptic winds. Figure 11 shows the model output of surface wind vectors for an opposing geostrophic wind input of 5 m s^{-1} from 220° at 1500, 1700, and 1900 hours. The surface wind vectors exhibit temporal CW turning from the coast to 30 km seaward and minimal effects to 50 km seaward. Starting 20 km inland from the coast a temporal CCW turning of the surface wind vector is seen due to the approach of the sea-breeze front. Figure 9 with a geostrophic wind from 227° exhibits similar characteristics for at-sea and land stations.

Figure 12 is the model output of wind velocity versus height 5 km offshore (simulating conditions on the Jones Islands) for the same conditions as in Fig. 11. A speed decrease at the inversion layer top coincides with the layer of most drastic wind directional change. The output shows CW turning of the wind with height up to 250 m, which compares well with actual data (Fig. 10) and shows a 65° CW rotation of the surface wind vector in 4 hours compared to 70° CW shown by the real data.

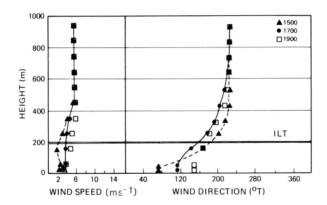

FIGURE 12. Model profiles of wind speed and direction 5 km seaward from the coast, for 1500, 1700, and 1900 ADST (V_G of 5 m s^{-1} from 220°). Compare actual measurements in Fig. 10. ILT is inversion layer top.

C. Storm-induced Leads and Effects on Surface Winds

When gale-class easterly winds lasting 3 days (Fig. 13) hit Alaska's north coast on 10 November 1981, the sea ice north of the fast ice zone was fractured, opening a lead 60 km wide parallel to the coast (Fig. 14). After the storm, a weak synoptic wind field existed. The large lead represented a source of moisture and heat 20°C above typical ambient air temperatures, causing an unstable boundary layer, and greatly reduced visibility due to steam fog.

The surface air temperature, wind speed, and wind direction at Deadhorse are presented in Fig. 15 before, during, and after the gale. Two criteria for major sea ice movement were met (Agerton and Kreider, 1979): The wind speed threshold of 13 m s^{-1} was exceeded and stress direction alternated from 240° to 80°. The thermal contrast between the

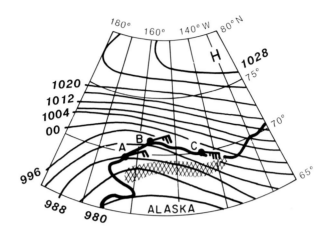

FIGURE 13. Section of a NWS synoptic chart for 12 November 1981 (0000 GMT). The Brooks Range is crosshatched. The wind speeds at A (Point Lay), B (Barrow), and C (Barter Island) are 10 m s^{-1}, 12.5 m s^{-1}, and 17.5 m s^{-1} respectively.

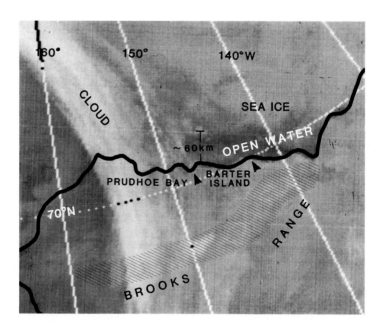

FIGURE 14. Satellite infrared image of Alaska on 17 November 1981. The ice in the Beaufort Sea was fractured by the gale-force storm shown in Fig. 13, producing a 60-km-wide lead north of the coast. Dark tones represent warmer areas north of the coast.

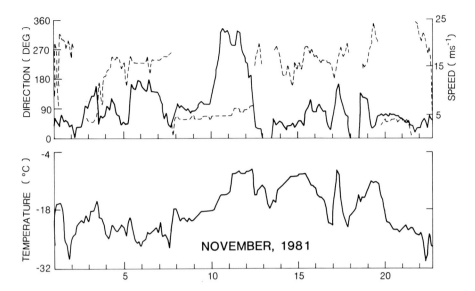

FIGURE 15. Wind speed (solid) and direction (dashed) and air temperature at Deadhorse for 1-22 November 1981.

snow-covered land, sea ice, and open-water lead induced southwesterly surface winds. This sub-synoptic event was not accounted for in the NWS forecasts for the area, thus predictions for visibility, wind direction, and temperatures were extremely poor. Temperature and wind direction returned to seasonal values in a week as the lead refroze.

IV. MESOSCALE EFFECTS INDUCED BY OROGRAPHY

A. Corner Effect

The Brooks Range, with a mean height of 1525 m, has a considerable orographic effect on the cyclonic scale winds for much of the study area (Arctic Forecast Guide, 1962). Instances of surface winds faster than 14 m s^{-1} at weather stations near mountains once were usually attributed to drainage of cold air (katabatic winds) down steeply sloping terrain, but studies have shown that most strong winds at such sites are due to above-normal pressure gradients and that the winds are super-gradient (Dickey, 1961). However, these winds do not blow straight downslope and are accompanied by rising temperatures as often as by falling temperatures in the same season. Hence topographically induced effects other than purely katabatic ones had to be examined.

Winter arctic inversions in the lower atmosphere lead to deflection of large-scale horizontal flows around topographic barriers, resulting in high wind speeds at certain locations near mountain ranges and large variations in surface wind directions. This corner effect (Dickey, 1961) has been noted in the vicinity of Barter Island (Fig. 16), where strong surface winds have a preference for west or east directions and can exceed the calculated geostrophic wind by 50%. Dickey (1961) used the 600-m elevation contour as the upper boundary to horizontal flow and approximated the Brooks Range by a vertical cylinder (Fig. 16). The expressions for the velocity distribution around a cylinder in steady, horizontal, irrotational, frictionless flow of an incompressible fluid are as follows:

Radial velocity component

$$U_r = U(1 - a^2/r^2)\cos \phi \tag{5}$$

Directional velocity component

$$U_\phi = -U(1 + a^2/r^2)\sin \phi \tag{6}$$

Magnitude

$$V = (U_r^2 + U_\phi^2)^{1/2} \tag{7}$$

where a is the radius of the cylinder, r the distance to a measuring station, and ϕ the smallest angle from the head of the basic wind vector U to r. Figure 16 shows the effects at Barter Island upon a westerly wind U from 285°: a is 274 km, r is 322 km, and ϕ is 110°. From Eqs. (5-7), $U_r = -.095U$, $U_\phi = -1.619U$, and $V = 1.62U$ from 269°.

Dickey's model (1961) agreed well with observations and showed that the corner effect has an omnidirectional influence as much as 350 km from Barter Island during storm conditions. This effect can therefore modify winds at Prudhoe Bay to the west or far north on the polar pack.

B. Mountain-barrier Baroclinity

Mountain-barrier baroclinity, a predominantly winter phenomenon, is responsible for 180° surface wind shifts along the Alaskan coast between

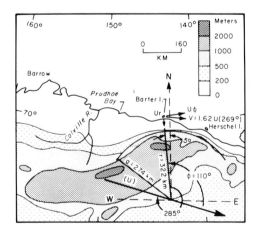

FIGURE 16. Wind U from 285° and resultant surface wind V at Barter Island due to orographic modification (Dickey, 1961). The 600-m contour is approximated by a circular arc of radius 274 km.

Barrow and Barter Island during moderate wind conditions. The mechanism was first described theoretically by Schwerdtfeger (1974) using data from stations north of the Brooks Range as evidence. He stated that a stable air mass moving toward a mountain range (Fig. 17) without heating from below induces baroclinity by causing a tilting of isobaric and isothermal surfaces away from the obstacle. This results in a thermal wind parallel to the range horizontal axis and a mesoscale pressure-gradient force perpendicular to this axis from B to A with a maximum value near the inversion layer top (level 1, Fig. 17). For the case of Barter Island, northerly flow induces a strong westerly wind component near the bottom of the cold air layer (shaded area, Fig. 17) over the North Slope. Therefore, flow is toward the reader between level 1 and 0.

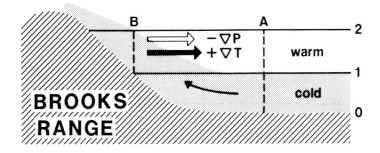

FIGURE 17. *Cross section of a stable air mass moving toward a mountain range (Schwerdtfeger, 1974).*

The effects of mountain-barrier baroclinity were experimentally examined through use of a pressure data network. Figure 18 is a section of a NWS synoptic chart for 11 March 1979 (1200 GMT) with surface wind and pressure data locations indicated. Only data from S, F, and H were used to construct contours. The numbers next to sites are surface pressures minus 1000 mb. The winds at S and F and the pressure at B13 (1037.2 mb) are reasonable for the offshore high pressure system shown. However, the winds at G and H are in opposite directions to the anticyclonic flow indicated, and actual surface pressure at B14 is 5 mb less than the chart estimate. The pressure change from B13 to B14 was 15.3 mb (map contours indicate 8 mb) while the change from F to H, a comparable distance, is 5.6 mb. Simultaneous Barter Island (H) rawinsonde data (Fig. 19) show a strong temperature inversion extending up to 400 m and approximately 150° of wind turning within the inversion layer. The computed North Slope geostrophic wind for this time using pressure data from buoys B13, B14, and Barrow was from 36° at 15 m s^{-1}.

The above data and other evidence (Kozo, 1980) demonstrate the existence of an additional pressure gradient force away from the Brooks Range axis since frictional wind turning usually accounts for only $20-30^\circ$ CCW change from upper free-stream direction to the 10-m level.

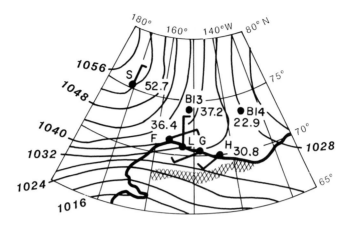

FIGURE 18. *Weather Service synoptic pressure chart section for 11 March 1979 with the Brooks Range crosshatched. F, L, G, H, and S show surface wind data locations at Barrow, Lonely, Prudhoe Bay, Barter Island, and a Soviet ice station, respectively. B13 and B14 are pressure (only) buoys which have an accuracy of ±1.5 mb. (Pressure contours in mb, wind flag = 1.5 to 3.5 m s^{-1}.)*

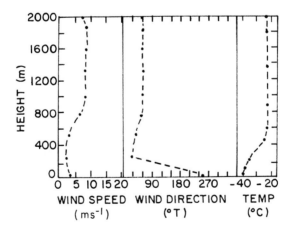

FIGURE 19. *Profiles of wind speed, direction, and temperature at Barter Island for 11 March 1979 (1200 GMT).*

Figure 20A shows three-hourly surface wind directions at Barter Island versus Barrow for March 1979. Enclosed by the solid line are simultaneous winds with an average directional difference of 140°. Since Barrow is more than 300 km north of the Brooks Range, and not subject to its orographic effects, the surface wind should vary less than 30° from the geostrophic wind direction.

To compare surface winds at Barter Island with geostrophic winds, a least-squares technique was applied on a three-hourly basis to pressure data from B13, B14, and Barrow. The Barter Island data were excluded because the pressure is modified by mountain-barrier baroclinity. The station combination used provided a good fit to the measured surface wind directions at Barrow (scattergram not shown) and was representative of the 800-2000-m flow in the Barter Island rawinsonde data (Fig. 19).

Figure 20B shows a three-hourly measured surface wind versus computed geostrophic wind directions at Barter Island for March 1979. It is similar to Fig. 20A since surface wind directions at Barrow should be within 30° of the geostrophic wind directions for the same periods. The upper level flow in the Barter Island rawinsonde data also fits the calculated geostrophic winds. Hence the data points enclosed by the solid line in Fig. 20B are evidence for mountain-barrier baroclinity, causing an average of 140° of turning from the geostrophic wind level to the surface. This is one reason for predominant surface westerly winds at Barter Island in the winter months, while Barrow has a predominance of easterly winds (Brower et al., 1977).

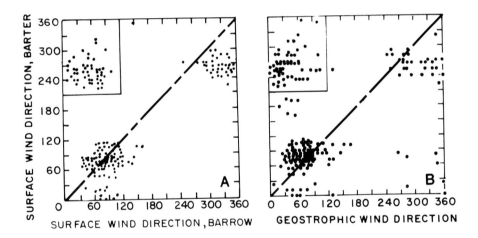

FIGURE 20. A, Simultaneously measured surface wind directions at Barter Island and Barrow for March 1979. Boxed values represent directional differences averaging 140°. B, Barter Island surface wind directions and simultaneous geostrophic wind directions. Boxed values occur during mountain-barrier baroclinity events averaging 140° differences.

Barter Island soundings were used to provide an estimate of the thermal wind (V_T) due to piling up of cold air on the North Slope. The thermal wind can be calculated as (Dalrymple et al., 1966)

$$V_T = \frac{g}{f} \frac{\Delta T}{\overline{T}} \, G \times k \tag{8}$$

Here g is the acceleration of gravity, f is the Coriolis parameter (1.37 x 10^{-4} s^{-1} at 70°N), ΔT the temperature difference between the bottom and top of the inversion layer, k is the vertical unit vector, $\overline{T}$ is the mean inversion layer temperature, and G is the slope of the cold air layer. Figure 20 shows $\Delta T \simeq 16°C$, and $\overline{T} \simeq 241$ K. If G is assumed to be 50% of the terrain slope (0.007), then using Eq. (8), $V_T \simeq 17$ m s^{-1} from the east. Thermal winds of this magnitude flowing parallel to the mountains, when coupled with surface friction effects, account for the wind direction changes seen in Barter Island soundings from 800 m to the surface.

Figure 21, derived from 20-year historical compilations (Brower et al., 1977) of monthly wind direction frequency histograms, is a plot of the percentage difference $\Delta\%$ between total surface wind direction frequencies in the east-northeast quadrant and west-southwest quadrant for three coastal stations. These are Lonely, Oliktok, and Barter Island (Fig. 1), approximately 275 km, 165 km, and 64 km, respectively, from the foothills of the Brooks Range. Positive $\Delta\%$ implies more east winds than west winds while negative $\Delta\%$ implies more west winds than east winds. The months October through April show a great disparity between Lonely (farthest from the mountains) and Barter Island (closest to the mountains), or further evidence of mountain-barrier baroclinity during winter when the atmospheric boundary layer is most stable.

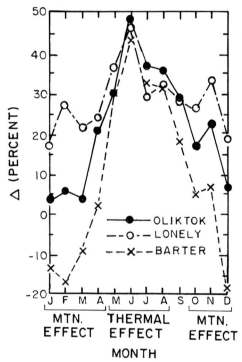

FIGURE 21.   Monthly percentage difference ($\Delta\%$) between total surface wind direction frequencies in the east-northeast quadrant and the west-southwest quadrant for data locations at Barter Island, Oliktok, and Lonely.

In May through August, when insolation exceeds 20 hours and there is little or no snow cover, the stability of the boundary layer over land decreases and conditions at the three coastal stations become similar. During this time the frequency of east winds at all stations averages 35% greater than the frequency of west winds. Mountain-barrier baroclinity is minimized and the thermal effect of the coastline becomes a major influence on coastal surface winds (Kozo, 1982a).

In Fig. 22 a record of the average daily albedo over the arctic tundra has been separated into four characteristic time periods (Maykut and Church, 1973):

(1) the *winter stationary period* (mid-October to early June), characterized by a completely snow-covered tundra,

(2) the *spring transitional period* (mid-June), which shows a rapid drop in albedo from 75% to 10% over the tundra in 3-5 days,

(3) the *summer stationary period* (June to August), characterized by a lack of snow and ice cover on land, and

(4) the *autumn transitional period* (September), which shows an increase in albedo with large, rapid fluctuations as the land surfaces freeze and become snow covered. The precise dates of the transition period may change from year to year but not the general shape of the curve nor the magnitudes of albedo.

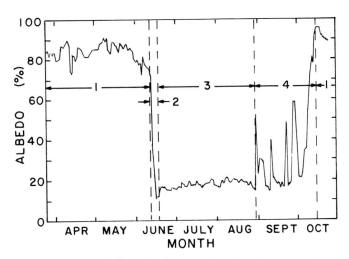

FIGURE 22. *Average daily albedo over the tundra near Barrow from 25 March to 25 October 1966 (Maykut and Church, 1973).*

The importance of a stable air mass moving toward a mountain range without heating from below can now be seen. Barter Island has a surface wind regime with a significant amount of west-southwest winds during the winter stationary period. But surface wind directions are similar to the other stations during the summer stationary period when air masses traversing the snowless tundra are heated from below, thereby decreasing the static stability. September is a transition month on both Figs. 21 and 22. The data (Fig. 21) in the month of May show an apparent inconsistency for Barter Island, but it is a 20-year average for a month where insolation reaches 24 hours duration, atmospheric stability diminishes, and the tundra can be devoid of snow in its last week.

Mountain-barrier baroclinity is a major physical process responsible for the wintertime abundance of west-southwest coastal winds from Barter Island to Prudhoe Bay along the Alaskan arctic coast. The phenomenon requires a stable atmospheric boundary layer, which is related to high surface albedo in winter. It disappears when the albedo is low in summer. The horizontal extent of approximately 120 km and 25% winter occurrence rate of this effect indicate that winter surface wind measurements for a coastal zone from east of Barter Island to Prudhoe Bay may be poor indicators of the stress exerted on nearshore ice. Instead, the geostrophic wind should be calculated for a position farther seaward to determine stress transmitted to the inshore ice.

V. CONCLUDING REMARKS

Mesoscale meteorological wind effects can periodically be the sole driving force for summer surface currents or winter sea ice movement in coastal areas. Recognizing the existence of such effects and predicting their occurrence can only be a starting point and guide for deriving ultimate locations of pollutants spilled at sea. However, these locations cannot be defined without information from sub-synoptic networks, which did not exist in the area before the Outer Continental Shelf Environmental Assessment Program. The activities of exploration and development of energy resources will increase. The operational efficiency of industry, government, and the native peoples will be greatly enhanced by continuing maintenance of these weather networks.

REFERENCES

Agerton, D.J., and Kreider, J.R. (1979). In "Proceedings POAC '79," Vol. 1, p. 177. Norweg. Inst. Technology, Trondheim.
Albright, M. (1978). In "AIDJEX Bull." No. 39, p. 111. Univ. of Washington, Seattle.
Albright, M. (1980). In "A Symposium on Sea Ice Processes and Models," p. 402. Proceedings of the International Commission on Snow and Ice/Arctic Ice Dynamics Joint Experiment (R. S. Pritchard, ed.), Univ. Washington Press, Seattle.
Arctic Forecast Guide (1962). U.S. Navy Weather Research Facility, Norfolk, VA.
Borisov, A.A. (1959). In "Climates of the U.S.S.R." p. 106. Aldine, Chicago.
Brower, W.A., Diaz, H.F., Prechtel, A.S., Searby, H. W., and Wise, J.L. (1977). "Climatic Atlas of the Outer Continental Shelf Waters and Coastal Regions of Alaska." NOAA, Asheville, NC.

Dalrymple, P., Lettau, H., and Wollaston, S. (1966). *In* "Studies in Antarctic Meteorology" (M.J. Rubin, ed.), Vol. 9, p. 13. Am. geophys. Union Antarctic Research Series.

Defant, F. (1951). *In* "Compendium of Meteorology," p. 655. American meteor. Soc., Boston.

Dickey, W.W. (1961). *J. Met. 18*, 790.

Dzerdzeevskii, B.L. (1945). "Tsirkuliatsionnye skhemy v troposfere tsentral noi Arktiki." *In* Izdatelstvo Akad. Nauk., p. 228. (English translation in Sci. Report #3 under Contract AF 19 (122), Univ. Cal. Los Angeles).

Gonella, J. (1972). *Deep-Sea Res. 19*, 833.

Halpern, D. (1979). *Mon. Wea. Rev. 107*, 1525.

Kozo, T.L. (1979). *Geophys. Res. Letters 6*, 849.

Kozo, T.L. (1980). *Geophys. Res. Letters 7*, 377.

Kozo, T.L. (1982a). *J. appl. Met. 12*, 891.

Kozo, T.L. (1982b). *J. appl. Met. 12*, 906.

Krumbein, W.C. (1959). *J. geophys. Res. 64*, 823.

Maykut, G.A., and Church, P. E. (1973). *J. appl. Met. 12*, 620.

Moritz, R.E. (1977). *Arctic Alp. Res. 9*, 427.

Neumann, J. (1977). *J. atmos. Sci. 34*, 1914.

O'Brien, J.J. and Pillsbury, R.D. (1974). *J. appl. Met. 13*, 820.

Rogers, J.C. (1978). *Mon. Weath. Rev. 106*, 890.

Roll, H.U. (1965). "Physics of the Marine Atmosphere." Academic Press, New York, p. 1972.

Rossi, V. (1957). *Mitt. Met. Zentralans, Helsinki, 41*, 1.

Sater, J.E., Walsh, J. E., and Wittmann, W.I. (1974). *In* "The Coast and Shelf of the Beaufort Sea" (J.C. Reed and J.E. Sater, eds.), p. 85. Arctic Inst. of North America, Arlington.

Schmidt, F.H. (1947). *J. Met. 4*, 9.

Schwerdtfeger, W. (1974). *In* "Proceedings of the 24th Alaskan Science Conference," p. 240. Geophys. Institute, Univ. Alaska, Fairbanks.

Walsh, J. E. (1974). *J. atmos. Sci. 31*, 2012.

THE BEAUFORT UNDERCURRENT

Knut Aagaard

School of Oceanography
University of Washington
Seattle, Washington

I. INTRODUCTION

The southern Beaufort Sea is generally portrayed as a region of mean westward water and ice motion, corresponding to the southern edge of the anticyclonic gyre of the Canadian Basin. However, except over the inner continental shelf the average subsurface motion is actually in the opposite direction: over the slope and shelf seaward of about the 50-m isobath there is a strong mean eastward motion. This eastward flow is by far the most conspicuous feature of the regional circulation, and I will refer to it as the Beaufort Undercurrent.

II. HYDROGRAPHY

It has been clear for some years that an eastward flow exists on the shelf at least occasionally, since in summer the most prominent hydrographic feature on the shelf is a subsurface temperature maximum, generally found seaward of about 40-50 m depth. This temperature maximum is associated with the eastward flow of water originating in the Bering Sea. The influx was first described by Johnson (1956), and it has since been discussed by Hufford (1973), Mountain (1974), Paquette and Bourke (1974), and others.

The warm water that enters the Beaufort Sea has come through the eastern Bering Strait and followed the Alaskan coast around Point Barrow. This intrusion is in fact composed of two water masses, called Alaskan coastal water and Bering Sea water by Mountain (1974). The former can have summer temperatures west of Barrow as high as 5-10°C, but the salinities are low, being less than 31.5 parts per thousand. The Bering Sea water is more saline and is contained in the density range 25.5 (1.0255 g cm^{-3}) to slightly over 26.0 in σ_t, as has been demonstrated in the detailed analyses of Mountain (1974) and Coachman *et al.* (1975). Figures 1 and 2, adapted from Mountain (1974), show the temperature on density surfaces associated with the two intruding water masses. The restriction of the warm eastward flow to the outer continental shelf and the slope is clear, as

is the difference in extent of influence of the two water masses. The
Alaskan coastal water mixes rapidly with the ambient surface water as it
moves eastward and is not clearly identifiable east of about 147-148°W.
On the other hand, the Bering Sea water, with its deeper temperature
maximum (in the σ_t range 25.5-26.0), can be traced at least as far as Barter
Island at 143°W.

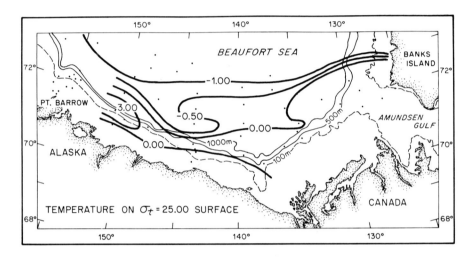

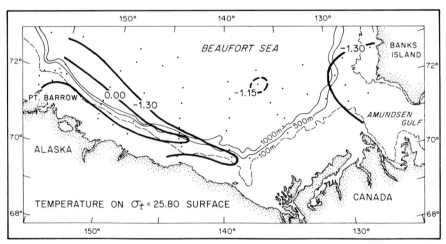

FIGURE 1. Temperature on density surface corresponding to Alaskan
coastal water (σ_t = 25.00) and Bering Sea water (σ_t = 25.80), August –
September 1951. Dots represent station positions. Adapted from Mountain
(1974).

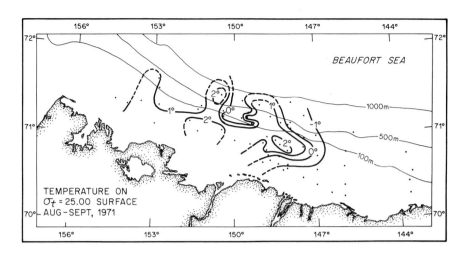

FIGURE 2.   Summer temperature on density surface σ_t = 25.00.   Dots
represent station positions. Adapted from Mountain (1974).

The temperature maximum over the shelf (as contrasted to off-shore
over the deep basin) is primarily a summer phenomenon. Figure 3 shows
the temperature–salinity correlation at two stations on the shelf near
153°W. Station W 25-19 was occupied in early November 1976 and W 27-1,
at the same location, the following March. The temperature maximum of
about -0.9°C at 43 m depth at station 19 occurred at σ_t = 25.8 and
represents Bering Sea water having rounded Point Barrow earlier in the
year at a higher temperature. Somewhat later in the winter, the Bering
Sea temperature signal is effectively erased on the shelf, as shown by the
temperature–salinity correlation in the upper 50 m at station W 27-1; down
to about 40 m, where the density is 26.5 in σ_t, the water is at the freezing
point. Such low temperatures extending into or past the density range of
the core of Bering Sea water are typical on the shelf in winter, and in fact
these conditions can readily be found already in November. Therefore one
probably cannot generally use temperature to trace the Bering Sea water
on the Beaufort shelf much past freeze-up in the fall. Also, after that time
the new water entering the Beaufort Sea from the Chukchi Sea is itself
near the freezing point after being cooled in its fall and winter transit
northward.   Delineation of the eastward flow under these conditions
therefore depends on hydrographic parameters other than temperature or
on direct current measurements.
    The patchiness of the temperature distribution in Fig. 2 is of particular
interest. Mountain (1974) attributed such features to interruptions in the
influx of warm water onto the Beaufort shelf, and in fact the appropriate
variability has been observed upstream in Barrow Canyon (Mountain et al.,
1976). On the other hand, similar temperature structures commonly seen in
frontal regions elsewhere in the ocean are frequently attributed to flow
instabilities (see Wadhams and Squire, 1983, for a recent example from an

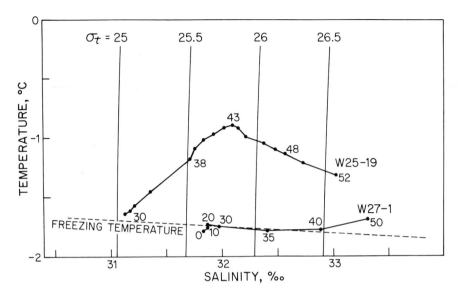

FIGURE 3.   Temperature-salinity correlations at a fall (W 25-19) and winter (W 27-1) station near 153°W.

ice-covered area).   However, except perhaps during density upwelling events (Hufford, 1974) and near the major rivers during peak discharge, there do not appear to be marked density fronts over the Beaufort shelf. While there are sizable temperature gradients in summer associated with the warm waters being carried by the Beaufort Undercurrent, there is not normally a strong summer density front (Fig. 4).   During winter the temperature maximum disappears and persistent strong fronts of any kind are apparently absent over the shelf (Fig. 5).   While it is therefore doubtful whether frontal dynamics are of major importance in the shelf regime, temporary or weak fronts exist.   Hunkins (1979), in fact, calculated an e-folding time for baroclinically unstable current perturbations north of Point Barrow of 28 days.   This is a very slow growth rate and points toward the inefficacy of baroclinic instabilities in producing mesoscale features over the Beaufort Shelf.

Although the strong Bering Sea hydrographic signal is predominant, the eastward flow probably also involves water masses other than those from the Bering Sea.   This would be particularly true seaward of the shelf break.

III. CURRENT MEASUREMENTS

Over a four-year period beginning in March 1976 we obtained a total of 2335 record days of moored current measurements on the Beaufort Sea shelf between 146° and 152°W (Table I).   The mooring techniques that were

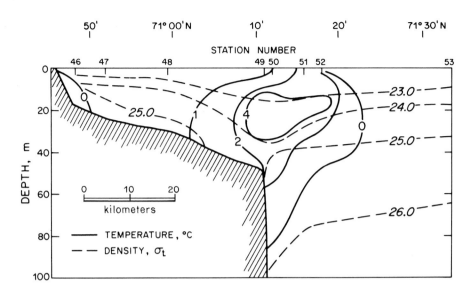

FIGURE 4.   Summer temperature and density section across the shelf and
upper slope at 150°W. Adapted from Mountain (1974).

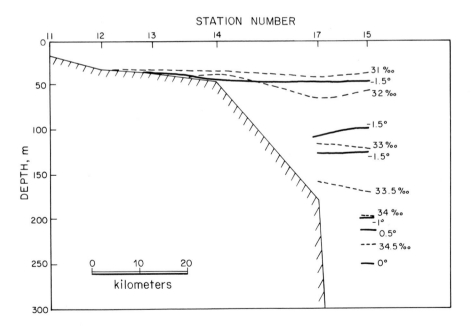

FIGURE 5.   Winter temperature and density section across the shelf and
upper slope at 147°W.

developed for these ice-covered waters have been described by Aagaard *et al.* (1978).

The measurements are summarized in Fig. 6 in the form of current roses. At each indicated mooring site, the current rose for a particular instrument (instrument depth given adjacent to the rose) represents both the mean speed and the direction, the latter in terms of the frequency of occurrence of the current in each 20° sector. For example at mooring LO-5 on the outer shelf, which recorded from 13 March to 7 October 1978 (Table I), the current was predominantly toward true azimuth 100 ± 10°. It registered in this sector 43% of the time, and the mean speed over the nearly 7 months was about 15 cm s^{-1}. At the same location the next most commonly observed current was in the reciprocal direction, and in the mean it was about 2 cm s^{-1} slower. Note that Fig. 6 by itself provides no information on observational time spans, and must therefore be examined together with Table 1. For example, at site LO-1 the rather round current rose at 78 m represents just over a week of measurements, whereas the elongated elliptical rose at 152 m represents nearly 7 months and is therefore in a probabilistic sense far more representative of the current conditions.

IV. THE INSHORE REGIME

It is useful to consider the Beaufort Undercurrent against the contrasting background of the circulation on the inner shelf. A number of investigators have found evidence for a wind-driven circulation there in the summer (Barnes and Reimnitz, 1974; Barnes *et al.*, 1977; Drake 1977; Hufford and Bowman 1974; Hufford *et al.*, 1974; Hufford *et al.*, 1977; and Wiseman *et al.*, 1974). In general there appears to be a westward water motion driven by the prevailing easterly winds, but the circulation responds rapidly to changing wind conditions, such that under westerly winds (which is the secondary wind mode) the motion is eastward. More recently, Matthews (1981a) and Barnes and Reimnitz (1982) concluded, primarily from drifter data, that significant flow on the inner shelf is primarily a summer phenomenon, when the winds drive near-surface currents at about 3% of the wind speed. Bottom currents were deemed considerably slower.

Little work has been done on the winter circulation. However, during March-April 1976 we obtained two 3-week winter current records from the inner shelf: NAR-1 and NAR-2, from water respectively 27 and 38 m deep (Fig. 6). Both instruments were under fast (or grounded) ice, NAR-2 near the outer edge and NAR-1 14 km farther inshore. It is apparent from Fig. 6 that the flow was relatively slow and that there was little net motion. Examination of the individual records shows that the currents never

FIGURE 6. *Current roses at Beaufort Sea mooring sites. Each vector represents the mean current in a sector of 20°, the vector length being proportional to the speed (see scale). The number at the end of each vector is the frequency of occurrence of a current within that 20° sector. Depth of measurement shown adjacent to each rose. Heavy dots show mooring locations.*

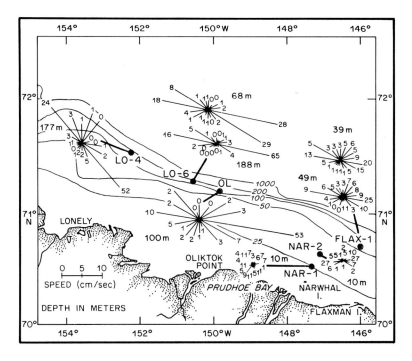

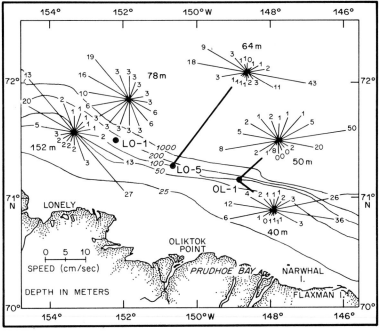

TABLE I. Beaufort Sea Current Records

Mooring	Lat.	Long.	Water depth	Instrument depth	Valid record Begins	Valid record Ends
	(N)	(W)	(m)	(m)	(GMT)	
NAR-1	$70^o32.2'$	$147^o20.0$	27	10	28/3/76	22/4/76
NAR-2	$70^o38.7'$	$147^o09.0'$	38	10	28/3/76	19/4/76
OL	$71^o12.6'$	$149^o53.0$	225	100	27/5/77	1/9/76
LO-1	$71^o31.1'$	$152^o11.3'$	192	78	29/3/77	6/4/77
				152	29/3/77	20/10/77
LO-4	$71^o31.8'$	$152^o15.3$	192	177	13/11/77	30/10/78
LO-5	$71^o17.0'$	$150^o44.1'$	99	64	13/3/78	7/9/78
LO-6	$71^o17.7'$	$150^o37.9'$	203	68	11/3/78	4/9/78
				188	11/3/78	14/9/78
FLAX-1	$70^o43.6'$	$146^o00'$	59	39	22/2/79	22/7/79
				49	22/2/79	22/7/79
OL-1	$71^o10.0'$	$148^o52.7'$	60	40	21/2/79	6/3/80
				50	21/2/79	6/3/80

exceeded 10 cm s^{-1} and were generally less than 5 cm s^{-1}. The mean flow during the 3 weeks was essentially negligible (0.1 cm s^{-1} at NAR-2 and 0.3 cm s^{-1} at NAR-1).

The low-frequency variability at the two sites was quite similar. Moreover, comparison with the simultaneous wind record from nearby Narwhal Island suggests that much of the variability was wind-driven. The progressive vector diagrams (Fig. 7) show that the initial motion was northeastward at both sites; this was near the end of a period of southwest winds of 5-7 m s^{-1}. The current then set southwest from about 1 to 6 April, during which time the wind blew from the northeast at 5-6 m s^{-1} before weakening the last two days. The wanderings of the current vectors during the following week (with a net north or northeast displacement) were under variable winds. From 13 to 15 April the wind was west-southwesterly at 5-8 m s^{-1} and the current followed within one-half day, setting eastward. Finally, from late on 15 April the winds blew from the northeast, with episodes to 8 m s^{-1}, during which time the current vectors were directed westward. Generally these displacements were much smaller (perhaps one-half as great) at NAR-1 than at NAR-2, consistent with greater frictional damping at the latter site, which was in shallower water farther inshore under the fast ice. The NAR-1 record does not significantly lag NAR-2 in phase, as would be expected if the current inshore of the fast ice edge were generated by lateral entrainment. It is therefore probable

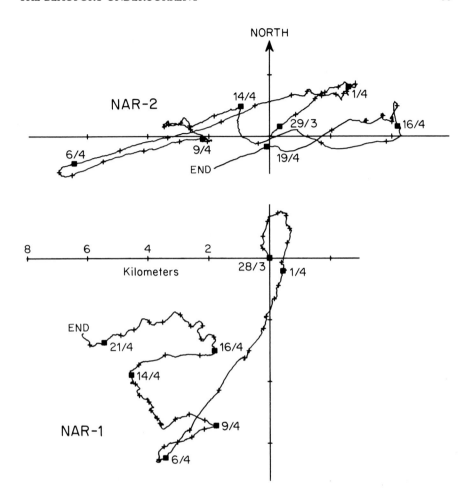

FIGURE 7. *Progressive vector diagrams for the current at 10 m depth during March-April 1976 at moorings NAR-1 (27 m water depth) and NAR-2 (38 m water depth). The moorings were separated by 13.9 km. The time ticks are 12 hr apart.*

that the wind-induced currents were in fact responses to coastal wind setup (or conceivably directly to atmospheric pressure gradients). The pertinent offshore length scale over which the effects of set-up should be observable (the barotropic deformation radius) is 125 km for a water depth of 30 m. (This can be contrasted with the baroclinic radius of 2.5 km, calculated from a STD section done the previous month.)

The clear suggestion of these winter flow measurements on the inner shelf is that even though the kinetic energy levels are suppressed in winter (at least under the fast ice), there is a wind-driven circulation component throughout the year. How far inshore under the fast ice the wind effects persist is not known, but there is evidence for at least a 15-km extension shoreward of the fast or grounded ice edge. Indeed, if the response to the winds is through coastal set-up, wind effects would be expected to extend to the coast (or to the barrier islands).

There is undoubtedly also a thermohaline circulation on the inner shelf, although the present data cannot address this issue. Most notably, we should expect a nearshore buoyancy-driven flow during the summer runoff and a density-driven circulation during winter associated with brine drainage. The latter has been discussed by Matthews (1981b), but the evidence for it is inconclusive (contrast Barnes and Reimnitz, 1982).

Finally there is the question of possible effects of the fast ice edge itself. The role of an ice edge in promoting characteristic horizontal and vertical circulations is a topic of increasing interest (see Røed and O'Brien, 1981 and Niebauer, 1982 for physically different recent examples). However, it is doubtful that the fast ice edge is of much consequence in this regard in the Beaufort Sea. Normally there is essentially no open water seaward of the edge, nor is there strong and prolonged shear at the edge. Instead, the close ice pack seaward of the fast ice moves rather slowly, so that at the edge the ocean probably does not have strong lateral discontinuities at its upper surface.

V. THE BEAUFORT UNDERCURRENT

Seaward of about the 50-m isobath, the flow is substantially different, being characterized by relatively strong flow throughout the year which is locally aligned with the isobaths (Fig. 6). The consistent orientation of the velocity vectors parallel with the isobaths resembles the situation on narrow shelves elsewhere (Hickey, 1979). While the bathymetry portrayed in Fig. 6 suggests the local isobath trend at LO-4 to be about 120°, other more detailed bathymetric constructions for the area indicate a southwestward topographic indentation in this area. The local isobath orientation may therefore well be more southeast-northwest, as the current measurements at both LO-1 and LO-4 suggest. Because of the regional variation in isobath trend, the actual orientation of the prevailing current varies by about 45° over the entire mooring area, but for simplicity we shall consider the octant east to southeast as easterly and west to northwest as westerly. Note further in Fig. 6 both that the statistically prevailing current direction at each site is easterly and that this sector contains the strongest currents (except at FLAX-1). The net motion is therefore everywhere toward the east (see also Table II).

A. Extent and Dynamics of the Mean Flow

Examination of the FLAX-1 records, in 60 m of water, shows a flow regime that is less constrained directionally than at the other offshore moorings and also has a slower mean eastward motion. This suggests that while the Beaufort Undercurrent was still observed at that location, the instruments were located near the in-shore edge of the current. As a rule

TABLE II. Outer Shelf Mean and Maximum Currents

Mooring	Instrument depth (m)	Record length (days)	Mean velocity Speed (cm s^{-1})	Mean velocity Direction	Maximum speed (cm s^{-1})
OL	100	95	12.8	099°	56
LO-1	78	9	Record too short		
	152	205	2.3	154°	61
LO-4	177	352	4.8	142°	66
LO-5	64	208	3.8	100°	66
LO-6	68	172	6.4	112°	60
	188	188	6.9	102°	57
FLAX-1	39	149	1.0	067°	34
	49	149	1.3	073°	33
OL-1	40	380	7.0	089°	73
	50	380	9.2	082°	70

of thumb we can therefore probably assign the Beaufort Undercurrent to lie seaward of about the 50-m isobath.

The northern edge of the current is at present far less clearly defined, but the eastward flow probably extends seaward a considerable distance. The deepest moorings (OL and LO-6) were at depths of 225 and 203 m respectively, with the former mooring recording the fastest mean flow of any site and the latter the third fastest. Furthermore, the current at the upper instrument at LO-6 was about 40% stronger than at LO-5, which was moored inshore at the same time in water 99 m deep. These measurements show no diminution of the current seaward, but rather a strengthening.

The hydrography also points to a considerable seaward extent of the Beaufort Undercurrent. For example, Figs. 1 and 2 suggest eastward flow seaward of the 1000-m isobath. Likewise, although there is only a hint of it in Fig. 5, the vast majority of the STD sections across the shelf and over the slope clearly show a sinking of the isopycnals seaward. A good example is given in Fig. 8. This density distribution corresponds to a geostrophic shear in which the eastward current increases with depth. The effect extends at least to the end of the sections, which was generally in water deeper than 600 m. Such a shear is in fact in agreement with that observed at all three moorings where currents were recorded at two different depths; at LO-6, FLAX-1, and OL-1 the mean current in each case increase downward (Table II). Therefore, not only does the Beaufort

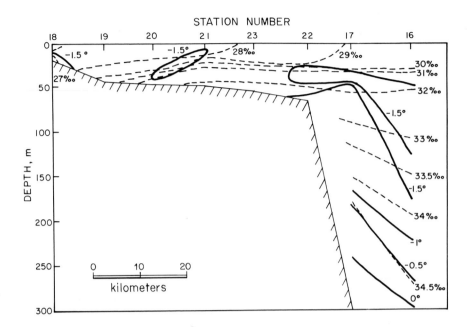

FIGURE 8. Fall temperature and salinity section across the shelf and upper slope near 142°W.

Undercurrent extend seaward a considerable distance and increase downward, but at least over the slope it also transports water masses in addition to those from the Bering and Chukchi Seas. A deep flow of Atlantic water eastward along the slope was in fact suggested some years ago by Newton and Coachman (1974). Their conclusion was based in part on core layer analysis and in part on finding a trough in the deep dynamic topography over about the 2500-m isobath.

The implication of these various lines of evidence is that the Beaufort Undercurrent typically extends from the near-surface to the bottom between the 50- and 2500-m isobaths, a horizontal distance of 60-70 km. If this view proves to be correct, then the eastward flow over the shelf is the inshore manifestation of a major boundary current which is part of the large-scale circulation of the Canadian Basin. The driving force for such a large-scale boundary current is at present unknown, but Sverdrup-type calculations using the curl of the wind stress show only westward flow in the southern Beaufort Sea (Newton, 1973).

If the flow over the shelf is part of the large-scale circulation, then it is not surprising that attempts to understand the dynamics of the shelf flow in more local terms have met at best limited success. For example, the mean winds over the southern Beaufort Sea are easterly, so that local wind driving is not a viable mechanism for the mean eastward flow. Mountain (1974) argued that a mean eastward motion might be driven by the momentum flux of the upstream flow, *i.e.*, the flow through Bering Strait, which continues northward along the Chukchi coast of Alaska; a similar opinion was advanced by Hufford (1975). In this view the outer shelf flow in the Beaufort Sea would be strongly inertial, and the local momentum balance might be maintained by the advective influx of momentum, the retarding effect of bottom friction, and the augmenting or retarding effect of winds, depending on whether they were westerly or easterly. Mountain (1974) showed through a scale analysis that in the absence of favorable (westerly) winds, such an inertial current likely would be restricted by the bottom friction to lie west of about 143-145°W, which represents a propagation of 500 km or less. Since the prevailing winds in fact are easterly and the wind effects are very important to the momentum balance, we should expect an inertially driven flow to diminish significantly toward the east, and probably to disappear in the Canadian Beaufort Sea, even if Mountain's scaling had greatly overestimated the effects of bottom friction.

Neither expectation is in fact borne out observationally. For example, comparison of the mean eastward current strength at mooring sites with records longer than 3 months shows that with the exception of the FLAX-1 measurements at 146°W, which appear to have been very near the inshore edge of the Beaufort Undercurrent and are therefore anomalous in this connection, there is no diminution toward the east of current speed. On the contrary, the flow appears to increase toward the east, although the mean speed variation at any given meridian is nearly as large as the trend, so that such a conclusion is highly tenuous.

There is also very clear evidence for an extension of the mean eastward flow along the Canadian Beaufort Sea shelf. In particular, moored current measurements made on the outer shelf between 129° and 137°W by Huggett et al. (1975) showed mean flow toward the northeast of up to 8 cm s^{-1} over 3-1/2 months during 1975.

The conclusion must therefore be that the Beaufort Undercurrent extends along the entire Beaufort Sea shelf and slope, and that an inertial explanation of the mean eastward flow is contrary to observation. On the other hand, there is evidence (discussed below) of inertial effects contributing to the low-frequency variability of the flow over the western part of the Beaufort shelf.

The discussion so far has been restricted to the subsurface flow. We do not have current measurements on the outer shelf above about 40 m depth, so that we cannot be sure of the motion near the surface. It is possible that the mean surface motion is westward rather than eastward. Not only are the prevailing winds easterly, but the baroclinic structure on the outer shelf is also appropriate. For example, in about 80% of the 17 STD sections we took across the shelf during 1975-77, the dynamic topography over the outer shelf was such as to sustain a geostrophic shear in which the motion at the surface was westward relative to that at 40 m. A mean geostrophic velocity difference between these levels (excluding the 20% of cases with reversed shear) was about 7 cm s^{-1}. Examination of Table II shows that this roughly corresponds to the mean easterly flows calculated from the current records, suggesting little or no mean geostrophic motion at the surface. Note also in Table II that at the shallowest of the paired measurements made in the strong eastward flow, at OL-1, the observed long-term mean shear (2 cm s^{-1} over 10 m) is just about what we calculated above as the mean baroclinic effect (7 cm s^{-1} over 40 m). The geostrophically balanced baroclinicity in the upper layer may therefore by itself be sufficient to offset the deeper mean eastward flow. Furthermore, this baroclinic structure results from an upper-layer salinity that, except during summer, decreases seaward. Such a salinity distribution might result from salinization of the shelf water during freezing or from upwelling of saline water onto the shelf. The interesting possibility thus arises that if there is a mean westward motion in the upper layer, it may be due as much to the presence of dense water on the shelf as to direct wind effects. Appropriate models would in that case have to be both thermohaline and wind-driven.

B. Low-Frequency Variability

While both the mean and the strongest motion are eastward on the outer shelf, at least below the surface layer, Fig. 6 shows the secondary mode to be westerly and to contain the next-fastest flow. Figure 9, showing the daily mean current over a year at LO-4, illustrates the characteristics of the motion and can be taken as representative of the low-frequency longshore flow on the outer shelf. The flow alternates sharply in azimuth between about 140° (the primary mode) and 320° (the secondary mode). The nominally easterly flow (actually southeast) is both the fastest and the most persistent of the two principal flow directions, so that we are seeing a mean easterly motion upon which is superimposed a strong reciprocating low-frequency signal with a typical period of 3-10 days.

Figure 10 shows the spectral characteristics for the first 6 months of the LO-4 record. The reference coordinates for the current have been rotated 50° clockwise, such that the U-component is oriented in the azimuth 140°, parallel with the principal current axis (and presumably also parallel with the isobaths). The energy is seen to be almost entirely contained in the longshore mode (U-component), and then principally at the lowest frequencies. By contrast, the already low energy level in the cross-shelf (V) component drops off even further at frequencies less than about 0.5 cycle per day. Note also in Fig. 10 the absence of obvious tidal peaks. This relatively low tidal energy level is typical for the Beaufort shelf. For example, semidiurnal amplitudes range from less than 1 cm s^{-1} to perhaps 3 cm s^{-1}, depending upon location.

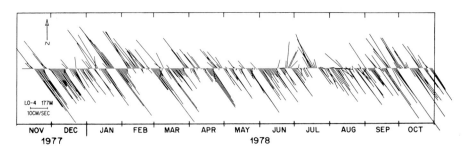

FIGURE 9. *Daily mean current at 177 m depth at mooring LO-4. Water depth is 192 m. The speed scale is indicated by the bar.*

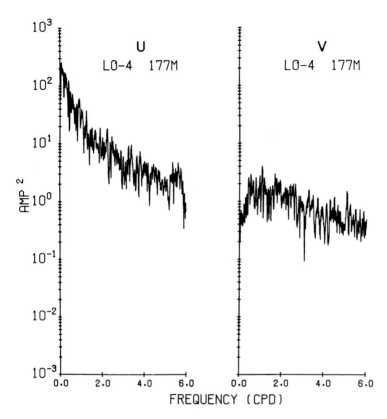

FIGURE 10. Energy-conserving spectra for first 6 months of LO-4 current record.   U-component is longshore (azimuth 140°) and V-component offshore (50°).

A more detailed examination of the frequency band between about 0.1 and 0.4 cycle per day reveals considerable spectral structure, and comparison with the second 6 months of the LO-4 record shows notable differences between the two portions of the record. The low-frequency statistics are thus non-stationary. In effect, this current regime appears to have a rather broad low-frequency forcing and admittance, such as, for example, would be expected for a wind-driven regime.

While a locally wind-driven regime would force current events that are essentially synoptic over a large portion of the Beaufort shelf, there is also evidence that current fluctuations can propagate over such distances at rates comparable to the mean flow speed. These events point toward the importance of inertial effects in contributing to the low-frequency variability over the western part of the Beaufort shelf. From 25 August 1976 to 10 August 1977, a current meter was moored west of Cape Lisburne in the Chukchi Sea at 68°55'N, 167°21'W. Designated NC-7 (Coachman and

Aagaard, 1981), it was located in the core of the northerly flow along the Chukchi coast. It overlaps in time the LO-1 record during the period 29 March–10 August 1977. Comparison of these two records initially suggested no relation. However, if the time scale is shifted about 106 days, so that 14 December 1976 in the NC-7 record coincides with 29 March 1977 in the record from LO-1 (Fig. 11), then there is a remarkable similarity. Specifically, at the beginning of the LO-1 record, there were two long periods of nearly uninterrupted westward flow, 24 and 28 days in duration, separated by 6 days of eastward flow. Each major period of reversed (westward) flow had four or five subordinate oscillations within it. These major reversals are unique in our Beaufort Sea records. Similarly, beginning 11 December 1976 there were two 26-day periods of southward flow at NC-7, again interrupted by 6 days of normal flow. Each major reversal (southward) had four or five embedded subordinate oscillations. There were no other reversals of comparable duration at any time during the year-long record, nor anywhere else at a large number of mooring sites farther west in the Chukchi Sea, in Bering Strait, and in the northern Bering Sea, other than at the mooring immediately west of NC-7 (NC-6), which was also in the core of the mean northward flow along the Chukchi Sea coast.

The NC-7 and LO-1 record portions which contain the reversals in longshore flow are well correlated (0.58 with a lag of 106 days, which is significant at the 95% level), while the remaining portions are uncorrelated. Furthermore, there is nothing in the Barrow wind records to suggest local wind forcing as the cause of the prolonged LO-1 current reversals. Winds during the period 30 March–25 May 1977 were from the easterly sector about 75% of the time but were very light, reaching 5 m s^{-1} on only one day, and were usually in the range 1–3 m s^{-1}. The similarity between the two current records with respect to these unusual events

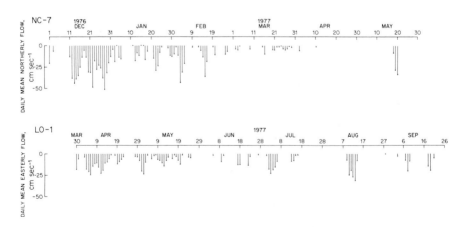

FIGURE 11.   Southerly flow events at NC-7 near Cape Lisburne in the Chukchi Sea during December 1976–May 1977, and westerly flow events at LO-1 during March–September 1977. Where no daily mean flow is shown, the current was in the normal northerly (NC-7) or easterly (LO-1) direction.

therefore suggests the absence of eastward momentum input during a period that was determined by upstream events in the Chukchi Sea. Moorings NC-7 and LO-1 were separated by about 650 km, suggesting a signal propagation rate of 7 cm s^{-1}, roughly comparable to the long-term mean flow in the area. However, a propagation distance of 650 km exceeds that predicted by Mountain's (1974) scale analysis by as much as a factor of 2, even in the absence of opposing easterly winds. Alternatively we might therefore more appropriately consider major upstream flow events to be forced synoptically along the entire Chukchi coast, as Mountain et al. (1976) did. In order to explain the prolonged reversals, this only requires the flow to persist inertially from Point Barrow eastward to the LO-1 site, a distance less than 200 km, which is easily supportable by scaling arguments. We do not know how much farther eastward on the shelf momentum advection from the Chukchi Sea might contribute to the flow, since LO-1 was the only mooring deployed at the time.

The Beaufort Sea current records in general prove to be correlated both vertically and over intermediate horizontal distances. For example, Fig. 12 shows the coherence and phase for the U-component at the two LO-6 instruments. The low-frequency band is seen to be both coherent and nearly in phase vertically over a depth of 120 m. Furthermore, examination of both the daily means and the energy spectra for the two instruments shows that while there are minor differences, the low-frequency energy levels are not substantially different (Table II). The suggestion is therefore that the entire water column between these levels responds nearly as a single unit. Since the largest density stratification occurs above the level of the upper instrument, this is dynamically reasonable.

At intermediate separations on the shelf the mooring records are on the whole fairly well correlated. For example, Fig. 13 shows the daily mean longshore component at the LO-4 and deep LO-6 instruments, separated by 63 km. It is clear that very many of the low-frequency events are both coherent and nearly in phase, and in fact the 35-hr filtered data from the two instruments yield a correlation of 0.64 with a lag of only 6 hours. For the given station separation, this implies a phase propagation of nearly 300 cm s^{-1}. While we cannot definitively preclude these events as representing the propagation of wave-like disturbances (an internal Kelvin wave has a phase velocity of about 65 cm s^{-1} under typical winter shelf conditions), comparison of the current records with the wind field suggests that wind forcing is a principal cause of the low-frequency variability.

C. Wind Forcing

Wind records suitable for comparison with the current measurements are available in the form of calculated geostrophic winds at 72°N, 150°W during 18 February–31 July 1979, a nominal point about 100 km from OL-1 and 200 km from FLAX-1; the calculations were made by T. Kozo. Surface winds and pressures are also available from Barter Island, over 100 km to the east of FLAX-1, during 23 February–4 May 1979 and 30 May–22 July 1979. I shall refer to these latter two periods as parts 1 and 2 of the surface wind records.

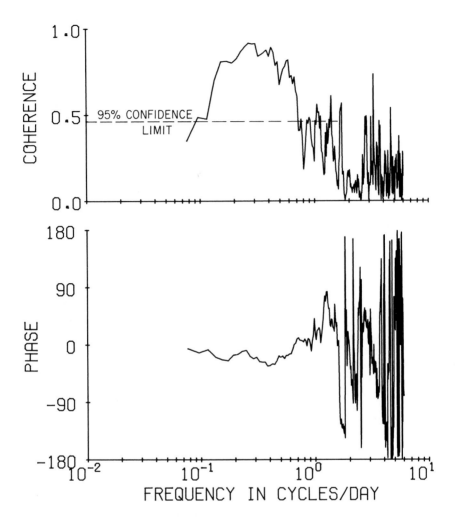

FIGURE 12. Coherence and phase relations for the upper (68 m) and lower (188 m) LO-6 current records, U-component (longshore).

Statistically, the wind and current records are correlated at low frequencies. For example, Fig. 14 shows the coherence and phase spectra between the U-components of current at the upper FLAX-1 instrument and of the surface wind at Barter Island during part 1. Likewise, Fig. 15 shows the coherence between the U-component at the upper OL-1 instrument and the geostrophic wind. The records from the lower instruments at each mooring are very similar to those from the upper instruments. While the correlation with the geostrophic wind is somewhat poorer than the correlation with the surface wind, the overall suggestion of the two figures is of a current reasonably coherent with the wind at time scales longer than about 2-3

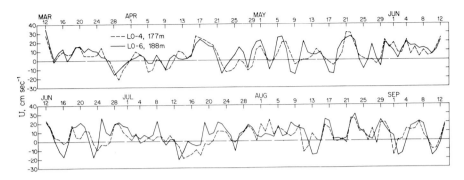

FIGURE 13. *Daily mean U-component of velocity (longshore) at the LO-4 and lower LO-6 instruments. The moorings were separated by 63.4 km.*

days. Figure 14 shows no statistically significant phase difference, while Fig. 15 suggests a changing phase relation, with the current lagging the wind at the lowest frequencies, but leading at frequencies between about 0.2 and 0.4 cycle per day. Except close to the coherence and phase peaks near 0.3 cycle per day, however, the positive phase difference is not statistically significant. (The 95% confidence limit on phase is about ± 39°.) There was no significant positive phase difference between the current and the surface wind during part 2 of the Barter Island wind record.

Except for a significant but slightly smaller correlation between the east component of current and the north component of wind (zero phase difference), no other wind and current combinations prove coherent. The implication of these calculations is therefore that a significant portion of the low-frequency variability in the longshore current is wind-driven, primarily by the longshore wind component, with which it is nearly in phase. The governing mechanism is probably a coastal (or fast-ice edge) Ekman layer divergence, giving rise to a geostrophic longshore flow; *i.e.,* the low-frequency wind driving is local and acts through coastal set-up. Seaward of the baroclinic deformation radius (2-3 km) the response would be barotropic, thus uniform with depth (see Fig. 12). Wind-driven events would essentially be in phase with the wind (the response time scale is the inverse of the Coriolis parameter, which is less than 2 hours) and coherent in the long-shore direction over the coherence scale of the wind itself (see Fig. 13). These expectations are in agreement both with our observations and with the general results of wind-driven coastal circulation models (Allen, 1973; Hamilton and Rattray, 1978).

Examination of individual events in the current records also suggests wind forcing as contributing to the low-frequency variability. For example, comparison of the Barter Island winds with the OL-1 and FLAX-1 current records shows that under strong easterly winds the current normally reverses. This appears to hold in about 75-80% of the instances when the easterly winds exceed 8-10 m s^{-1}. On the other hand, current reversals also occur that are not obviously related to the wind. Figure 16 shows an example of both kinds of events, depicting surface winds and OL-1 and FLAX-1 currents during 22 February-8 March 1979. Initially, winds and

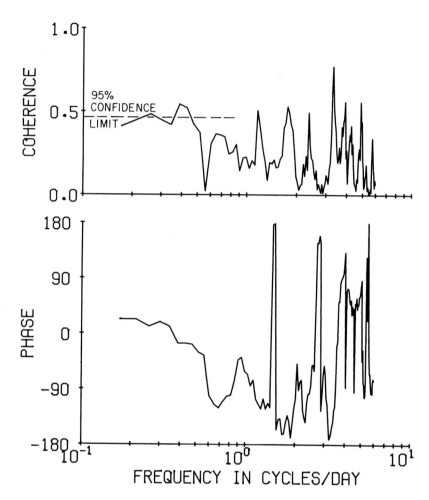

FIGURE 14. *Coherence and phase relations for the upper FLAX-1 (39 m) current record and the Barter Island surface winds, 23 February–4 May 1979, U-component (longshore).*

currents were in the same direction (eastward), but during the period centered around 28 February, the currents were weak and variable (and included some westward motion), while the winds continued moderately strong toward the east. About 4 March the currents entered an increasingly pronounced reversed mode which peaked on 6 March. The winds at Barter Island, meanwhile, had begun blowing toward the west at midnight on 4 March; they reached their maximum strength on March 6, coincident with the current peak.

The conclusion from these considerations is that the longshore wind plays an important, but not always prevailing, role in the low-frequency

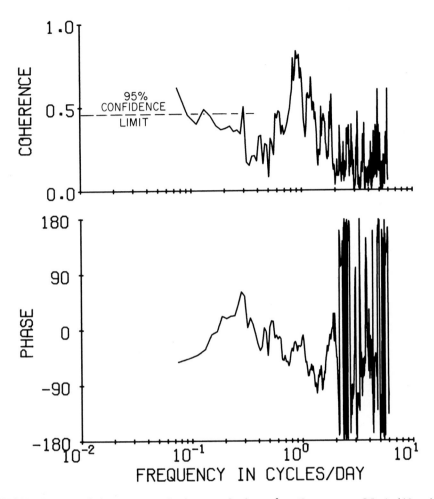

FIGURE 15. Coherence and phase relations for the upper OL-1 (40 m) current record and the geostrophic wind at $72^{O}N$, $150^{O}W$, 18 February-31 July 1979, U-component (longshore).

variability of the subsurface longshore flow. The wind appears to be particularly important in flow reversals. Specifically, when the winds are easterly at 8-10 m s^{-1} or more, the currents over this part of the shelf will normally reverse and set westward. Ascertaining the dynamics of the wind-driven flow with some certainty and detail will require special experiments, the most difficult part of which may well be to obtain good measurements of the offshore wind field. Probably much of the noise in the present wind and current coherence calculations is due to an inadequate definition of the wind stress.

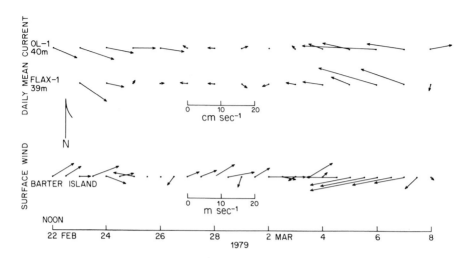

FIGURE 16. Daily mean current at the upper OL-1 and FLAX-1 instruments, together with surface wind at Barter Island, 22 February-8 March 1979.

D. Long-Term Variability

Two moorings, LO-4 and OL-1, were deployed for a year and the records are therefore long enough to provide an indication of variability on at least the seasonal time scale. Examination of the 3-week mean longshore currents at these moorings shows no seasonal signal, however, and in fact both the strongest eastward (OL-1) and westward (LO-4) 3-week mean currents occurred in late January to early February. A distribution-free runs-test indicates that the 3-week means in fact constitute random sequences. There is, however, some tendency toward the lower significance limit on the number of runs, *i.e.*, there is some indication of a persistence of currents weaker or stronger than the mean, on time scales several multiples of the 3-week time base.

The annual mean flow at the lower OL-1 instrument in 1979-80 was nearly twice that at LO-4 in 1977-78. While this suggests that there may be large interannual variations in current strength, the difference in means could conceivably be due largely to difference in location of the moorings. However, OL-1 was a shallow mooring, probably located close to the inner edge of the Beaufort Undercurrent, where the NAR-1, NAR-2, and FLAX-1 records all suggest reduced flow, whereas LO-4 was close to the shelf break. Interannual variability is therefore a very plausible explanation for the difference between the OL-1 and LO-4 records.

Overall, the impression is of a long-term mean easterly flow which, although variable on extremely long time scales, does not exhibit any regular seasonal variations. Rather, there are random fluctuations on a time scale of 1-2 months or more, probably including significant interannual variations.

VI. CONCLUSIONS

We have seen that substantially different circulation regimes prevail on the inner and outer shelf. The demarcation zone between the regimes lies near the 50-m isobath. The inner shelf is strongly wind-driven in summer, but it undergoes large seasonal changes and is far less energetic in winter, although wind effects probably persist even under the fast ice close to shore.

In contrast, the outer shelf circulation is energetic at sub-tidal frequencies throughout the year. Its dominant feature is the Beaufort Undercurrent, a bathymetrically steered mean eastward flow extending seaward from the 40-50-m isobaths at least to the base of the continental slope. The flow apparently increases with depth. This relatively strong (of order 10 cm s^{-1}) deep-reaching boundary current is probably part of the large-scale circulation of the Canadian Basin, and is thus not locally driven. However, the portion of the Beaufort Undercurrent overlying the shelf can be modified by local forcing, most importantly by the wind. In the western part of the shelf, momentum flux from the Chukchi Sea through Barrow Canyon also appears to be dynamically important.

Much of the significance of the energetic outer shelf circulation, and of the Beaufort Undercurrent in particular, is that it provides an efficient dispersal and transport mechanism. This means not only that events on the Alaskan shelf can propagate long distances into the Canadian Beaufort Sea, but it also means that events in other places affect the north Alaskan shelf. For example, water from the Bering Sea has a major impact on the Beaufort shelf hydrography and undoubtedly also on the biota.

Furthermore, near the inshore edge of the Beaufort Undercurrent there are in fact frequent cross-shelf motions which are capable of transporting materials between the inner and outer shelf regimes. For example, Fig. 17 shows the daily mean current at FLAX-1. Flow events directed seaward were more common than landward ones (compare also the current roses in Fig. 6); cross-shelf flow was relatively strong (daily means frequently exceeded 5 cm s^{-1}); and such events typically had a duration of

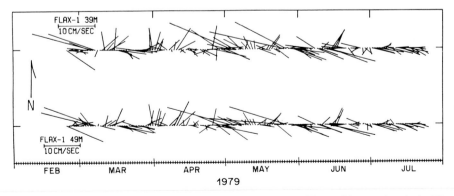

FIGURE 17. Daily mean current at mooring FLAX-1. Water depth is 59 m. The speed scale is indicated by the bar.

3 days or more. (A particle being moved offshore at 5 cm s^{-1} for three days will undergo a total excursion of 13 km.) Offshore flow occurred most frequently during the first 3 months, although there is no obvious reason why this should be so. There appears to be no significant correlation between the cross-shelf flow and either wind component, though this may conceivably be due to an inadequate representation of the local wind field. The important point for present purposes is that cross-shelf flow is frequent and links the Beaufort Undercurrent with the nearshore environment.

ACKNOWLEDGMENTS

I am indebted to C. Darnall, S. Harding, F. Karig, and R. Tripp for their efforts in the field, where we often worked under difficult conditions. J. Schumacher performed a number of the statistical calculations. This work was supported by the Bureau of Land Management through interagency agreement with the National Oceanic and Atmospheric Administration, under which a multi-year program responding to needs of petroleum development of the Alaskan continental shelf is managed by the Outer Continental Shelf Environmental Assessment Program (OCSEAP) office. This is Contribution No. 1293 from the the School of Oceanography, University of Washington, Seattle, WA 98195.

REFERENCES

Aagaard, K., Darnall, C., and Karig, F. (1978). *Deep-Sea Res. 25*, 127.

Allen, J.S. (1973). *J. Phys. Oceanogr. 3*, 245.

Barnes, P., and Reimnitz, E. (1974). *In* "Coast and Shelf of the Beaufort Sea" (J.C. Reed and J.E. Sater, eds.), p. 439. Arctic Institute of North America, Arlington, VA.

Barnes, P., and Reimnitz, E. (1982). U.S. Geol. Surv. *Open-file Rept. 82-717.*

Barnes, P., Reimnitz, E., and McDowell, D. (1977). U.S. Geol. Surv. *Open-file Rept. 77-477.*

Coachman, L.K., and Aagaard, K. (1974). *In* "Marine Geology and Oceanography of the Arctic Seas" (Y. Herman, ed.), p.1. Springer-Verlag, New York.

Coachman, L.K., and Aagaard, K. (1981). *In* "The Eastern Bering Sea Shelf: Oceanography and Resources" (D.W. Hood and J.A. Calder, eds.), Vol. 1, p. 95. Office of Marine Pollution Assessment, distributed by University of Washington Press, Seattle.

Coachman, L.K., Aagaard, K., and Tripp, R.B. (1975). "Bering Strait: The Regional Physical Oceanography." University of Washington Press, Seattle.

Drake, D. (1977). *Open-file Rept. 77-477.* U.S. Geol. Surv.

Hamilton, P., and Rattray, M., Jr. (1978). *J. Phys. Oceanogr. 8*, 437.

Hibler, W.D. III (1979). *J. Phys. Oceanogr. 9*, 815.

Hickey, B.M. (1979). *Prog. Oceanogr. 8*, 191.

Hufford, G.L. (1973). *J. Geophys. Res. 78*, 2702.

Hufford, G.L. (1974). *J. Geophys. Res. 79*, 1305.
Hufford, G.L. (1975). *J. Geophys. Res. 80*, 3465.
Hufford, G.L., and Bowman, R.D. (1974). *Arctic 27*, 69.
Hufford, G.L., Fortier, S.H., Wolfe, D.E., Doster, J.F., and Noble, D.L. (1974). *Oceanogr. Rept. CG-373*, U.S. Coast Guard, Washington, D.C.
Hufford, G.L., Lissauer, I.M., and Welsh, S.P. (1976). *Oceanogr. Rept. CG-D-101-76*, U.S. Coast Guard, Washington, D.C.
Hufford, G.L., Thompson, B.D., and Farmer, L.D. (1977). *Ann. Rept. RU 81 OCSEAP Arctic Project*, unpubl. ms.
Huggett, W.S., Woodward, M.J., Stephenson, F., Hermiston, W., and Douglas, A. (1975). *Tech Rept. No. 16, Beaufort Sea Project*, Dept. of Environment, Victoria, B.C.
Hunkins, K. (1979). *In* "Notes on Polar Oceanography," Tech. Rept. No. 79-84, Vol. 1, WHOI, Woods Hole, MA.
Johnson, M.W. (1956). *Arctic Instit. N. America, Tech. Pap. No. 1*, Washington, D.C.
Matthews, J.B. (1981a). *J. Geophys. Res. 86*, 6653.
Matthews, J.B. (1981b). *Ocean Manage. 6*, 223.
Mountain, D.G. (1974). "Bering Sea Water on the North Alaskan Shelf." Unpubl. Doctoral Dissertation, University of Washington, Seattle.
Mountain, D.G., Coachman, L.K., and Aagaard, K. (1976). *J. phys. Oceanogr. 6*, 461.
Muench, R.D. (1970). "The physical Oceanography of the Northern Baffin Bay Region." Doctoral Dissertion, University of Washington, Seattle.
Newton, J.L. (1973). "The Canada Basin; Mean Circulation and Intermediate Scale Flow Features." Ph.D. thesis, University of Washington, Seattle.
Newton, J.L., and Coachman, L.K. (1974). *Arctic 27*, 297.
Niebauer, H.J. (1982). *Continental Shelf Res. 1*, 49.
Paquette, R.G., and Bourke, R.H. (1974). *J. Mar. Res. 32*, 195.
Røed, L.P., and O'Brien, J.J. (1981). *J. Geophys. Astrophys. Fluid Dyn. 18*, 263.
Searby, H.W., and Hunter, M. (1971). *Tech Memo. NWS-AR-4*, NOAA, Anchorage.
Wadhams, P. and Squire, V.A. (1983). *J. Geophys. Res. 88*, 2270.
Wiseman, W.J., Suhayda, J.N., Hsu, S.A., and Walters, C.D. (1974). *In* "The Coast and Shelf of the Beaufort Sea" (J.C. Reed and J.E. Sater, eds.), p. 49. Arctic Institute of North America, Arlington, VA.

OBSERVATIONS AND ANALYSES OF
SEDIMENT-LADEN SEA ICE

Thomas E. Osterkamp
Joan P. Gosink

Geophysical Institute
University of Alaska
Fairbanks, Alaska

I. INTRODUCTION

Sea ice and seabed sediments interact strongly in the nearshore areas of Alaska's coasts, either by direct contact or through the water column; as a result of this interaction, seabed sediments can be incorporated into the sea ice cover. Matthews (1982) published a photograph of a block of fast ice containing a high concentration of fine-grained sediments in the top 1 m. The processes that incorporate such high sediment concentrations in the sea ice cover are poorly understood, but they may include direct contact (gouging by ice keels, flotation by anchor ice, seabed freezing and subsequent flotation of the ice cover), and interactions involving the water column, mainly entrainment in frazil ice and derived forms of frazil ice under conditions of high turbulence. Sediments affect the light transmittance, structure, mechanical strength and other parameters of sea ice. Because sediment in sea ice can be concentrated by several orders of magnitude over its presence in seawater, motion of the sediment-laden sea ice, as during freeze-up and breakup, may be a significant sediment transport agent in the nearshore areas of the Arctic Ocean (Naidu, 1979 and Barnes *et al.*, 1982). This process could be equally significant in transporting pollutants such as drilling mud, cuttings, oil, or dredge spoils.

There have long been reports of sediments entrained in sea ice in Alaskan nearshore areas; early descriptions include those of Tarr (1897), Kindle (1909), Sverdrup (1935), and Polunin (1949). More recently Sharma (1974) described occurrences in the Bering Sea, and Barnes and Reimnitz (1974), Naidu (1979), and Barnes *et al.* (1982) reported them in the Beaufort Sea. Reimnitz and Dunton (1979) reviewed the results of their diving observations and several related miscellaneous observations of sea ice and sediments by other investigators. A summary of discussions of sediment-laden sea ice at a meeting held in Seattle on February 7–8, 1980, and a related synthesis of the winter studies in the Beaufort Sea were given in Schell (1980).

THE ALASKAN BEAUFORT SEA:
ECOSYSTEMS AND ENVIRONMENTS

73

These sources indicate that sediment-laden sea ice is widely distributed, having been observed in the Bering, Chukchi, and Beaufort Seas. Sediment concentrations in the ice cover appear to be higher in some years, with fall storms during the freeze-up period apparently producing the highest concentrations.

Two distinct types of sediment-laden sea ice have been observed. The first type contains material that is usually coarse, consisting of gravel, kelp, mud, or other material that appears to have been lifted off the sea bed. The second contains fine-grained materials such as clay, silt, sand, and organics that usually appear distributed in a layer near the top of the ice cover, as shown in the cores in Fig. 1. Occurrences of the first type (mostly coarse-grained material) are probably the result of ice gouging, anchor ice flotation, seabed freezing in shallow waters with subsequent flotation, an influx of ice-bearing sediments from rivers (Benson and Osterkamp, 1974), and other processes (see Larsen, 1980). A number of processes that could give rise to fine-grained sediment entrainment have been proposed by Larsen (1980), Naidu (1980), and Osterkamp and Gosink (1980). All of these processes involve frazil ice formation and sufficient turbulence to suspend sediments in the water column during the freeze-up period.

We report here the effects of sediment in ice on light transmission through the sea ice cover, the results of observations and analyses of sediment-laden sea ice cores and ice samples, and theoretical analyses of the scavenging and filtration processes for entraining sediment in sea ice.

II. OBSERVATIONS

A. Reduction of Light Intensity

Sea ice cores were obtained from Harrison Bay during April and May 1971 (Osterkamp, 1972). The reduction of light intensity with depth in the ice cover was measured by suspending selenium photocells at various depths in some of the core holes; current output of the photocells is proportional to the light intensity. Weller and Schwerdtfeger (1967) showed that the errors introduced by this borehole installation are small. If it is assumed that the results follow Bouger's Law for absorption, then the absorption coefficients can be determined from a graph of photocell current versus depth in the ice cover. Figure 2 shows graphs of photocell current versus depth for two different sites. At the site where core F-1 was taken, the ice was clear, and at the site where core F-2 was taken, small amounts of sediment and organic debris were found from the surface to 0.35 m depth, although no measurements of sediment volume were made. Of 27 cores

FIGURE 1. Two sediment-laden sea ice cores taken from the ice cover offshore from Lonely. The cores were separated laterally by 0.5 m. A thin-section analysis showed that the sediment occurs in association with frazil ice and that the bottom of the sediment layer corresponds to the transition from frazil ice to columnar ice. Core 1 is at the bottom, core 2 is at the top.

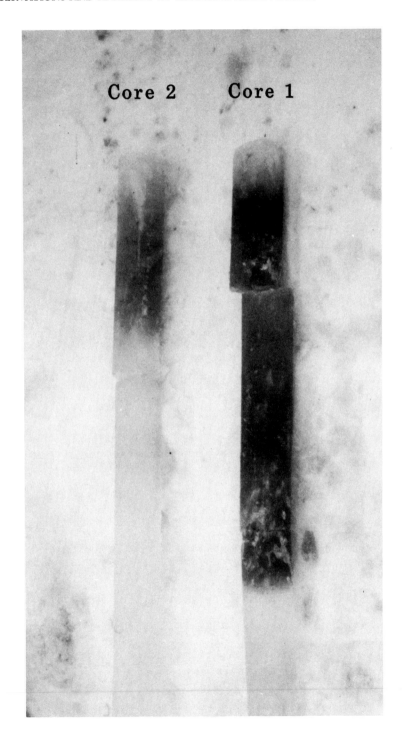

from Harrison Bay, 10 contained visible sediment and organic debris. Figure 2 shows that the reduction of light intensity (absorption coefficient)for sea ice containing small but visible amounts of sediments and organic debris was about 10 times that for clear sea ice. Thus relatively small but visible amounts of sediments entrained in a sea ice cover can drastically reduce the light intensity beneath it, a fact also noted by divers (Reimnitz and Dunton, 1979).

B. *Dive Site #11*

Dive site #11 (DS-11) was established north of Foggy Island Bay for the purpose of conducting interdisciplinary winter studies during the 1978-1979 and 1979-1980 winters (Schell, 1980). Dives made at DS-11 after freeze-up in 1978 showed that the ice cover had a slightly rough surface with solid ice about 0.5 m thick and "slushy" ice extending down to nearly 3 m from the ice surface (Reimnitz and Dunton, 1979). The "slush" ice consisted of plate-like ice particles that were loosely bonded together. When this ice was disturbed by the divers it contributed fine suspended particles to the water column with the largest particles having noticeable settling velocities. Observations made when the diving hole was cut also showed that the ice contained a considerable amount of sediment.

A visual examination of "slush" ice from DS-11 showed that this ice was very porous and consisted of millimeter- to centimeter-sized platelets. It was concluded that this "slush" ice was frazil ice.

Four ice cores obtained from the sea ice cover near DS-11 were analyzed for sediment concentrations with the results shown in Table I. The sediments consisted of clay and silt-sized particles with occasional fine sand grains and significant amounts of plant debris. It has been suggested (S. Naidu, personal communication) that the plant debris originated from peat sources.

Horizontal and vertical thin sections from all four cores showed that the ice in these cores was primarily frazil ice. Grain sizes were in the millimeter range. A few crystals in all four cores approached 10 mm in length. The grains were coarser near the center of core A and near the top of core B, where sediment concentrations were lowest; however, in cores C and D there was no apparent correlation of crystal size with sediment concentrations.

The sediments in the thin sections and in other pieces of ice from DS-11 were examined under a low-power binocular microscope. Figure 3 is a photomicrograph of sediment in a thin section. This sediment was generally found at grain boundaries, intersections of three or more grains, and in association with air bubbles. In a few cases, the sediment appeared to be inside an ice crystal, but this could not be verified since the three-dimensional structure of the ice crystals was not studied.

The sediment was flocculated with floc sizes usually a few tenths of a millimeter long, although a few were nearly 1 mm long. This flocculation appears to be similar to that observed when wastewater sludge is frozen unidirectionally (Osterkamp, 1974).

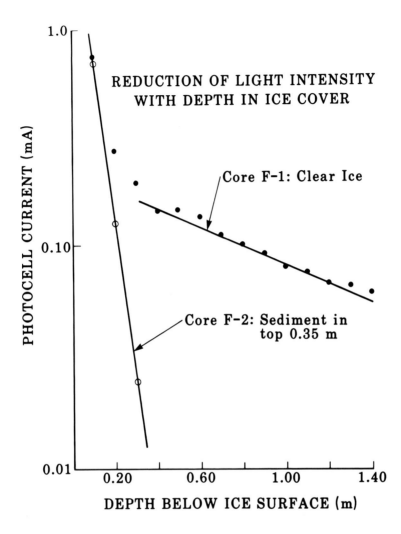

FIGURE 2. Reduction of light intensity (photocell current) with depth in the sea ice cover at two sites. Core F-1 was clear ice and core F-2 contained small amounts of sediment and organic debris in the upper 0.35 m, which substantially reduced the amount of light penetrating the ice cover.

C. Lonely

The two cores shown in Fig. 1 were obtained from the fast ice offshore from Lonely and were separated by 0.5 m. Core 1 had about 7 cm of relatively clear ice at the top underlain by ice with a high concentration of sediment down to 0.64 m depth and relatively clear ice below this level. Core 2 had about 8 cm of relatively clear ice underlain by sediment-laden ice down to 0.27 m depth and relatively clear ice below this level. Core 1(1.42 m in length) and the top 0.82 m of core 2 were returned to the laboratory for sediment and thin-section analyses. The sediment concentration in core 1 decreased sharply at 0.64 m depth (Table I), which coincided with the boundary between frazil ice and columnar ice. In core 2 the sediment concentration decreased sharply at 0.27 m depth, again coinciding with the transition from frazil to columnar ice. Wang (1979) also found that sediment-laden sea ice was composed of small granular crystals and that the boundary between sediment-laden ice and the clear ice below it was marked by a transition to columnar crystals (e.g. Weeks and Hamilton, 1962). These observations show that the sediment concentrations occur in the frazil ice and that the underlying columnar ice is relatively free of sediment. They also show that the bottom of the frazil ice is very uneven as reported by Reimnitz and Dunton (1979) and Barnes et al. (1982).

D. Other Sites

Ice cores, ice samples, and observations of sediment-laden sea ice in nearshore areas have been obtained in Norton Sound (Bering Sea), in the Chukchi Sea near Barrow, and in the Beaufort Sea near Prudhoe Bay. These data will not be reported here, but they suggest that the occurrence and distribution of sediment-laden sea ice is widespread. A core from Norton Sound had the highest sediment concentration observed in these investigations, 3.1×10^4 mg L^{-1}.

III. INTERPRETATION

The above field observations and laboratory studies of sediment-laden sea ice can be used to develop criteria for the processes responsible for sediment entrainment in the sea ice cover. Table I and the sediment concentration studies of Naidu (1979) and Barnes et al. (1982) show that the local sediment concentrations in sea ice can be several orders of magnitude greater than sediment concentrations in seawater during summer. Therefore the sediment entrainment process must account for these high concentrations. Barnes et al. (1982) suggested that the most likely mechanism for resuspending nearshore bottom sediments is storms during the freeze-up period. Sea ice containing sediments has a relatively clear layer about 10 cm thick at the surface (see also Barnes et al., 1982), as shown in Fig. 1. The two cores shown in Fig. 1 were very close laterally and so demonstrate the local variability of the entrained sediments and of the bottom topography of the frazil ice. While sediment concentrations in the ice are locally variable, their wide-ranging occurrences imply that any sediment entrainment mechanism must be of general applicability.

TABLE I. Sediment Concentrations in Sea Ice at Dive Site 11 and Lonely

DS 11			Lonely	
Core depth (mm)	Concentration (mg L^{-1})		Core depth (mm)	Concentration (mg L^{-1})
Core A			Core A	
0-70	209		0-70	250
80-150	123		70-140	1290
170-240	92		140-210	710
260-330	98		210-280	322
350-420	221		280-350	353
430-500	110		350-420	528
			420-490	606
			490-560	591
Core B			560-635	725
10-70	37		635-710	128
90-150	66		710-780	60
170-240	44		780-850	52
260-330	43		850-900	51
340-400	281		900-975	68
410-490	127		975-1025	57
510-590	192		1025-1095	82
			1095-1175	71
			1175-1245	61
Core C			1245-1315	59
20-80	260		1315-1385	47
100-160	352		1385-1420	46
190-240	52			
260-320	42		Core B	
340-400	41		0-70	158
420-470	57		70-140	526
Core D			140-210	437
20-80	142		210-285	455
100-170	732		285-355	158
190-250	596		355-450	113
280-350	87		450-520	64
370-420	53		520-590	68
430-500	96		590-633	85
			633-703	63
			703-733	73
			773-818	43

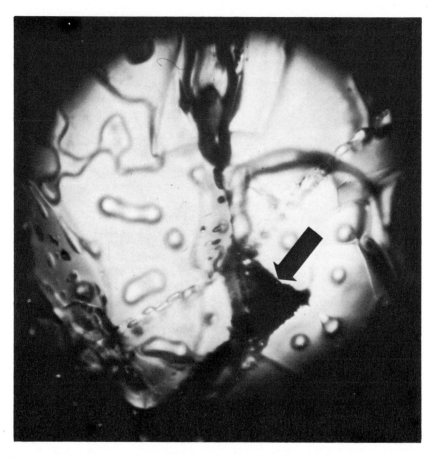

FIGURE 3. A photomicrograph of a sediment inclusion (arrow) in a thin section made from a sea ice core. The width of the field of view is about 1.8 mm. Flocculated sediment inclusions were generally found at grain boundaries, intersections of three or more grains, and in association with air bubbles.

A. Sediment Entrainment Processes

It is generally agreed that frazil ice formation under conditions of high turbulence during freeze–up is somehow involved in entraining fine–grained sediment, although there is no agreement on the detailed nature of the process or processes. Potential processes are described by Benson and Osterkamp (1974), Larsen (1980b), Naidu (1980), Osterkamp and Gosink (1980), and Barnes et al. (1982), and are summarized in Schell (1980). They include or involve the following:

1. Entrainment in anchor ice with subsequent flotation and incorporation into the overlying ice cover.
2. Entrainment by discharge of sediment-laden anchor ice from rivers.
3. Entrainment by ice gouging of the sea bed.
4. Entrainment by seabed freezing and subsequent flotation.
5. Entrainment during the initial ice cover formation from frazil ice or derived forms of frazil ice (e.g. grease ice, shuga, pans, floes, and brash ice) under conditions of high turbulence.
6. Entrainment and deposition under the ice cover by frazil ice formed in leads or in a broken ice cover.
7. Entrainment by "katabatic" flow of dense cold brine, formed nearshore under the ice, and draining downslope (offshore) where suspension of sediments and frazil ice formation could occur.
8. Entrainment by anchor ice formation of very short duration with subsequent flotation.
9. Entrainment by trapping in frazil ice when seawater is forced through it; that is, filtering.

The first four processes are generally agreed to be responsible for entrainment of coarse-grained material in the ice. They may occasionally cause entrainment of fine-grained sediments; however, they would not account for all the observations of fine-grained sediments as shown in Fig. 1. A more detailed discussion of the other processes is given below.

Process 5. Formation of a sea ice cover under turbulent conditions normally involves frazil ice nucleation, frazil growth, and evolution of the derived forms of frazil ice into an ice cover. A host of frazil ice nuclei are available under the normal environmental conditions found at the time of freeze-up (Osterkamp, 1977). Once formed, frazil ice crystals grow in the slightly supercooled water and, depending on the degree of turbulence, evolve into a variety of forms. The sequence is roughly: frazil slush or grease ice, shuga, frazil plumes, pancakes, pans, floes, and brash ice, proceeding from conditions of low turbulence to high turbulence. The resulting ice cover develops from these derived forms of frazil ice, usually when the turbulence decreases as after a storm or windy period. We have proposed that sediment is entrained or scavenged from the seawater when sediment particles adhere to individual frazil ice crystals or to derived forms of frazil ice and when seawater containing sediments is trapped in the congealing frazil ice cover (Osterkamp and Gosink, 1980). The observed clear, but somewhat rough ice at the surface would be the portions of the frazil ice floating above the water surface and the undulating bottom surface of the ice would be the underwater extensions of the frazil. Sediment particles would be flocculated by surface effects and by rejection from the growing ice crystals in a manner somewhat similar to that described by Osterkamp (1974) as the frazil crystals grow and as the derived forms of frazil ice solidify by downward freezing. This process appears to be capable of explaining most of the observations of sediment-laden sea ice. If the collection efficiency is great enough, it could increase the sediment concentration in the ice by several orders of magnitude over that in the seawater.

A somewhat similar scavenging mechanism was proposed by Ackley (1982) to account for the high concentrations of algae observed in sea ice of the Weddell Sea. In principle, the analyses given below should be applicable to the problems of algae concentration in a sea ice cover.

Process 6. When open water or leads occur in a recently formed ice cover, or in a broken ice cover during and after a storm or high winds, frazil ice may be produced in large quantities by the high heat transfer rates at the open water surface. When frazil crystals form, turbulence, density-driven salt plumes, and local currents may carry them down into the water column and subsequently deposit them on the underside of the nearby ice cover, producing a sort of underhanging frazil ice formation somewhat similar to those occurring in rivers. We have proposed (Osterkamp and Gosink, 1980) that sediment entrainment would occur by scavenging, interstitial trapping, and filtering. Thin sections of a sediment-laden ice core obtained from Norton Sound suggest that the sediment may have been entrained in the ice by process 6. This process probably could not produce the observed widespread distribution of sediment-laden sea ice but it may be important in some situations.

Modeling efforts (Gosink and Osterkamp, 1983) have focused on determining how deep frazil ice crystals could be carried under a lead and the shape of the underhanging frazil deposit. This work has shown that the frazil ice crystals should be deposited in a skewed normal distribution downstream from the lead, assuming an under-ice current is present. The depth of penetration by frazil crystals has not been theoretically determined, but Dayton *et al.* (1969) observed penetrations to 33 m under the Antarctic ice sheet.

Process 7. "Katabatic" flow of dense cold brine, formed nearshore and draining downslope (offshore), could cause suspension of sediments and formation of frazil ice. Salinity plumes associated with surface heat loss and ice growth may be expected to form in near-shore areas and to drain downslope (offshore). Conditions for this "katabatic" drainage would be most favorable in polynyas (*e.g.* in Norton Sound) or in open water along the coast during freeze-up. Initial estimates (Gosink and Osterkamp, 1983), assuming a heat loss of 100 Cal $m^{-2}s^{-1}$, indicate that downslope salinity-plume speeds may reach 0.1 m s^{-1} at a distance of 3-1/2 km offshore. Frazil crystals in the plume would grow until buoyant forces cause them to surface or be deposited under the ice cover. Sediment entrainment would be by scavenging, interstitial trapping, and filtering. This process provides the heat sink at depth necessary for frazil growth, and sufficient bed friction to suspend recently deposited fine-grained sediments. Large variations in sediment-laden sea ice from year to year, the apparent connection with storms during freeze-up, and observations of sediment-laden ice at distance of 20-30 km offshore are difficult to explain with this model.

Process 8. Frazil formation under conditions of high turbulence may cause frazil crystals to contact the sea bed for a short time. Larsen (1980) proposed that a transient form of anchor ice may develop with frazil crystals adhering to the seabed sediments. As the ice grows and its buoyancy increases, the frazil mass would lift off the sea bed carrying the sediments with it, to become part of the floating frazil ice which congeals and entrains the sediments. This process is nearly identical to process 1 above, except that this anchor ice formation is envisioned to be of much shorter duration. It is based on observations of anchor ice formation on a sandy portion of the river bed made by Arden and Wigle (1972) in the Niagara River. Their observations and others (Osterkamp, 1975) suggest that only a small portion of the frazil ice crystals formed in rivers evolve into floating anchor ice carrying entrained sediments. The sediment-laden sea ice cores, however, have sediments entrained through the whole thickness of frazil ice, except near the surface. It is difficult to envision how most of the frazil ice would evolve into the transient anchor ice. Also, published photographs of anchor ice containing river-bed sediments (Arden and Wigle, 1972) show that the sediment distribution in the ice is very uneven, much more so than in sediment-laden sea ice. Nevertheless, process 8 cannot be totally discounted without additional experimental observations.

Process 9. Formations of frazil ice consist of porous masses of ice crystals bonded together in varying degrees: sometimes it is easily possible to force an arm through it and at other times it may require some effort to force a meter stick into it. Naidu (1980) proposed that seawater containing suspended sediment may flow through the porous formations of frazil ice before, during, and after freeze-up and that sediments adhere to the ice crystal surfaces.   Over time, this process would allow very high concentrations of sediment to accumulate in the ice.   This filtration process is analyzed in greater detail below.

The above observations and the results of Barnes *et al.* (1982) suggest that process 5, proposed by Osterkamp and Gosink (1980), when combined with the filtering process 9 proposed by Naidu (1980), is the most likely mechanism for sediment entrainment in the sea ice cover and that process 6 may be an occasional alternate method.   Additional theoretical calculations relating to these processes are presented below.

B. Scavenging of Sediments by Frazil Ice Crystals

Scavenging of sediment particles by frazil ice crystals is a many-body problem. However, when sediment and frazil crystal concentrations are sufficiently small the scavenging may be analyzed on the basis of collisions between single frazil ice crystals and single sediment particles (Pruppacher and Klett, 1980). These concentrations may be considered sufficiently small when the distance between particles is large enough for two-body interactions to dominate. Typical frazil crystals about 1 mm in diameter and 10 μm in thickness have a volume of about 10^{-11} m^3. In a well-mixed water column, we expect concentrations of 10^4 to 10^6 frazil crystals m^{-3} (Osterkamp and Gosink, 1982). Hence the volume of frazil crystals in a cubic meter of water is about 10^{-6} m^3 and the ratio of frazil crystal diameter to distance between frazil crystals about $(10^{-6})^{0.33}$, or 0.01. Similarly, when a nearshore sediment concentration of 1 mg L^{-1} is assumed for sediment particles with a specific gravity of 2.6, the specific sediment volume (m^3 of sediment per m^3 of water) is 3.8×10^{-7}, and the ratio of sediment diameter to distance between sediment particles is about $(3.8 \times 10^{-7})^{0.33}$, or 0.007. Therefore, when frazil ice crystals are well mixed in the water column, an analysis based on two-body interactions should be adequate. After the frazil crystal concentration becomes high, sintering of the crystals (Martin and Kauffmann, 1981) produces complex agglomerations of frazil crystals or flocs, which are permeable to the surrounding water. The flocs accumulate near the surface, raise the local concentration of frazil crystals and reduce the distance between frazil crystals. However, as long as the frazil ice is well mixed in the water column, the criteria for applying the two-body model are satisfied.

This model follows the models of two-body collisions as described by Pruppacher and Klett (1980) and Twomey (1977). These models define a collection efficiency E based on the hydrodynamic flow pattern that describes the grazing trajectory of a small sphere around another larger sphere. Since neither ice crystals nor sediment particles are spherical, the model is not completely appropriate, but it should yield reasonable first-order estimates of E. Clearly the hydrodynamics of a permeable frazil floc will differ appreciably from that of a single crystal and significantly affect E. The implications on E of the flow pattern around and through a permeable frazil floc are discussed later.

C. Collection Efficiency

The collection efficiency E is the percentage effective cross-section for collision of two approaching spheres. More precisely, E is defined as the ratio of the actual collision cross section πy_c^2 and the geometric cross section $\pi (a_1 + a_2)^2$, where a_1 and a_2 are the radii of the larger and smaller spheres respectively and y_c is the maximum offset from the initial approach path of the two spheres such that a grazing trajectory will result. Thus

$$E = y_c^2 (a_1 + a_2)^{-2} \qquad (1)$$

Experimentally and theoretically, E is generally found to be less than 1 due to the pattern of flow lines around a sphere. The approaching smaller sphere will be deflected outward on a stream line offset from the larger sphere, and thus, only spheres approaching from within a small radius from their centerline will actually collide. E can be determined by experimental and numerical methods when a simple velocity field is assumed for the particles. The simplest velocity fields are determined by gravity, and the subsequent collision process is referred to as gravitational coagulation. According to Friedlander (1957) the collision efficiency is given by

$$E = (a_1/a_2)^2 / 2(1 + a_2/a_1)^2 \qquad (2)$$

and according to Twomey (1977) a good estimate of collision efficiency for gravitational coagulation is given by the Stokes number Stk:

$$E \simeq Stk = 2\,\rho_s v_0 (a_2)^2 / 9\,\mu\,a_1 \tag{3}$$

where ρ_s is the density of the sediment particle, v_0 is the terminal relative velocity between particles, and μ is the viscosity of the fluid. However, the Stokes number approximation is useful only for Reynolds numbers below about 10, and a previous study of frazil crystal rise velocity (Gosink and Osterkamp, 1982) determined appropriate Reynolds numbers as high as 70. The collision efficiencies using the two formulae, as given by Friedlander (1957) and Twomey (1977), have been computed in Table II for two values of sediment radii ($a_2 = 1\ \mu m$ and 10 μm) and constant relative velocity ($v_0 = 0.01$ m s^{-1}). Collection efficiencies for gravitational coagulation as measured by Wang and Pruppacher (1977) and Kerker and Hampl (1974) are also listed in Table II.

Turbulent velocity fields have been simulated in experimental studies by Woods and Mason (1964), and Jonas and Goldsmith (1972), and in theoretical studies by de Almeida (1975, 1976). The importance of wake capture and other inertial effects have been studied by Klett and Davis (1973), Lin and Lee (1975), and Schlamp et al. (1976). Most of these studies have involved small Reynolds numbers, corresponding to collector particles of radii less than 100 μm, in comparison to frazil crystal radii of about 500 μm. However, there is a clear trend of enhanced collision efficiency in these investigations, often as high as several orders of magnitude over that measured with simple gravitational coagulation. A few of the calculated and experimental collision efficiencies determined by Schlamp et al. (1976), Klett and Davis (1973), Kinzer and Cobb (1958) and Beard and Pruppacher (1971) are listed in Table II for comparison.

For turbulent velocity fields, de Almeida's (1975, 1976) numerical calculations show that E is increased due to the increased probability of interaction between particle and collector. However, his studies suggest that there may be a particular turbulence level for maximum E, beyond which E decreases. This is due in part to the decreasing scale of turbulent eddies associated with high–Reynolds–number turbulence, and the subsequent limit on trajectory of the individual particle. In particular, he found that E increases as turbulent dissipation ε increases from 0 to 10^{-4} m^{2}s^{-3}, and then decreases as ε increases to 10^{-3} m^{2}s^{-3}. The turbulent dissipation rate associated with a severe storm can be estimated if it is assumed that production of turbulent eddies is balanced by dissipation:

$$\varepsilon = \overline{-u'v'}\ \partial U/\partial y = u_*^2\,\Delta U/\Delta y$$

where $\overline{-u'v'} = u_*^2$ is the friction velocity stress due to wind drag, ΔU is the current speed, and Δy for the nearshore environment is the water depth. Then, assuming a wind speed of 20 m s^{-1} and a current speed of about 0.5 m s^{-1}, with a water depth of 6.0 m,

$$\varepsilon \simeq 7.5 \times 10^{-5}\ m^2 s^{-3}$$

In general, in the nearshore region $\Delta U/\Delta y$ is always less than about 0.1 s^{-1}, and u_* is less than 0.07 m s^{-1}; therefore, $\varepsilon < 5 \times 10^{-4}$ m^{2}s^{-1}. Thus it should be expected that the value of E for sediment particles scavenged by frazil

TABLE II. Determinations of Collision Efficiency E

Coagulation mechanism	Investigator	Determination	$a_1(\mu m)$	$a_2(\mu m)$	E
Gravity	Friedlander	Theoretical	500	10	$2 \cdot 10^{-4}$
		Theoretical	500	1	$2 \cdot 10^{-6}$
	Twomey	Theoretical	500	10	$6 \cdot 10^{-4}$
		Theoretical	500	1	$6 \cdot 10^{-6}$
	Wang & Pruppacher	Measured	300	0.25	$2 \cdot 10^{-3}$
		Measured	500	0.25	$9 \cdot 10^{-3}$
	Kerker & Hampl	Measured	1000	0.36	$8 \cdot 10^{-3}$
		Measured	254	0.4	10^{-4}
Turbulence or inertia	Schlamp et al.	Theoretical	305	10	0.9
		Theoretical	305	2	0.07
	Klett & Davis	Theoretical	70	10	0.7
		Theoretical	70	3	0.1
	Kruger & Cobb	Measured	250	8	0.8
	Beard & Pruppacher	Measured	250	5.5	0.6

ice crystals is always enhanced by turbulent processes in the nearshore region.

D. Sediment Scavenging Model

An individual frazil crystal moving through sediment-laden water will scavenge K sediment particles, where

$$K = C \times E \times V \tag{4}$$

C is the number of sediment particles suspended in the water per unit volume, and V is the volume traversed by the frazil crystal. C can be found from estimates of oceanic suspended sediment load in nearshore regions. Buss and Rudolfo (1972), Drake et al. (1972), and Drake (1977), suggested oceanic sediment concentrations of about 1 mg L^{-1}. Higher concentrations that may occur during storms would enhance the proposed mechanisms. If a specific gravity of 2.6 is assumed for this sediment, then the volume fraction of sediment in the water column is 3.8×10^{-7}. The number of sediment particles depends upon the approximate size of the particle. For example, if the sediment radius is 10 μm, then the individual particle volume is about 4.2×10^3 μm^3, and $C \approx 9 \times 10^{-11}$ μm^{-3}. If the sediment radius is 1 μm, then $C \approx 9 \times 10^{-8}$ μm^{-3}.

The volume V traversed by the frazil crystal is assumed to be a cylinder of length L with a diameter equal to the frazil crystal diameter. In quiescent water, frazil crystals rise perpendicular to their plane (Gosink and Osterkamp, 1982b), and a regular cylinder is a good representation of the traversed volume, whereas in a turbulent fluid the frazil crystal will have a random orientation to its trajectory. L is thus the effective trajectory distance of a frazil crystal until it becomes entrained into the floating surface ice or becomes part of a frazil floc or other derived form of frazil ice. It could also be considered the effective collection distance of the frazil crystal, and be written as $L = v\tau$ where v is the relative velocity of the crystal and τ is a scavenging time scale. If a typical frazil diameter is 1 mm, then $V = 7.85 \times 10^5 L$ (μm^3) where L is in μm.

Therefore, the number of sediment particles K scavenged by a single frazil crystal is

$$K_{10} = 0.7 \times 10^{-4} EL$$

for sediment particles with radius 10 μm and

$$K_1 = 0.7 \times 10^{-1} EL$$

for sediment particles with radius 1 μm.

This may be compared with the number of sediment particles N found trapped in an ice core per unit frazil crystal. N is equal to the total number of sediment particles in an ice sample, say T, divided by the number of frazil crystals in the ice sample M, or $N = T/M$. The total number of sediment particles in the ice sample is equal to the sediment volume fraction divided by the volume of a single sediment particle V_s. High sediment concentrations in "dirty" ice cores are near 10^3 mg L^{-1}. For a sediment specific gravity of 2.6, this suggests a sediment volume fraction of 3.8×10^{-4}, and $T = 3.8 \times 10^{-4}/V_s$ particles μm^{-3}. The number of frazil crystals in the ice sample M is given by

$$M = (1-e)/V_b \qquad (5)$$

where e is the porosity of the ice and V_b is the volume of an individual frazil crystal. Assuming that $e = 0.5$, the frazil diameter is 1 mm and the thickness is 10 μm, then $V = 7.8 \times 10^6$ μm^3, $M = 6.4 \times 10^{-8}$ frazil crystals μm^{-3}, and N is given by

$$N = T/M = 5.9 \times 10^3/V_s$$

For a sediment particle with a radius of 10 μm, $V_{s10} = 4.2 \times 10^3$ μm^3, and $N_{10} = 1.4$. For a sediment particle with radius 1 μm, $N_1 = 1408$.

Setting the number of sediment particles scavenged by the individual frazil crystal, K, equal to N, the number found in dirty ice per frazil crystal, gives

$$K_{10} = N_{10} \qquad \text{and} \quad K_1 = N_1$$

or $0.7 \times 10^{-4} EL = 1.4$ \qquad or $0.7 \times 10^{-1} EL = 1408$

$EL = 2 \times 10^4$ $\mu m = 0.02$ m \qquad $EL = 2 \times 10^4$ $\mu m = 0.02$ m

When collection efficiencies typical for gravitational coagulation are assumed, the collection distance is found to be about 100 m for 10-μ m particles and 10 km for 1-μ m particles. If collection efficiencies typical of turbulent coagulation are assumed, the collection distance is only 0.1 m. Conversely, if collection distances about the depth of the water column are assumed (L = 6.0 m), then the collection efficiency must be as high as 0.003 in order for individual frazil crystals to capture the number of sediment particles generally found in dirty ice.

In summary, the scavenging model indicates that sea ice sediment concentrations several orders of magnitude greater than that found in the water may be accounted for by the process of scavenging by individual frazil crystals. If the process is very efficient, as in a turbulent flow, the necessary collection distance for the frazil crystal is as low as 0.1 m. If the efficiency is an intermediate value of about 0.003, which is sufficient to account for the observed sediment concentrations in the ice, then the collection distance required is the same as the water depth. Scavenging by individual frazil crystals may dominate until sintering of frazil crystals and the resultant buoyant forces produce surface accumulations of other frazil forms and stratification of the well-mixed flow into a two-layer frazil and water flow.

E. Filtration Process

When frazil crystals have increased in size and number to the point where the turbulence cannot maintain a well-mixed flow, the frazil ice crystals tend to collect near the surface. If surface heat loss persists, the frazil crystals form into flocs and other forms of frazil, possibly by the sintering process described by Martin and Kauffman (1981). When this occurs, the two-body scavenging model is no longer appropriate. One process by which sediment particles could continue to be trapped by frazil flocs may be called a filtration mechanism (Naidu, 1980) whereby the flow of water through the frazil filters out the sediment. This mechanism is discussed with particular emphasis on the determination of the flow rate or discharge through the frazil flocs.

The porosity of the frazil flocs and pans suggests that the flow of water through them may be modeled by application of Darcy's Law for flow through a porous media, with the individual frazil crystals in the frazil flocs representing the grains in the porous media. Darcy's Law is

$$q = K \, d\phi/ds \tag{6}$$

$$\text{and } K = k\rho g/ \mu \tag{7}$$

where q is the flow rate, K is the hydraulic conductivity, $d\phi/ds$ is the hydraulic gradient or pressure head (to be defined more precisely below), ρ is the fluid density, μ is the fluid viscosity, g is the gravitational constant, and k is the permeability of the matrix. The constants ρ, g, and μ are determined from standard handbooks, and the determination of q, then, depends upon the estimates of k and $d\phi/ds$.

Permeability has been found to be a function of the shape and size of the grains in the matrix, and various formulae exist for the definition of k (see Bear, 1972). For lack of better information regarding the permeability of frazil pans, the simple formula given by Bear (1972) is adopted:

$$k = 0.617 \text{ x } 10^{-3} d^2 \qquad (\mu m^2) \qquad (8)$$

where d is the grain diameter in micrometers ($d = 10^3$ μm for frazil crystals). This implies that $k = 617.0$ μm^2.

The hydraulic gradient $d\phi/ds$ is the driving force for the flow, effectively the difference in pressure and potential energies across the frazil pan. In the following, two interpretations of the hydraulic gradient, corresponding to horizontal and vertical flow, are assumed.

For horizontal flow, a frazil pan blocks the water motion, giving rise to a pressure differential across the pan. Bernoulli's equation gives the change in pressure head across the pan:

$$d\phi/ds \simeq \Delta P/\rho gH = \Delta U^2/2gH \qquad (9)$$

If the current velocity U is about 0.10 m s^{-1}, and H, the pan width in the direction of flow, is about 1.0 m, then

$$d\phi/ds \simeq 5 \text{ x } 10^{-4}$$

and $q_h = 1.6$ μm s^{-1} is the horizontal flow rate through the pan.

Substantially higher vertical flow rates are found when a wave splashes over floating frazil ice. For example, a 0.1-m overflow on 1.0-m-thick frazil implies $d\phi/ds \simeq 0.1$ and a vertical flow rate, $q_v = 0.3$ x 10^{-3} m s^{-1}.

Using these horizontal and vertical flow rates through the frazil ice, the effectiveness of the filtration process can be estimated by determining the time required to filter out the observed sediment concentration. The flux of sediment delivered to a frazil pan with cross section area A is given by $C_w qA$, where C_w is the suspended sediment concentration in the water. If this flux continues for a time t, depositing sediment in the frazil matrix with efficiency E', then the sediment deposited in the frazil pan at the end of time t is equal to $C_w E'qAt$. Equating this sediment to that found in the ice samples gives the mass balance. The pan volume equals AH, so the sediment found in the pan is $C_i AH$, where C_i is the measured sediment concentration in the ice, and the mass balance is

$$C_w E'qt = C_i H \qquad (10)$$

Assuming that $C_i \simeq 10^3$ mg L^{-1} and $C_w \simeq 1$ mg L^{-1} implies

$$E'qt = 10^3 H \qquad (11)$$

An estimate of the minimum time required to filter out the observed sediment concentrations in the dirty ice may be found by considering cases involving maximum efficiency and maximum flow rate through ice.

Therefore choosing a vertical flow through the ice pan, and a filtering efficiency of 1.0, the minimum time becomes

$$t = 1000 \ H/q_v \simeq 3 \times 10^6 \ s$$

when a frazil thickness H of 1 m is chosen. This is a long deposition time scale (about 36 days); however, recalling that $q \propto H^{-1}$, which implies $t \propto H^2$, if smaller frazil ice thicknesses are assumed, say 0.1 m, then

$$t \simeq 3 \times 10^4 \ s$$

This deposition time scale is considerably closer to the period of a storm event capable of raising waves sufficient to overtop the frazil (about 8 hours). This suggests that the filtration mechanism is substantially more effective in the presence of small frazil flocs or pans with diameters much less than 1 meter. After the frazil pans reach a meter in size, the porous flow rate through the frazil matrix becomes too small to admit further sediment transport. Coincidentally, the suspended sediment in the interstitial water in the matrix is effectively trapped and prohibited from settling out when the turbulent intensity in the water lessens.

In summary, it appears that the filtration process is capable of concentrating sediment in the ice matrix in the observed concentrations. The process is most effective when frazil ice crystals begin to cluster into small flocs less than a meter across. As the flocs gather into larger pans, the interstitial water and sediment load becomes trapped within the matrix. The efficiency of the filtration process is unknown, as is the permeability of the frazil ice matrix. A systematic program of laboratory and field studies is required to quantify these factors. The present model may serve as a framework for further studies of the filtration process.

SUMMARY

Sea ice near the Alaskan coast has been found to contain concentrations of fine-grained sediment up to several orders of magnitude higher than the concentrations normally found in sea water. Measurements show that relatively small but visible sediment concentrations can drastically reduce the light intensity incident on the water column under the sea ice cover. Studies of thin sections and ice samples from sea ice cores show that the sediments are clay and silt-sized particles with occasional fine sand and organic debris. The sediment incorporated into the sea ice was flocculated with floc sizes on the order of a few tenths of a millimeter in maximum dimensions. Sediment flocs were found at grain boundaries, intersections of three or more grains, in association with air bubbles, and occasionally with in ice crystals. High sediiment concentrations were found to be restricted to the frazil ice in the sea ice cover, suggesting that the entrainment processes involve turbulence and frazil ice production during the freeze-up period. An evaluation of potential sediment entrainment processes suggests that those involving turbulence, frazil ice formation, sediment scavenging by individual frazil crystals and filtration by derived forms of frazil ice (e.g. grease ice, shuga, flocs, pans, floes) are the most viable.

A tentative sediment scavenging model for sediment entrainment in sea ice including the above components is proposed which is a combination and elaboration of the models of Naidu (1980) and Osterkamp and Gosink (1980). The proposed model assumes sufficiently windy conditions during freeze-up to produce the turbulence required for resuspension of fine-grained sediments and frazil ice formation and their entrainment in the water column.

A tentative filtration model for sediment entrainment in sea ice shows that the filtration process is also capable of concentrating sediment in the sea ice in the observed concentrations. This process will be most effective when the frazil ice has evolved into flocs, shuga or small pans with dimensions up to a few tenths of a meter. As the flocs evolve into larger forms of frazil ice the model suggests that the interstitial water and sediment load becomes trapped within the matrix.

When the derived forms of frazil ice congeal into an ice cover, it is proposed that sediment particles are flocculated by surface effects and by rejection of sediment particles from the growing frazil ice crystals, which forces them into flocs in the interstices between crystals and into association with air bubbles.

ACKNOWLEDGMENTS

We thank K. Dunton for obtaining ice samples for us at DS-11 and A. S. Naidu for measuring sediment concentrations in several sea ice cores. V. Gruol, J. Hanscom, and C. Stephens helped in the preparation of sea ice thin sections.   This research was partially supported by the Outer Continental Shelf Environmental Assessment Program.

REFERENCES

Ackley, S. F. (1982). *EOS 63, No. 3.*
Arden, R. S. and Wigle, T. E. (1972). *In* "The Role of Snow and Ice in Hydrology," Vol. 2,   Proc. of Banff Symposia, UNESCO-WMO-IAHS, Banff, Alberta.
Barnes, P. W., Reimnitz, E. and Fox, D. (1982). *J. sed. Pet.* 52, No. 2, 493.
Barnes, P. W. and Reimnitz, E. (1974). *In* "The Coast and Shelf of the Beaufort Sea" (J. C. Reed and J. E. Sater, eds.), p. 439. Arctic Institute of North America, Arlington.
Bear, J. (1972). "Dynamics of fluids in porous media." Elsevier, New York.
Beard, K. V. and Pruppacher, H. R. (1971). *J. atmos. Sci.,* 28, 1455.
Benson, C. S. and Osterkamp, T. E. (1974). *In* "Oceanography of the Bering Sea" (D. W. Hood and E. J. Kelley, eds.). Institute of Marine Science, University of Alaska, Fairbanks.

Buss, B. A. and Rudolfo, K. S. (1972). In "Shelf Sediment Transport: Process and Pattern" (D. B. Duane and O. H. Pilkey, eds.). Dowden, Hutchinson, and Ross, Stroudsburg, PA.

Dayton, P. K., Robilliard, G. A. and Devries, A. L. (1969). Science 163, 273.

de Almeida, F. C. (1975). "On Effects of Turbulent Fluid Motion in the Collisional Growth of Aerosol Particles." Research Rept. 75-2, Dept. Meteor., University of Wisconsin, Madison.

de Almeida, F. C. (1976). J. atmos. Sci. 33, 1571.

Drake, D. W. (1977). U.S. Geol. Surv. Open-File Report 77-477.

Drake, D. E., Kolpack, R. L., and Fischer, P. J. (1972). In "Shelf Sediment Transport: Process and Pattern" (D. P. Swift, D. B. Duane and O. H. Pilkey, eds.). Dowden, Hutchinson, and Ross, Stroudsburg, PA.

Friedländer, S. K. (1957). Am. Inst. chem. Eng. J. 3, 43.

Gosink, J. P. and Osterkamp, T. E. (1982). "Measurements and Analyses of Velocity Profiles and Frazil Ice Crystal Rise Velocities during Periods of Frazil Ice Formation in Rivers." Presented at International Glaciological Society, Second Symposium on Applied Glaciology, August 1982.

Gosink, J.P. and Osterkamp, T.E. (1983). "An Evaluation of Potential Mechanisms for the Incorporation of Sediment in Sea Ice." Geophysical Institute, University of Alaska, Fairbanks. in press.

Jonas, P. R. and Goldsmith, P. (1972). J. fluid Mech. 52, 593.

Kerker, M. and Hampl, B. (1974). J. atmos. Sci. 31, 1368.

Kindle, E. M. (1909). Am. J. Sci. 29, 175.

Kinzer, G. D. and Cobb, W. E. (1958). J. Meteor. 15, 138.

Klett, J. D. and Davis, M. H. (1973). J. atmos. Sci. 30, 107.

Larsen, L. (1980). In "Beaufort Sea Winter Watch" (D. M. Schell, ed.). OCSEAP Arctic Project Office, University of Alaska, Fairbanks.

Lin, C. L. and Lee, S. C. (1975). J. atmos. Sci. 32, 1412.

Martin, S. and Kauffman, P. (1981). J. Glaciol. 27, No. 96, 283.

Matthews, J. B. (1983). "Seasonal Circulation in Some Alaskan Arctic Lagoons." Oceanologica Acta (in press).

Naidu, S. (1979). Annual report, Outer Continental Shelf Environmental Assessment Program, Juneau.

Naidu, A. S. (1980). In "Beaufort Sea Winter Watch" (D. M. Schell, ed.). OCSEAP Arctic Project Office, University of Alaska, Fairbanks.

Osterkamp, T. E. (1972). Rept. No. R72-3, 49-56. Institute of Marine Science, University of Alaska, Fairbanks.

Osterkamp, T. E. (1974). J. Glaciol. 13, No. 67, 155.

Osterkamp, T. E. (1975). "Tanana River Ice Cover." Proc. Third Int. Symp. on Ice Problems, IAHR, Hanover, NH, 1975.

Osterkamp, T. E. (1977). J. Glaciol. 19, No. 81, 619.

Osterkamp, T. E. and Gosink, J. P. (1980). In "Annual Report to BLM/NOAA OCSEAP" (T. E. Osterkamp and W. D. Harrison, eds.). Boulder.

Osterkamp, T. E. and Gosink, J. P. (1982). "Selected Aspects of Frazil Ice Formation and Ice Cover Development in Turbulent Streams." Presented at Workshop on Hydraulics of Ice-covered Rivers, Edmonton, Alberta, June 1-2, 1982.

Polunin, N. (1949). "Arctic Unfolding." Hutchinson, New York. 348 pp.

Pruppacher, H. R. and Klett, J. D. (1980). "Microphysics of Clouds and Precipitation." D. Reidel, Dordrecht, 714 pp.

Reimnitz, E. and Dunton, K. (1979). In "Annual Report to BLM/NOAA OCSEAP" (P. W. Barnes and E. Reimnitz, eds.). April, 1979.

Schell, D. M., ed. (1980). "Beaufort Sea Winter Watch Ecological Processes in the Nearshore Environment and Sediment Laden Sea Ice: Concepts, Problems and Approaches." Special Bulletin #29 OCSEAP Arctic Project Office, University of Alaska, Fairbanks. 73 pp.

Schlamp, R. J., Grover, S. N., Pruppacher, H. R. and Mamielec, A. E. (1976). J. atmos. Sci. 33, 1747.

Sharma, G. D. (1974). In "Oceanography of the Bering Sea" (D. W. Hood and E. J. Kelley, eds.), p. 517. Occ. Publ. 2, Inst. of Marine Science, University of Alaska, Faribanks.

Sverdrup, H. U. (1935). Am. J. Sci. 35, 370.

Tarr, R. S. (1897). Am. J. Sci. 3, No. 15, 223.

Twomey, S. (1977). "Atmospheric Aerosols." Elsevier, Amsterdam. 302 p.

Wang, Y. S. (1979). "Crystallographic Studies and Strength Tests of Field Ice in the Alaskan Beaufort Sea." Proc. Conf. Port and Ocean Eng. Under Arctic Conditions, Norwegian Institute of Technology, Trondheim.

Wang, P. K. and Pruppacher, H. R. (1977). J. atmos. Sci. 34, 1664.

Weeks, W. F. and Hamilton, W. L. (1962). Am. Mineral. 47, 945.

Weller, G. and Schwerdtfeger, P. (1967). "Radiation Penetration in Antarctic Plateau and Sea Ice." Polar Meteorology, Tech. Note No. 87, WMO, No. 211, TP 111, 120 p.

Woods, J. D. and Mason, B. J. (1964). Quart. J. roy. meteor. Soc. 90, 373.

BEAUFORT SEA ICE MOTIONS

Robert S. Pritchard

Research and Technology Division
Flow Industries, Inc.
Kent, Washington

I. INTRODUCTION

Motions of the Beaufort Sea ice cover are of general scientific interest and of engineering importance because of ongoing petroleum development along the North Slope of Alaska. This paper addresses the question of where an oil spill in ice would be transported. The overall analysis of oil spill behavior is presented by D. R. Thomas in this volume. He determined that because the oil would be trapped in the bottomside relief and frozen into the ice cover, ice motion would be the prime cause of oil motion. Although most of the oil would be released at spring break-up to move relative to the ice, the ice motion would still be the dominant motion component. The general information needed by regulatory agencies and oil company engineers includes some statement of the probability that oiled ice will be transported to a specific location. Since the lifetime of North Slope oil fields can be expected to be 30 years, or more, estimates of the range of behavior must include interannual variability.

During the past decade, intensive scientific effort has been aimed at developing a mathematical modeling capability for simulating the behavior of the sea ice cover of the Beaufort, Chukchi, and Bering Seas. One such effort, the Arctic Ice Dynamics Joint Experiment (AIDJEX), was a coordinated program involving many investigators. A summary of research during AIDJEX is given in the numerous papers collected and published at the conclusion of that program (Pritchard, 1980). The Outer Continental Shelf Environmental Assessment Program (OCSEAP) has also carried on a model development program, focusing its attention on areas nearer to shore where petroleum development is occurring. During both of these programs, data on ice motions were obtained to gain better knowledge of actual ice motions and to validate the performance of the models being developed.

The model that has evolved from these efforts is the most sophisticated, complete model developed to date for simulating the behavior of arctic sea ice. I presented the mathematical formulation of this "complete" ice dynamics model earlier (Pritchard, 1981) and also outlined the historical development of the model. In the model, quadratic drag laws describe the tractions applied to the top and bottom of the ice cover by air and water (winds and currents). The gravitational force acting on the ice due to differences in the sea surface level is also considered. In addition, the internal ice stress can exert an important force on the ice. The divergence of the internal ice stress can cause large localized events or deformations whose extent is dependent on existing ice conditions, such as ice strength and concentration. The strength of the ice, or its resistance to deformation, depends inversely on the amount of open water and thin ice present at each location. Thus, in the model, ice conditions are described by the ice thickness distribution. An elastic-plastic constitutive law has been developed to describe the relation between stress and deformation (Pritchard, 1975). Hibler (1980a, b) developed a viscous-plastic constitutive law for the same purpose. The formulations of yield surface, flow rule, and ice strength (and changes in strength as hardening or softening occurs) have evolved over the last eight years through a series of theoretical analyses, conceptual experiments, and ice dynamics simulations. An excellent review is presented by Rothrock (1979).

The complete ice dynamics model is designed to describe average values over length scales of tens of kilometers. Over such distances, leads, ridges, and other features can be ignored to allow approximation as an isotropic, homogeneous medium. When the average daily velocity or daily displacement is determined, the inertia of both the ice and the water column is negligible. Justification for these assumptions can be found in Rothrock (1979), Pritchard (1981), or the many references cited in those works. Pritchard and Kollé (1981) and Kollé and Pritchard (1983) showed that the complete model simulates average daily ice velocities to within about 3.5 cm s^{-1} in the Beaufort Sea. This gives errors in daily displacement of about 3 km day^{-1}. Spatial resolution in simulations of the Beaufort Sea published to date has been on the order of 40 km. D. R. Thomas (personal communication) has considered resolution as small as 5 km in simulations of nearshore behavior.

A free-drift model has also been used to simulate pack ice motions. This model is a simplified version of the complete ice dynamics model in which internal ice stress is neglected. Numerous authors have taken the free-drift approach. Zubov (1943) offered his now-famous rule that ice moves at 2% of the surface wind speed in the direction parallel to the sea-level pressure isobars. To first-order accuracy, this rule serves well. Sater et al. (1974) presented a study using Zubov's rule and historical wind data to estimate the historical range of Beaufort Sea ice motions. More recently, McPhee (1980) and Thomas and Pritchard (1979a) presented results of a similar study but also considered effects of nonlinearities on the motion. McPhee focused efforts on validating model performance during summer. Thomas and Pritchard considered the entire year and estimated errors each month during the AIDJEX main experiment in 1975-76. Thorndike and Colony (1982) made a similar comparison between simulated free-drift ice motion and data collected during 1979 and 1980 as part of the First Global Atmospheric Research Program (GARP).

In this paper, the free-drift model is used to simulate ice motions in the Beaufort Sea, and these simulated motions are compared with observed motions. The ice motions described here are representative only of pack-ice motions seaward of the stamukhi zone. Over the inner continental shelf, ice stress becomes a more important factor. Due to this limitation, the ice motions described here provide estimates of oil trajectories only after the spill has been incorporated into the moving pack ice.

This report has two primary objectives:

(1) To demonstrate that a free-drift model in which ice stress is neglected is adequate for describing ice motion.

(2) To use the free-drift model and historical wind and current data to estimate the historical range of ice motions.

For the first objective, simulations are made using the free-drift model with ice motion data obtained during AIDJEX. Quantitative comparisons are then made between the simulated and observed motions to verify the accuracy of the model. For the second objective, historical wind data from a 25-year period are used with long-term geostrophic currents to estimate the range of behavior of ice motions in the Beaufort Sea. Average monthly motions determined using the free-drift model are presented.

II. FREE-DRIFT MODEL

In the free-drift model, the motion of the ice cover at a point is calculated from the local balance of air stress, water stress, Coriolis force, and sea surface tilt. Inertial forces are ignored for periods of a day or longer. The force balance is

$$\tau_a - \tau_w - mf\, k \times v - mg\, \nabla H = 0 \tag{1}$$

where

m	is the mass per unit area of ice,
v	is the ice velocity in the horizontal plane,
τ_a	is the air stress, or the traction exerted by the atmosphere on the upper surface of the ice,
τ_w	is the water stress, or the traction exerted by the ocean on the lower surface of the ice,
f	is the Coriolis parameter,
k	is the unit vector in the vertical direction,
g	is acceleration due to gravity, and
H	is the height of the sea surface.

The mass per unit area m is found from the ice density and the mean thickness of the ice, which is assumed constant. The mean thickness value used for each monthly simulation is taken from Thorndike et al. (1975).

The air stress is determined from the geostrophic wind, U_g, in the atmosphere as

$$\tau_a = \rho_a c_a \,|U_g|\, B_a U_g \tag{2}$$

Monthly average values of air density ρ_a are used, which range from 1.29 kg m^{-3} in summer to 1.47 kg m^{-3} in winter (Leavitt et al., 1978). The air drag coefficient c_a is found from the 10-m drag coefficient c_{10} (Leavitt et al., 1978), since the ratio of the wind speed at 10 m to the geostrophic wind is nearly a constant (Albright, 1980). The rotation operator B_a describes the counterclockwise turning angle between the geostrophic wind and the surface stress. The turning angle used is 24° (Albright, 1980).

The water stress is computed from

$$\tau_w = \rho_w c_w \ |\,G\,|\,B\,G \tag{3}$$

where

$$G = v - v_g \tag{4}$$

is the difference between the ice velocity and the geostrophic ocean current v_g; G also represents the relative drift velocity caused by winds. A drag coefficient c_w of 0.0055 and an Ekman turning angle β in the rotation operator B of 23° (counterclockwise) are used (McPhee, 1980).

The sea surface tilt is related to the ocean geostrophic current as follows:

$$mg\,\nabla H = -mf\,k \ \times \ v_g. \tag{5}$$

The sea surface tilt and Coriolis force terms may thus be combined and the force balance equation rewritten as

$$\tau_a - \tau_w(G) - mf\,k \ \times \ G = 0 \tag{6}$$

This equation can then be solved using the local geostrophic wind to determine the relative drift velocity G.

Using the free-drift ice motion equation (6), relative ice drift values are calculated for a range of geostrophic wind velocities. In Fig. 1, relationships are shown between (1) the nondimensional geostrophic wind speed $N_a^{1/2}\,|\,U_g\,|\,/V_R$ and the turning angle δ and (2) the nondimensional geostrophic wind speed and the relative ice speed, $|\,G\,|\,/V_R$. Dimensionless variables have been introduced to allow all results to be presented with only the angle β remaining as an independent parameter.

The reference speed used to scale the dimensionless plots is

$$V_R = mf\,/\rho_w c_w.$$

For the parameters used in this study, this reference speed is about 3 cm s^{-1}. Therefore, a dimensionless relative ice speed of 5 corresponds to an actual relative ice speed of 15 cm s^{-1}. The wind is further scaled by the square root of the ratio of drag coefficients,

$$N_a = \rho_a c_a/\rho_w c_w.$$

FIGURE 1. Free-drift ice response to winds. The nondimensional relative ice speed and the geostrophic turning angle, in degrees clockwise, are plotted against the nondimensional geostrophic wind for several values of the oceanic turning angle β.

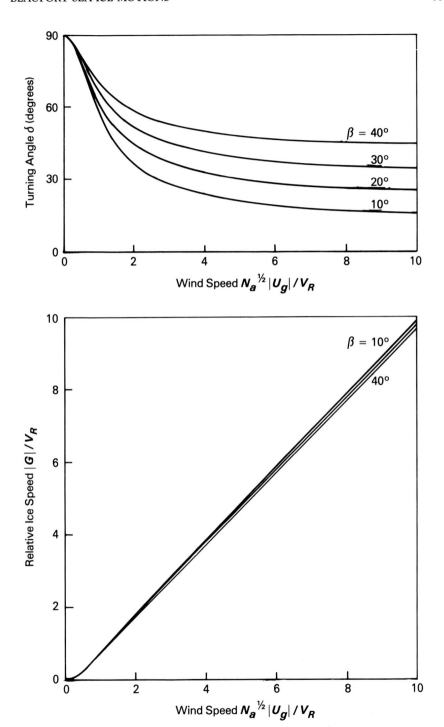

Again, for typical parameter values, $N_a^{1/2}$ = 0.014, thus the relative ice drift speed is about 1.4% of the geostrophic wind speed. This result may be expressed in terms of the surface wind speed if it is assumed that the ratio of the geostrophic to the surface wind speed is a constant. Following Albright (1980), assume

$$| U_g | / | U_{10} | \simeq 1.6.$$

Then

$$N_a^{1/2} | U_g | \simeq N_a^{1/2} (| U_g | / | U_{10} |) | U_{10} | = 0.22 | U_{10} |,$$

and Zubov's 2% percent rule is recovered to within 10%.

The relationships between $N_a^{1/2} | U_g | (V_R)^{-1}$, $| G | (V_R)^{-1}$, and δ given in Fig. 1 display free-drift ice behavior for all conditions. The relative ice speed is nearly linear for all but the weakest winds (which cause little ice motion). The turning angle varies greatly with wind speed, however, making the vector relationship between U_g and G nonlinear. The angle β has little effect on the speed but does change the turning angle δ substantially at all speeds.

A final observation about the free-drift ice-motion equations is that it is possible to separate ice motions caused by the geostrophic ocean current, v_g, from those caused by winds, G, by using Eq. (4). The cumulative effects of winds and currents can then be considered when estimating the mean and variance of free-drift ice motions. Over a 1-month period, the average daily ice motion $\bar{v}$ is determined from the average geostrophic current $\bar{v}_g$ and the average relative drift caused by winds $\bar{G}$. Thus

$$\bar{v} = \bar{v}_g + \bar{G}.$$

The variance in daily motions, σ_v^2, is related to the variance in geostrophic current, $\sigma^2_{v_g}$, and the variance in relative drift, σ_G^2, by

$$\sigma_v^2 = \sigma^2_{v_g} + 2 \text{ cov } (v_g, G) + \sigma_G^2 \tag{8}$$

where cov (v_g, G) is the covariance between v_g and G.

If geostrophic currents are assumed to be nearly steady over the month so that their variance $\sigma^2_{v_g}$ is small compared to the relative drift variance σ_G^2, then $\sigma_v^2 \simeq \sigma_G^2$. If geostrophic currents are variable, then the contribution to variance in ice motion depends on the correlation between v_g and G. If this correlation is small, then

$$\sigma_v^2 \simeq \sigma^2_{v_g} + \sigma_G^2 \tag{9}$$

and the ocean current variance is added directly to the relative drift variance. Other assumptions may be made, but Eq. (9) provides a reasonable estimate for the Beaufort Sea. The geostrophic ocean currents are expected to change much more slowly than the winds.

III. MODEL PERFORMANCE

Using the free-drift model, relative drift velocities in response to winds are calculated to determine daily ice motions, which are then accumulated to provide monthly motions. The performance of the model is evaluated in terms of mean velocity errors and the standard deviation of daily errors during each monthly period. These two error measures are useful, in part, because of the decomposition into mean and variance given in Eqs. (7) and (8).

Thomas and Pritchard (1979a,b) studied the mean and variance of motions in the free-drift model to isolate the processes that cause each. Winds vary substantially from day to day as weather systems of 3 to 5 days in duration pass over an area. On the other hand, currents are steady in the Beaufort Sea gyre over these periods. Although currents are slow, under 5 cm s^{-1}, they contribute to monthly trajectories because of their persistence. In the free-drift model, the effects of these two driving forces have been separated because of the large uncertainty in ocean current estimates. Errors may also be separated, with mean velocity errors attributed primarily to errors in currents or mean winds and variances of errors attributed primarily to variances in winds and to other factors, such as neglecting internal ice stress. Since ocean currents contribute as much to ice motion as winds and have large uncertainty and unknown variability, the treatment used here is not ideal. However, errors in ocean currents are probably no larger than the year-to-year variability in the wind-driven ice-motion components.

Thorndike and Colony (1982) reported that more than 70% of the variance of daily ice motions is caused by geostrophic winds and that the long-term average ocean current contributes roughly half of the long-term motion of the ice pack. Residuals in this comparison show a standard deviation of 4.0 cm s^{-1}, a number that compares favorably with the results of Thomas and Pritchard (1979a,b) for summer motions but is lower than winter values. Since Thorndike and Colony (1982) optimized all parameters in the free-drift model to determine the best fit, their residuals should be smaller.

Thomas and Pritchard (1979a,b) used the free-drift model to calculate 7 months of daily ice displacements (June through December 1975). In addition, free-drift simulations have been made for parts of April and May 1975 and parts of January and February 1976 (Pritchard and Kollé, 1981; Kollé and Pritchard, 1983). The data used in these studies were obtained by manned camps and drifting buoys located in the Beaufort Sea during the AIDJEX program. The data have been broken into monthly subsets; the mean monthly errors in displacement ϵ and the standard deviations of these errors σ_ϵ are shown in Fig. 2. The data are shown as a calendar year of values, though they actually cover the period from April 1975 through February 1976. The values are shown as being constant during the month, and the dashed lines that connect the data points suggest seasonal trends. As can be seen on the Figure, monthly mean errors are small. The standard error of the mean is estimated by dividing the standard deviation of the error for each month by the square root of the number of days in the set, $\sigma_\epsilon(N)^{-1/2}$.

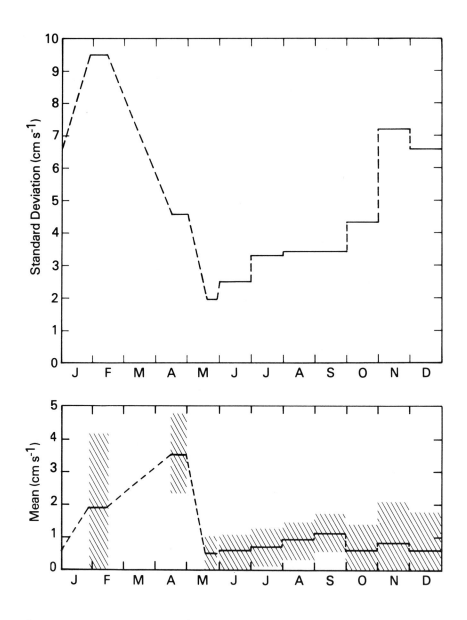

FIGURE 2. Mean errors and standard deviations in errors between simulated and observed ice motions. The shaded areas show the standard error of the mean, $(N)^{-1/2}$. The data cover the period from April 1975 through February 1976, though they are presented as a calendar year of values.

For the January-February results, N = 17; for April, N = 16; for May, N = 8; and for the June-December period, N is the number of days in each month. During almost the entire year, the mean errors are not significantly different from zero.

From June through September, average daily displacement errors are less than 1.0 cm s^{-1} and the standard deviation is less than 3.5 cm s^{-1}. During this time, the ice strength is low enough that it may be neglected. After September, ice growth and rapid freezing of open water cause the ice strength to increase. Effects of internal ice stress can then become important, even to the point of reducing ice motions to zero in strong winds (Pritchard, 1977). Errors in daily displacement gradually increase to a mean value of 2.0 cm s^{-1} and a standard deviation of nearly 10 cm s^{-1} for the January-February simulations. During April and May, simulation errors are smaller. This seems too early for ice decay and strength reduction. The seemingly smaller errors could be unreliable because of the small data set.

These comparisons were made for locations roughly 400 km north of land. We do not expect the free-drift model to perform as well near shore, where internal ice stress divergence is amplified. Although few comparisons are available for nearshore locations, Pritchard and Kollé (1981) and Kollé and Pritchard (1983) compared free-drift velocities with nearshore drifting buoy velocities during the period 27 January to 12 February 1976, for which a simulation was also performed using the complete ice-dynamics model. They found that the standard deviation of errors rose to over 10 cm s^{-1}, whereas it was roughly 8.0 cm s^{-1} in the pack. These errors are two to three times as large as those of the complete ice dynamics model. This is not a very precise simulation capability, but the difference is not dramatic. Therefore, it appears that the free-drift model can be used with this accuracy anywhere in the Beaufort Sea.

IV. ICE MOTIONS

Using the free-drift model and wind data for the years 1953 through 1977 (Jenne, 1979), a historical range of ice motions has been calculated and the range of motions for each month has been determined (Thomas and Pritchard, 1979a,b). In each of Figs. 3 through 14, the historical range of monthly ice motions is presented. Although data were calculated for 19 sites (see Thomas and Pritchard, 1979a,b), spatial variations are smooth enough that only three sites need to be shown for the present purposes. All sites are in deep water outside the stamukhi zone. Values at intermediate locations may be found by interpolation.

At each location, ice drift contributions from both winds and currents are indicated. The solid arrows pointing to the centers of the ellipses represent the components of ice motion caused by winds. Ice initially located at the tail of an arrow can be expected to move to the head of the arrow during the month. This is the mean for the 25-year simulation period. The three ellipses surrounding each mean monthly motion arrow represent the areas within which 50, 90, and 99% of all monthly

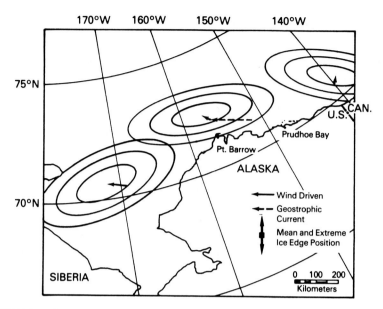

FIGURE 3. *Simulated range of ice motions for month of January.*

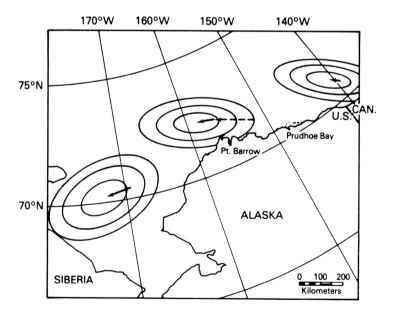

FIGURE 4. *Simulated range of ice motions for month of February.*

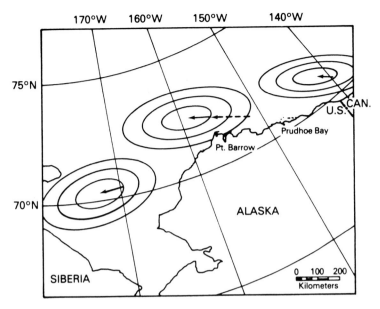

FIGURE 5. *Simulated range of ice motions for month of March.*

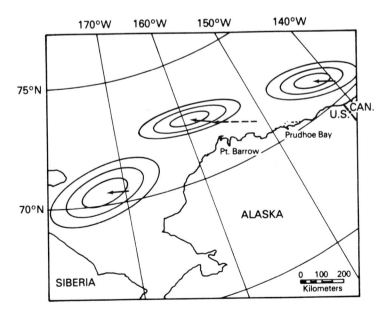

FIGURE 6. *Simulated range of ice motions for month of April.*

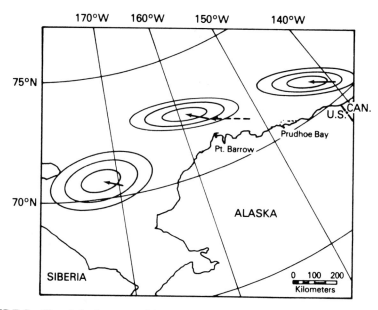

FIGURE 7. Simulated range of ice motions for month of May.

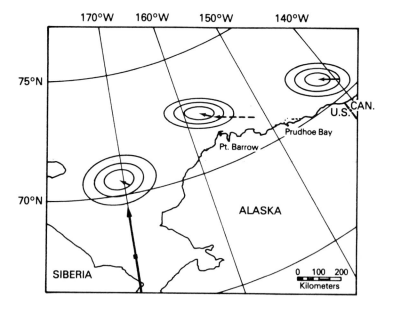

FIGURE 8. Simulated range of ice motions for month of June.

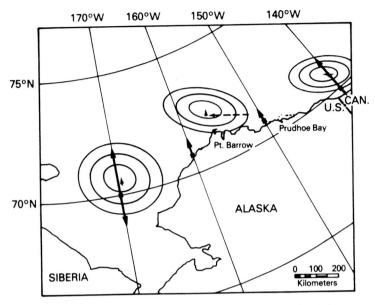

FIGURE 9. Simulated range of ice motions for month of July.

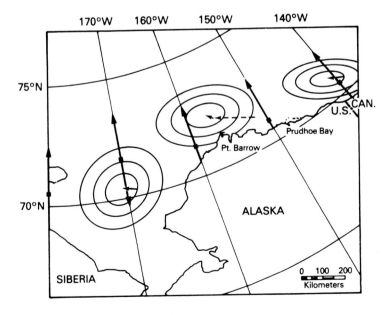

FIGURE 10. Simulated range of ice motions for month of August.

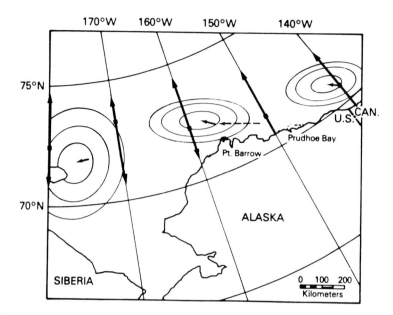

FIGURE 11. *Simulated range of ice motions for month of September.*

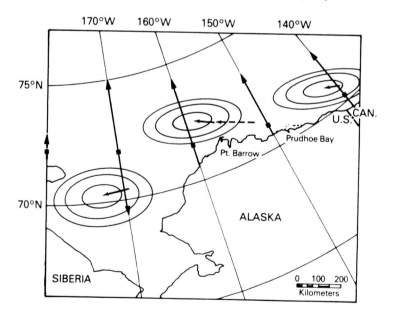

FIGURE 12. *Simulated range of ice motions for month of October.*

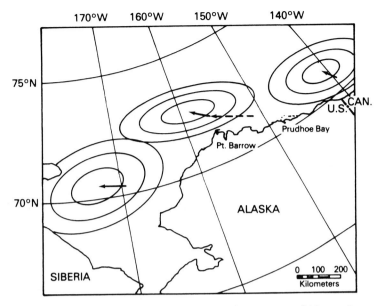

FIGURE 13. Simulated range of ice motions for month of November.

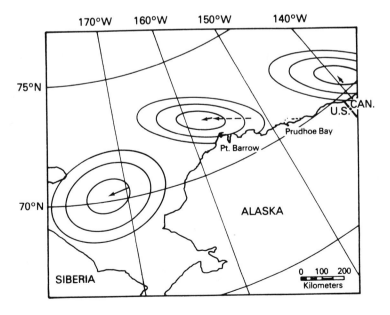

FIGURE 14. Simulated range of ice motions for month of December.

motions can be expected to occur. The most likely range of locations to which ice will move is defined as the area within the 50% ellipse. The 99% ellipse can be interpreted as the 100-year event because 1 event in 100 will fall outside it. The 95% confidence limits for the ellipses lie between 0.64 and 2.24 of the size of the 50% ellipse. The range of confidence limits is large because of the small 25-year sample size and because there is a large amount of variability from year to year.

The monthly contribution from geostrophic ocean currents (the same each month) is shown as a dashed arrow. A current vector is presented only at the central site between Prudhoe Bay and Point Barrow. There were no data in the Chukchi Sea to extend the current field, and values near the U.S.-Canadian border are near zero. Since the ocean current is simply added to the relative drift velocity, any additional ocean current data can be included in these results easily.

The simulated motions show clear westward ice drift, except during July when ice motion is northward and in the eastern Beaufort Sea in January. Variations are also generally concentrated along the east-west direction. There is an apparent dominant alongshore motion, especially near Point Barrow. Since the free-drift model is insensitive to the presence of land, this motion must be a function only of the driving forces, not the presence of shore.

In addition to ice motion data, Figs. 3-14 show the historical range of ice edge positions for each month estimated from Naval Oceanographic Office data for 1954 through 1970 (Brower et al., 1977). For each 10 degrees of longitude the mean monthly ice edge position is indicated by a solid square, although only from June to October do they lie in the map area. The range of observed ice edge positions shown by the arrowheads reflects the extremes of the 15-day average positions calculated for each year. Thus, for this value, the month is broken into two parts.

Since offshore winds are expected to drive the ice offshore, it is reasonable to assume that light ice years are associated with offshore winds. Thus, during years when the ice edge is far from shore, ice motions tend toward the seaward side of the ellipses. For those cases where an ellipse intersects the shoreline, the likelihood of oiled ice reaching shore is suggested. However, this is true only for oil originating at the indicated site. If a spill occurs farther offshore or closer to shore than shown, the ellipses must be translated seaward or shoreward, respectively. Alongshore interpolation is also required to describe drift from sites between the three locations shown. When extreme shoreward motions occur, the free-drift model is no longer expected to be accurate. In this case, the open water and thin ice are eliminated and the ice strength increases. This causes ice stress to build and resist further shoreward motion. Therefore, the equiprobability ellipses shown in Figs. 3-14 cannot be accurate under this condition.

For longer periods of time, trajectories are estimated by accumulating monthly mean free-drift motions. Monthly displacements are interpolated in space as the ice moves. The range of variability from year to year is found by assuming that variance grows as a square-root function and that monthly variance is constant. An example of a year-long ice trajectory is presented in Fig. 15. This trajectory begins near Prudhoe Bay on 1 October and continues for a year. Since geostrophic ocean current data were not

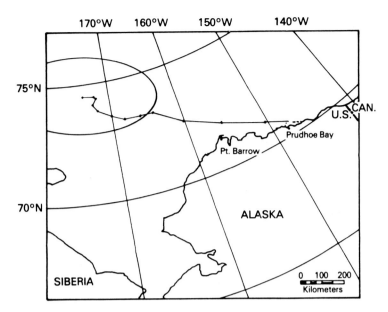

FIGURE 15.   Cumulative sea ice trajectory over one year beginning at Prudhoe Bay on 1 October.

calculated in the Chukchi Sea, zero values are assumed. This accounts for the reduction in monthly displacement after 4 months. The ellipse shown in the figure represents the accumulation of 50-percent equiprobability ellipses for the year and indicates the area to which the ice will likely drift after a year.

V. CONCLUSIONS

To determine the effects of an oil spill occurring in the Beaufort Sea, the probable motion of the ice cover in the area must be predicted. Complete ice dynamics models have been shown capable of simulating daily displacements with errors of 3.5 cm s^{-1} in the Beaufort Sea for both the winter and the summer. These sophisticated models have not been used here. Instead, a free-drift model that neglects ice stress has been used because of the lower cost of processing 25 years of wind data and because its performance has been judged adequate. The free-drift model has been shown to be as accurate as the complete model during summer months, but errors rise to a mean of 2.0 cm s^{-1} and a standard deviation of nearly 10 cm s^{-1} in the winter.

Errors in simulated ice motions arise from many sources: winds, currents, ice stress, and others. Since ocean current data remain sparse and geostrophic currents are not as accurate as the available wind data, the effects of winds and currents on ice motions are separated. This technique allows improvement of the modeled results as more and better current data become available.

Monthly ranges of ice drift calculated from historical winds over a 25-year period (1953-77) and long-term geostrophic ocean currents show an estimate of both the most likely and the extreme ice motions. Ice drift is typically westward everywhere along the Alaskan North Slope, in agreement with the gyral motion in the Beaufort Sea. It is concluded, therefore, that the ice will drift westward with the possibility of large year-to-year variability. The historical mean monthly motion and the variances from year-to-year variations have been quantified; monthly values can then be accumulated to estimate ice trajectories. These ice trajectories provide a reasonable estimate of trajectories that an oil spill would take.

ACKNOWLEDGMENTS

I thank Dr. Gunter Weller for his continued support of this work. Many years ago he saw that this model and the complete ice dynamics model would provide unique and valuable environmental information. His patience and persistence carried us through many times when outside support was not there. I also thank D. R. Thomas for his continuing help in this work. His contribution to the model and to the processing of data was essential to obtaining results, most of which have been presented in our jointly authored reports.

REFERENCES

Albright, M. (1980). In "Sea Ice Processes and Models" (R.S. Pritchard, ed.), p. 402. University of Washington Press, Seattle.
Brower, W.A. Jr., Diaz, H.F., Prechtel, A.S., Searby, H.W., and Wise, J.L. (1977). "Climatic Atlas of the Outer Continental Shelf Waters and Coastal Regions of Alaska," Vol. 3. Arctic Environmental Information and Data Center, Anchorage.
Hibler, W.D. III (1980a). Cold Reg. Sci. and Technol. 2, 299.
Hibler, W.D. III (1980b). In "Dynamics of Snow and Ice Masses" (S. Colbeck, ed.), p. 141. Academic Press, London.
Jenne, R.L. (1975). Technical Note 1A-III. National Center for Atmospheric Research, NOAA, Boulder.
Kollé, J.J., and Pritchard, R.S. (1983). J. Energy Resourc. Tech, 105, 346.
Leavitt, E., Albright, M., and Carsey, F. (1978). In "AIDJEX Bull." No. 39, p. 121. University of Washington, Seattle.
McPhee, M.G. (1980). In "Sea Ice Processes and Models" (R.S. Pritchard, ed.), p. 62. University of Washington, Seattle.
Pritchard, R.S. (1975). J. appl. Mech. 42, 379.
Pritchard, R.S. (1977). In "POAC 77 Proceedings" (D.B. Muggeridge, ed.), p. 494. Memorial University of Newfoundland, St. Johns.

Pritchard, R.S. (ed.) (1980). "Sea Ice Processes and Models." University of Washington Press, Seattle.

Pritchard, R.S. (1981). In "Mechanics of Structured Media, Part A" (A.P.S. Selvadurai, ed.), p. 371. Elsevier, Amsterdam.

Pritchard, R.S., and Kollé, J.J. (1981). "Flow Research Report No. 187," Flow Industries, Inc., Kent, WA.

Rothrock, D.A. (1979). J. Glaciol. 24, 359.

Sater, J.E., Walsh, J. E., and Wittmann, W. (1974). In "The Coast and Shelf of the Beaufort Sea" (J.C. Reed and J. E. Sater, eds.), p. 85. Arctic Institute of North America, Arlington.

Thomas, D.R., and Pritchard, R.S. (1979a). "Flow Research Report No. 133," Flow Industries, Inc., Kent, WA.

Thomas, D.R., and Pritchard, R.S. (1979b). In "The Physical Behavior of Oil in the Marine Environment," p. 5.17. National Weather Service, NOAA, Ashville, NC.

Thorndike, A.S., and Colony, R. (1982). J. geophys. Res. 87, 5845.

Thorndike, A.S., Rothrock, D.A., Maykut, G.A., and Colony, R. (1975). J. geophys. Res. 80, 4501.

Zubov, N.N. (1943). "Arctic Sea Ice," Translated by Naval Oceanographic Office and Am. Met. Soc. Naval Electronics Laboratory, San Diego.

STRUCTURE OF FIRST-YEAR PRESSURE RIDGE SAILS
IN THE PRUDHOE BAY REGION

Walter B. Tucker III
Devinder S. Sodhi
John W. Govoni

U.S. Army Cold Regions Research
and Engineering Laboratory
Hanover, New Hampshire

I. INTRODUCTION

Sea ice ridges constitute one of the major hazards to exploration and production of petroleum in the Beaufort Sea. They are of little concern to operations in fast ice near shore, but farther seaward the large number and size of ridges make them a serious problem, not least because normal surface transportation is virtually impossible in heavily ridged areas. Major construction projects thus are limited to the ice-free season, and winter transportation and resupply operations must be carried out by aircraft. A primary consideration in the design of a platform is that the proposed structure be able to withstand the impact of a free-floating multi-year or first-year ridge. Finally, possible damage to an undersea wellhead or pipeline by ridge keels must be considered.

It is well known that the study of sea ice mechanics can be enhanced by analyzing ridge features and onshore ice pile-ups: if we understand the mechanisms of ice failure in ridges and pile-ups, we can estimate the forces causing failure. With this knowledge, a designer may then decide whether or not the upper limit of the design load is governed by the forces required for ice failure or by environmental driving forces. However, many properties of ice relevant to failure are not well understood. We suggest that by studying existing features in which ice failure has occurred, we improve our understanding not only of ridges as potential hazards but also of those mechanisms that caused the ice to fail and form the ridge.

We know, for instance, that ice properties such as strength (flexural and compressive) and elastic modulus are dependent on the temperature and brine volume of the ice sheet (Schwarz and Weeks, 1977; Weeks and Assur, 1969). Temperature and brine volume vary with the thickness of the sheet (Cox and Weeks, 1974), so strength and elastic modulus are also a

ISBN 0-12-079030-0

function of thickness. We also know that in fast ice adjacent to the coast, the c axes of individual ice crystals are oriented in a nearly uniform direction thought to be controlled by ocean currents at the time of formation (Weeks and Gow, 1978, 1980; Kovacs and Morey, 1978). Preferred axis orientation has an effect on ice strength through the location of failure planes. These are all relatively small-scale properties, and although they are relevant to the design load problem, their effect is usually parameterized or excluded from models and calculations by assuming constant strengths and elastic moduli. The study of existing deformation features may throw light on the validity and applicable scale of these parameterizations. In addition, we may gain a better understanding of the processes involved in pressure ridge formation.

In most current models of ridging and formation of grounded features (Parmerter and Coon, 1973; Kovacs and Sodhi, 1980; Kry, 1980; Vivatrat and Kreider, 1981), other important properties of the deformation process are also assumed. Some of these are ice feature slope angles, ice density, rubble density, ice thickness, and the ice-to-ice and ice-to-beach coefficients of friction. Results of ridging models have been difficult to verify because so few data concerning ridges have been available (Parmerter and Coon, 1973). Ridge heights have been an exception because they can be measured near shore by laser profilometry (Tucker *et al.*, 1979), and a few have been profiled by manual surveying techniques (Weeks and Kovacs, 1970; Kovacs and Mellor, 1974; Wright *et al.*, 1978; Sisodiya and Vaudrey, 1981). Kovacs and Sodhi (1980) also reported on the heights and locations of major shore ice ride-up and pile-up features.

Ridges are formed by a combination of shearing and compressive forces on an ice sheet. Those in which the major formation forces are compressive are termed pressure ridges. This is not to imply that the ice fails in compression as the compressive strength is significantly higher than the flexural strength. Thus the ice is thought to fail in bending (Parmerter and Coon, 1973; Vivatrat and Kreider, 1981) or possibly by buckling (Kovacs and Sodhi, 1981). Ridges formed primarily through shear forces can be long (several kilometers) linear features typified by nearly vertical walls and thoroughly pulverized ice. On the other hand, while pressure ridges can be long, they may or may not be straight, and they contain ice blocks of a variety of sizes. Typical shear and pressure ridge features are shown in Fig. 1.

This report attempts to increase the small data base and understanding of morphological features of free-floating pressure ridges. Our field data include the geometry of ridges and measurements of the blocks incorporated in them. Some of these data and analyses were presented previously (Tucker and Govoni, 1981), but the data set has since been greatly supplemented and now includes measurements made during April of 1980 and 1981 on 84 free-floating pressure ridges.

As in Tucker and Govoni (1981), we examined ridge height and width as functions of the ice thickness. We also looked in more detail at slope angles, variations of ice-block thickness within a ridge, and the top-surface area of the measured blocks. In addition, knowing both ridge heights and ice thicknesses, we calculated the force required to generate the stored potential energy in the ridges. These forces are presented to show a typical range of values for free-floating pressure ridges.

FIGURE 1. *Examples of the two predominant types of sea ice ridges.*
A, shear ridge; B, pressure ridge.

II. DATA COLLECTION

Field measurements of ridges in the Prudhoe Bay region were made in April of 1980 and 1981. In 1980, ridges were chosen from five sampling sites located from 30 to 165 km north of Cross Island. Because no geographic variation was apparent (Tucker and Govoni, 1981), the April 1981 ridge sampling was restricted to within 65 km north of Cross Island. The well-known "shear zone" (Kovacs and Mellor, 1974; Reimnitz *et al.*, 1977; Tucker *et al.*, 1980) lies within this area, and large numbers of ridges were available for sampling.

We used the same sampling procedure both years. A typical pressure ridge (Fig. 1B) was identified from the helicopter. We selected, locally, the highest point along the ridge, usually less than 0.5 km from the landing site. At this point and at points 15 m to either side along the ridge, the angle and distance to the ridge crest were measured from the base of the ridge. The base was assumed to be at sea level because evidence of flooding was common there. An average height and width for each ridge was calculated using these values. Along this same section of the ridge, we measured six to ten ice blocks, selecting those on both sides of the ridge that had the largest measurable top-surface area. The thickness was normally easy to measure because it is usually the smallest dimension; otherwise it could be distinguished by observed layering in the ice caused by sediment or different crystal structures and by brine drainage features. We took the top-surface area to be the product of the longest and shortest axes of the top surface of the block.

In this study, only single ridges were selected for sampling; rubble fields and clustered parallel ridges were specifically avoided. The water depth at all sampling locations appeared sufficient to assure that the ridges were not grounded. This premise was based upon study of the local bathymetry and assuming that ridge sail heights were one-fifth the keel depths (Kovacs, 1972).

Ice conditions appeared to be similar in both years, with one exception. Whereas in 1980 no multi-year ice was observed near any of the sampling regions, during the 1981 field season a large amount of multi-year ice lay just off the barrier islands. Although less multi-year ice was evident farther seaward, some was observed near all sampling locations. Other than this difference, no year-to-year variations were apparent to us. The shear zone appeared to be similarly rough both years, and the number of leads was approximately the same both years.

III. RESULTS AND DISCUSSION

A. *Ridge Geometry and Ice Thickness*

In this analysis, a mean ice block thickness was calculated by averaging the thickness of the six to ten blocks measured in each ridge. We believe that this procedure provides a representative sampling of thickness contained within the ridges. Variations in block thickness within individual ridges are examined later in this report.

The relation between ridge height and block thickness was established by Tucker and Govoni (1981) using just 30 ridges. Figure 2 shows the combined 1980 and 1981 data, in which height is plotted as a function of thickness. This figure shows (1) that the relation between sail height and thickness is definite and (2) that there is no year-to-year variation in this morphological aspect of pressure ridges. Because none of our analyses detected differences between the two years, they are not discriminated in following figures. Similarly, a detailed examination of morphological differences with respect to sampling location gave no positive results. The only obvious difference is that the number of ridges varied with geographic location, an expected result when working in the vicinity of the shear zone.

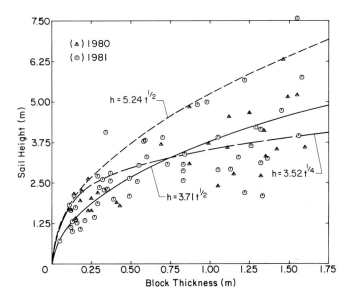

FIGURE 2. Sail height as a function of block thickness. Best fit curves as a function of $t^{0.25}$ (flexural failure) and $t^{0.5}$ (buckling failure) are shown. The $t^{0.5}$ function (h = 5.24 $t^{0.5}$), fitting heights that exceed the best fit curve (h = 3.71 $t^{0.5}$) by more than one standard deviation, is also shown.

Because we observed a great deal of multi-year ice in 1981 and none in 1980, we can infer that multi-year floes embedded in a first-year ice field do not affect the morphology of individual ridges. The presence of multi-year ice may stabilize or strengthen an ice sheet as a whole (Tucker et al., 1980) and thus limit the numbers of ridges formed over a given area. When the driving forces are sufficient to overcome this additional strength, the ridging processes appear to be the same for individual ridges whether or not multi-year ice is present. There is no doubt, however, that multi-year floes influence the location of ridges, as linear ridges skirt large multi-year floes or end at their edges.

Another deduction to be made from the lack of year-to-year variation is that the overall properties of the ice sheet that influence ridge form are relatively homogeneous. This hypothesis appears plausible when one considers those factors that may control the height of a ridge. The Parmerter and Coon (1973) ridging model limits ridge height on the basis of the thickness and strength of the parent ice sheet, which is loaded from above and below by ice blocks or rubble. When this sheet fails due to uneven vertical forces that exceed the flexural strength, the ridge begins to widen—to build laterally rather than vertically. The load that a parent sheet of certain thickness can bear depends upon its strength. If this hypothesis of limiting height is correct, our data indicate, by their similar scatters, that the strength for any given parent sheet thickness must have been similar for the two years. If the ice had greater strength for one of the years, we would expect generally higher ridges that year.

In making the above deduction we are assuming that the parent sheet thickness is the same as that of the ice found in the ridge, which appeared to be the case in our observations. However, if that is not the case, or if the Parmerter and Coon (1973) limiting height hypothesis is incorrect, whatever mechanism is responsible for the clear relation between sail height and ice thickness must still depend strongly on the strength of the ice. Thus, we believe that our hypothesis is valid regardless of the limiting height mechanism. The question of scale is also of interest. In examining many ridges, we are treating a phenomenon of geophysical scale in which year-to-year strength variations that may be detected on a small scale are averaged out.

A reasonable relation between ridge height in meters (h) and ice block thickness in meters (t) for the data in Fig. 2 is given by

$$h = 3.71 \, t^{0.5}. \tag{1}$$

This relation fits the data with a correlation coefficient of 0.77. The constant, determined by least-squares procedure, differs only slightly from Tucker and Govoni's (1981) value of 3.69. Because the empirical relation is based on a limited sample of data, the reader should be cautious when using Eq. (1) to extrapolate to large ridges where block thickness exceeds 1.5 m.

Tucker and Govoni (1981) presented a simple geometric derivation for sail height as a function of the square root of ice thickness, but some broad assumptions were required. It would be physically more satisfying if a height-thickness relation could be derived from energetic arguments. Functional relations can be established by examining the forces involved in the flexural and buckling modes of failure of the ice sheet along with the larger forces required in ridge-building: that required to generate the gravitational potential energy and that required to overcome friction. The buckling analysis results in sail heights being a function of the square root of thickness in exactly the same form as Eq. (1). Analysis of flexural failure gives a relation that is similar in form, only height is a function of the fourth root of thickness. The best-fit fourth-root relation is also shown in Fig. 2. Its constant, also calculated by least squares, is 3.52, and the correlation coefficient for the relation is 0.72, slightly less than that of the square-root relation. These functional derivations are carried out in the Appendix.

The relation strongly suggests that buckling may be a mechanism in pressure-ridge formation. Assumptions are necessary in this derivation, so the possibility exists that the resulting square-root function may be coincidental. Kovacs and Sodhi (1981) pointed out that buckling may be a possible failure mode in ice-beach and ice-structure interactions. Their analysis further shows that buckling is possible for thick ice, contrary to previous speculation that only thin ice buckles. They emphasized that failure does not occur uniformly along the entire edge of an ice sheet, but rather in small localized regions. Buckling creates cusps and jagged edges around the contact areas, thus promoting more buckling. We have observed this type of ice failure in small-scale experiments. We are inclined to believe that buckling (and, to a lesser degree, crushing) may also occur in ridge formation, considering a moving ice sheet in contact with the previously created rubble.

The number of ridges formed from thick ice is another point that can be gleaned from Fig. 2. Over 30% of the ridges sampled were composed of ice thicker than 1.0 m. This finding is significant in light of previous speculation that ridges formed only from the deformation of ice in leads. We contend that if driving forces are sufficient to deform the thicker ice, and if no thin ice is present, then ridging will proceed with the thick ice. This process may be more prevalent in the nearshore area where the amount of thick first-year ice is greater and the number of leads fewer than in the central Arctic Ocean.

Ridge height is much more variable in thick than thin ice. We believe that this reflects the amount of ice movement that has taken place at the ridging location. Ridge height reflects the volume of deformation taking place and hence the movement, which in turn is a function of the magnitude and duration of the force applied. The large scatter of heights on the right side of Fig. 2 probably indicates that the movement has not been sufficient for most of the ridges to reach their limiting heights. In contrast, there is much less height variation for the ridges composed of thin ice. These ridges are probably quite close to their limiting heights, as less movement is necessary to deform thin ice to its limiting height. Further movement would cause the ridge to widen into a rubble field or a cluster of ridges, which were excluded from our survey.

Equation (1) gives the best-fit square-root relation to the heights of all the sails, most of which probably had not reached their limiting heights. We have also attempted to define a relation to represent the limiting height curve, by fitting the square-root relation (by least-squares approximation) to those heights that exceeded the original relation (Eq. 1) by at least one standard deviation. This curve, in which the constant is 5.24, is also shown in Fig. 2. We emphasize that while this relation may approximate the limiting height for a given thickness, we have only a limited number of ridges, and by avoiding multiple ridges we missed ridges that had definitely reached their limiting height. As a result we have no way of knowing what confidence may be placed in this relation.

It is also interesting to examine ridge width with respect to ice thickness, as plotted in Fig. 3. Again, a square-root relation of the form

$$w = 16.35t^{0.5} \tag{2}$$

fits these data with a correlation coefficient of 0.71. The fact that the correlation is reasonable is not surprising because the ridges measured in this study had not started to build laterally. In addition, ridge slope angles, which are discussed in more detail below, are generally constant at about 25°. This implies that width varies proportionally to height and thus is also a function of ice thickness. Assuming a triangular shape of ridge sails and using Eqs. (1) and (2), the ridge slope angle is calculated to be 24.4°.

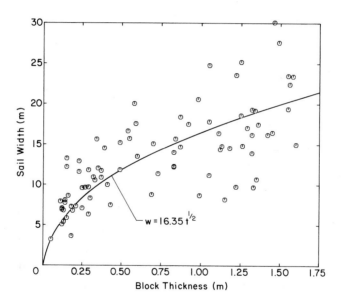

FIGURE 3. Sail width as a function of block thickness with best fit square root curve.

How do the constants relating height to thickness (Eq. 1) and width to thickness (Eq. 2) compare in our two years of data? In 1980, with 30 ridges measured, the constants were 3.69 and 15.64, respectively; in 1981, with 54 ridges, they were 3.73 and 16.83. The combined data resulted in constants of 3.71 and 16.35, as they appear in Eqs. (1) and (2) above. Both pairs of constants are in close agreement (less than 8% difference), showing that year-to-year variability is small in these aspects of ridge morphology.

Ridge slope angles averaged 26.1° for the 1980 data and 24.5° for the 1981 data. The mean slope angle for the combined total of 84 ridges is 25.1°, with a standard deviation of 5.0°. These slope angles are in close agreement with the 24° reported by Kovacs (1972) and the 25° reported by Wright et al. (1978). The actual slope angles for the 84 ridges varied from 7.0° to 56.4°, which is certainly a large range, but the relatively small standard deviation shows that most are close to the mean value. Parmerter and Coon (1973) also found a sail slope of approximately 25° in their model predictions.   They suggested that this slope is dictated by the limited amount of material (blocks) that is available to go into the sail because the majority of the deformed ice goes into the keel. The sail slope angle is less than the angle of repose, at which the ice blocks begin to slide back down the ridge.   Angles of repose in high shore-ice pile-ups have been measured by Kovacs (unpub. data) to be about 36°. We saw little evidence of ice-block backsliding, therefore we tend to agree with Parmerter and Coon (1973) in this respect.

B. Composition of Ridge Sails

So far we have been concerned only with the mean ice thickness and have not considered the variability of the thickness of ice incorporated into single ridges. In some ridges, more than one distinct thickness of ice was observed. This finding indicates that adjacent ice sheets of different ages were consumed in the ridging process.

The standard deviation of block thicknesses found in each ridge is plotted as a function of ridge height in Fig. 4. The magnitudes of standard deviation that approximately differentiate between two and three ice-thickness categories are also shown on this figure. A significant number of ridges (about 30) showed bimodal block thicknesses. The most important feature shown in Fig. 4 is that the higher ridges have the largest thickness variations. None of the ridges higher than 4.8 m contained ice of only one thickness.   Some of the large standard deviations may be attributed to variations in ice thickness due to differing snow depths on the surface of the ice. Kovacs et al. (1981) reported 10% thickness variations in 1.5-m-thick ice due to this effect, but that variation does not account for all of the standard deviation. The major variations in ice thickness must reflect ice of different ages (hence thicknesses) being deformed. This could happen, for instance, if a newly refrozen lead deforms, followed by the deformation of the thicker surrounding ice once the lead ice is consumed, assuming the driving forces remain sufficient.   A possible alternative is that the ridge resulted from two or more distinct deformation events during the season.

In contrast to the large standard deviations of the larger ridges, ridges less than 2.5 m high have very small standard deviations, which implies that the small ridges are formed of ice of nearly uniform thickness. Figure 2 shows that these ridges nearly always contain ice less than 0.5 m thick. Such thin ice is relatively new and tends to grow uniformly before snow covers it. Thus, we see very little variation in thickness due to growth

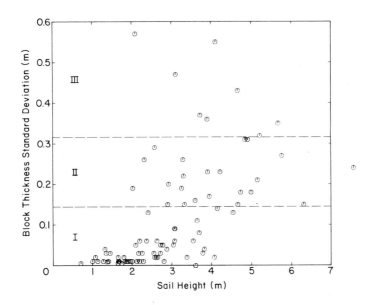

FIGURE 4. *Standard deviation of block thickness as a function of sail height. Divisions between areas I, II, and III represent approximate divisions in standard deviation between one, two, or three ice thickness categories contained in the ridge sail.*

differences. Small ridges appear either to have formed from ice in leads when forces were not sufficient to deform adjacent thicker ice or to have formed early in the season when all the ice was thin.

In four of the ridges surveyed, several blocks were observed that were too thick (one was 2.5 m) to have been formed during a single season. The blocks all occurred in ridges in which the majority of the blocks were of lesser thickness. Perhaps small multi-year floes were deformed and incorporated into the ridge. They may also have been the result of the deformation of previously rafted and refrozen blocks as described by Sisodiya and Vaudrey (1981).

We also examined the top–surface dimensions of the blocks we measured at each ridge. Block surfaces are generally not well–defined rectangles; nevertheless we took the top–surface area to be the product of the longest and shortest axis of the block. Figure 5 shows the mean top-surface areas as a function of mean block thickness. What is evident is simply that the largest blocks are also the thickest. As shown by Tucker and Govoni (1981), an exponential relation provides the best fit of area to thickness. We can find no physical basis for this relation, however, and a relation based on the square of thickness, also shown in Tucker and Govoni

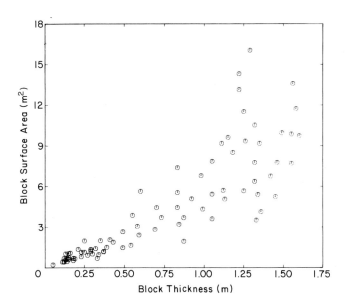

FIGURE 5. Block surface area of the largest blocks as a function of block thickness.

(1981), provides a poor fit at small thicknesses. Thus, no fit to the data is attempted here.

Of equal interest is the longest axis dimension because it may be related to the mode of failure. Figure 6 shows the mean block length (longest axis) versus mean block thickness for all the ridges sampled. Also shown is the best-fit linear regression line resulting in the equation

$$\ell = 2.04\ t + 0.72,\tag{3}$$

where ℓ represents the block length. This curve fits the data with a correlation coefficient of 0.91.

We may also examine the block length in terms of the characteristic length L, a property of ice defined in the Appendix and dependent in part upon its thickness, Poisson's ratio, and elastic modulus. Using typical values of elastic modulus (3 GPa), Poisson's ratio (0.3), and the specific weight of water (9.81 kN m^{-3}) we get

$$L = 12.94\ t^{0.75}\tag{4}$$

Using a least squares approximation, the best fit of block length to the three-fourths power of thickness is

$$\ell = 2.85\ t^{0.75}\tag{5}$$

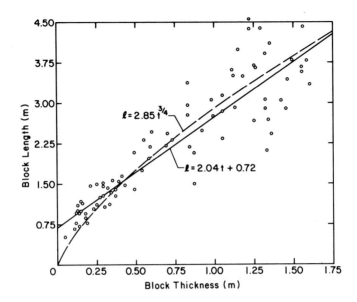

FIGURE 6.   *Block length as a function of block thickness. Best fit curves with length as a linear function of thickness and as a function of $t^{0.75}$ are also shown.*

This curve, also shown in Fig. 6, fits the data with a correlation coefficient of 0.90. The ratio of the observed block length to the characteristic length is 1:4.5.   As the largest uncertainty in the determining characteristic length for sea ice is the elastic modulus, it is useful to examine the effect of various elastic moduli.   For elastic moduli of 1 GPa and 10 GPa, the ratios are 1:3.4 and 1:6.3, respectively.

As the measured block lengths are less than the lengths predicted for flexural failure, so are they also less than those predicted for buckling failure.   Parmerter and Coon (1973) predicted that the stress maximum, and therefore cracking, occurs at ice sheet lengths of $\pi/4$ L, which is approximately 3.5 times our observed lengths.   Vivatrat and Kreider (1981) predicted a length-to-thickness ratio of 5 for flexural failure occurring at the toe of a grounded feature forming from 1.8-m-thick ice.   For this thickness, our observations show the ratio to be 2.8.   The ratio of 5 is only observed for thin ice (< 0.25 m), which is in agreement with the observations of Weeks and Kovacs (1970).

That blocks are shorter than flexural failure would predict does not necessarily indicate that flexural failure is not occurring or that the predictions are in error.   It is much more likely that other failure modes (shearing, buckling, crushing) and secondary failure also occur during ridge formation.   Secondary failure, such as cantilevered blocks breaking under their own weight or tumbling from the pile, will reduce block sizes regardless of the original failure mechanisms.

C. Force to Generate the Stored Potential Energy

The consensus among workers in ice mechanics is that the two largest forces involved in ridge building are that required to overcome friction and that required to generate the gravitational potential energy that is stored in the ridge (Parmerter and Coon, 1973; Kovacs and Sodhi, 1981; Vivatrat and Kreider, 1981). The cited investigators also reviewed other energy sinks which include fracture of the sheet, deformation or penetration of the rubble, and elastic deformation. Little is known about the forces involved, but magnitudes can be crudely inferred.

The frictional force is difficult to estimate because the ice-to-ice area of contact and the coefficient of friction are important quantities that must be assumed. Various assumptions and estimates for this force are summarized by Vivatrat and Kreider (1981).

In contrast, calculating the force required to generate the gravitational potential energy is very simple. The assumed variables in the calculation can also be estimated with reasonable accuracy. If the center of the ridge is assumed to be in isostatic equilibrium, the equation for the effective force per unit width (in $N\ m^{-1}$) is

$$F_{pe} = 0.5\ \rho_i g h t \tag{6}$$

where ρ_i is the ice density and g is gravitational acceleration (Parmerter and Coon, 1973). The major assumption here is the value of the ice density, and this is usually taken to be 920 kg m^{-3}.

Using this simple equation, the effective force to generate the stored potential energy for the 84 ridges has been calculated. The mean height and mean block thicknesses for each ridge were used for the calculations.

Figure 7 shows the effective force needed to generate the stored potential energy as a function of block thickness (Fig. 7A) and ridge height (Fig. 7B). One point of interest is the small magnitude of these forces; only one ridge required a force exceeding 50 kN m^{-1}. With the exception of Parmerter and Coon's (1973) predicted force levels, these values are very much less than the other energetic model predictions as summarized in Vivatrat and Kreider (1981). The models that were formulated for shore-ice pile-ups and grounded ridges generally include friction, which is expected to be of considerable magnitude. While our values are not directly comparable, it is interesting to note that Vivatrat and Kreider (1981), in their own grounded ridge model, predicted force levels for ridge deformation and sheet fracture (the formulations appear to be applicable to floating ridges) that are larger than our calculated forces for potential energy generation. This may indicate that these two forces can no longer be neglected.

A simple relation for F_{pe}, which is a function of ice thickness alone, can be derived using the best-fit empirical height to thickness relation Eq. (1). Substituting Eq. (1) into the force equation for height (Eq. 6) yields

$$F_{pe} = 0.5\ \rho_i g\ 3.71\ t^{1.5} \tag{7}$$

and combining all constants we arrive at

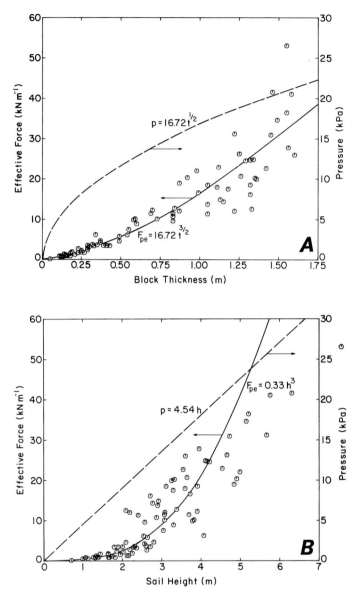

FIGURE 7. *Force to generate gravitational potential energy per unit width as a function of: A, block thickness; B, sail height. Curves represent derived expressions for effective gravitational potential energy and effective pressure as functions of block thickness and sail height. Scale for force and plotted values is on left and pressure is on right.*

$$F_{pe} = 16.72 \, t^{1.5} \tag{8}$$

The curve representing this function is plotted in Fig. 7A. As expected, it provides a reasonable fit to the calculated data. The pressure or stress in the ice sheet (in kPa) that is required to generate this potential energy force can be obtained by simply dividing Eq. (8) by thickness, which yields

$$p = 16.72t^{0.5} \tag{9}$$

This function is also plotted in Fig. 7A. Data values are not expressed in terms of pressure, however.

Similarly, F_{pe} (in kN m^{-1}) as a function of ridge height alone can be obtained by substituting into Eq. (6) for thickness obtained from Eq. (1). Combining constants reduces the equation to

$$F_{pe} = 0.33 \, h^3 \tag{10}$$

This relation is plotted in Fig. 7B. The agreement with the data is poor for ridges higher than 4.5 m. This is a result of the substitution made using Eq. (1), which requires that large ridges be constructed of very thick ice. These thicknesses required for higher ridges are inferred in Eq. (10). Consequently the force is unrealistically large for ridges higher than 4.0 m, and the curve should not be used for these larger ridges.

Likewise, the pressure (in kPa) required to generate this force as a function of height can be obtained by dividing Eq. (10) by thickness and again making a substitution for thickness from Eq. (1). This yields the linear relation

$$p = 4.54 \, h, \tag{11}$$

which is also plotted on Fig. 7B. A very similar relation can also be obtained more straightforwardly by simply dividing Eq. (6) by thickness.

Representative values of the effective force necessary to generate the stored potential energy in ridges have been presented. We believe these are representative because a variety of ridge heights and ice thicknesses have been sampled. One must keep in mind, however, that these values are applicable only to free-floating ridges. Expressions for the force as a function of height or thickness only are derived from these time- and space-averaged values. The forces necessary to generate the stored potential energy for ridges that may have reached their limiting heights may be estimated by following these same procedures using the upper limit curve shown in Fig. 2. Grounded features would be expected to generate larger forces because they can build to much greater heights.

IV. CONCLUSIONS

1. Ridge height and width have been shown to be a function of the thickness of the ice found in ridges. These results support Parmerter and Coon's (1973) height limit hypothesis in which the limiting height is

determined by the load that the parent ice sheet can withstand prior to bending failure, assuming that the parent sheet is of the same thickness as the ice being ridged. Other mechanisms may be responsible for limiting ridge height and width. Height and width are reasonable empirical functions of the square root of thickness. While better statistical fits may result from other empirical relations, the square-root dependence is physically satisfying because height and width go to zero with thickness.

2. Buckling may be a common failure mode in pressure ridge formation. An energetic argument of buckling results in height being related to the square root of ice thickness, while flexural failure results in a fourth-root relation. This is not to imply that buckling is a more prevalent failure mechanism than bending, only that it is possible to have some buckling mode of failure.

3. No apparent geographical or temporal variation affected the examined morphological features. A large range of both height and thickness was found each year. The lack of year-to-year variation is made especially obvious by the close agreement of the constants determined for the empirical relations of height and width to block thickness for both years. What is implied is that the properties of the ice sheet that influence the morphology of ridge sails (overall strength, density, etc.) can be considered homogeneous from year to year at this scale.

4. Many (> 30%) of the ridges sampled were formed of ice over 1 m thick. While our sampling procedure was not sufficiently random to state that 30% of all ridges in the nearshore area are formed of thick ice, it at least indicates the deformation of ice that is not in leads. We contend that if the stress state in the ice reaches high enough levels, and no thin ice is present to deform, then thick ice will form ridges. Thick-ice ridges may be peculiar only to the nearshore region. In the central Arctic Ocean more leads are present; thus we would expect fewer ridges to be composed of thick ice. The fact that there appears to be an upper limit to the thickness of the ice that deforms indicates either that sufficient thinner ice is present or that the driving force is limited.

5. Ice of several thicknesses is frequently contained in a single ridge sail. A likely explanation is that ridging begins with the ice in the lead and that the driving forces are strong enough to continue deforming the adjacent thicker ice once the lead is closed.

6. The top-surface areas and lengths of the ice blocks composing the ridges are reasonable functions of block thickness. Length is strongly correlated with the three-fourths power of thickness. This is significant because the characteristic length of ice is also dependent upon thickness to the three-fourths power. The ratio of observed block length to the characteristic length is 1:4.5.

7. Forces required to generate the potential energy stored in the ridges are small, the largest among our samples being 52 kN m^{-1}. The calculated values represent time- and space-averaged forces per unit length of a ridge. Expressions derived from the empirical relation of sail height to ice thickness can be used to give the force to generate the potential energy in terms of sail height or ice thickness only. The most important point here is that we have calculated these forces using actual data rather than assumed values. It is undoubtedly true that grounded

features can build to much greater heights, thus requiring a significantly greater force than is the case for free-floating pressure ridges.

V. ACKNOWLEDGMENTS

This work was supported by funding from the Bureau of Land Management through interagency agreement with the National Oceanic and Atmospheric Administration, under which a multi-year program responding to needs of petroleum development of the Alaskan continental shelf is managed by the Outer Continental Shelf Environmental Assessment Program (OCSEAP) Office. This study was conducted under Research Unit 88. We are also grateful for the valuable technical comments provided by A. Kovacs, W.F. Weeks, J.R. Kreider and others.

APPENDIX: ICE FAILURE MODE ANALYSIS

Consider the typical ridge-building case in which an ice sheet is moving against a line of rubble. At various locations along the linear feature the moving sheet attempts to penetrate the rubble, but the resistive forces, consisting of gravitational potential energy and friction, prevent penetration. Buckling failure may then occur at the toe of the ridge as shown in Fig. 8. The contact area along the edge of the ice sheet is assumed to be nonuniform along the length of the ridge. This assumption is based on the observations made during small-scale experiments.

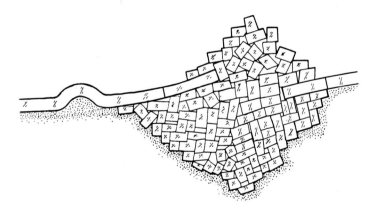

FIGURE 8. Scenario of possible buckling failure as it may occur in a localized region along the contact area between an advancing ice sheet and a ridge. The sheet attempts to penetrate the ridge but is halted by resistive forces. Buckling then occurs at the toe of the ridge.

The functional relation of the buckling force per unit width, F_{bu}, can be expressed as (Sodhi, 1979; Sodhi and Nevel, 1980; Sodhi *et al.*, 1983)

$$F_{bu} = \gamma \rho_w g L^2 \qquad\qquad (A1)$$

where $\rho_w g$ is the specific weight of water, γ a constant that depends on the geometry of the ice sheet and the boundary condition, and L the characteristic length of the ice, given by

$$L = \left(\frac{Et^3}{12 \, \rho_w g \, 1 - \nu^2} \right)^{0.25} \qquad\qquad (A2)$$

Where E is the elastic modulus and ν is Poisson's ratio of the ice sheet.
Substituting Eq. (A2) into Eq. (A1) we get

$$F_{bu} = C_1 \, t^{1.5}, \qquad\qquad (A3)$$

where C_1 results from combining various constants.
We assume that sufficient environmental forces exist to cause localized buckling of the ice sheet. The resistive force per unit length R_p, offered at the toe of the ridge sail, may be expressed in the following functional form:

$$R_p = 1/2 \, \rho_i g h t \beta, \qquad\qquad (A4)$$

which is the force required to generate the gravitational potential energy multiplied by a factor β (>1) to include the resistance in penetrating the rubble. Including penetrating resistance in this manner, it is implicitly assumed that the resistance pressure encountered by an ice sheet penetrating the block rubble of a ridge is proportional to the height of the sail.
Combining constants,

$$R_p = C_2 h t. \qquad\qquad (A5)$$

By equating F_{bu} in Eq. (A3) with R_p in Eq. (A5), we get

$$h = C_3 t^{0.5}, \qquad\qquad (A6)$$

where C_3 represents the combination of constants C_1 and C_2.
Similarly, the functional form of the crushing force is

$$F_{cr} = C_4 \sigma_c t \qquad\qquad (A7)$$

where σ_c is the ice crushing strength and C_4 depends on parameters related to ice–ridge contact. Equating F_{cr} in Eq. (A7) and R_p in Eq. (A5), we get

$$h = C_5. \qquad\qquad (A8)$$

A model for possible bending failure of an ice sheet advancing against a pressure ridge is shown in Fig. 9. Instead of penetrating the rubble, the

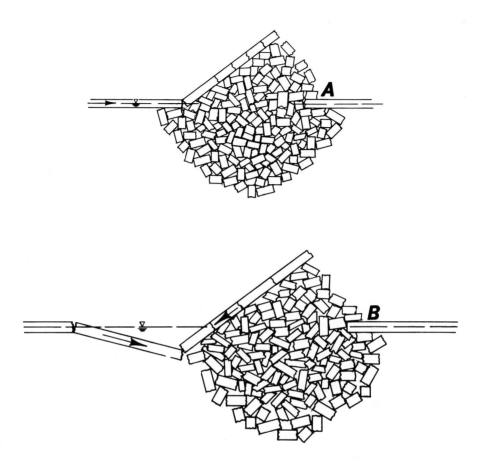

FIGURE 9. Model for possible bending failure mechanism. The advancing
ice sheet rides up over the ridge until its weight is great enough to cause
its horizontal part to submerge and fail in bending at the toe of the ridge.

ice sheet is pushed up (or down) against the side of the ridge. As the ice
sheet is pushed up, it may fail in bending at the toe of the ridge. Unless
the advancing ice sheet is very thick, the forces required to break it in
bending are small. But the ice sheet may still maintain its integrity to
keep advancing up the side of the ridge. As the weight of the climbing
sheet increases, part of its floating counterpart submerges, eventually
causing bending failure at the toe of the ridge. At the loss of support from
underneath, the lifted ice may tumble down to generate rubble.

The downward force, R_b, that acts at the edge of the advancing ice
sheet is proportional to the weight of the ice sheet pushed up on the ridge,
i.e. the product of sail height h and ice thickness t. Thus we have

$$R_b = C_6 ht \tag{A9}$$

The load per unit width, F_b, required for the ice sheet to fail in downward deflection is proportional to $\sigma_f\, t^2\, L^{-1}$ (Kerr, 1975), where σ_f is the flexural strength of the ice sheet:

$$F_b = C_7\, \sigma_f t^2\, L^{-1} \tag{A10}$$

Equating F_b in Eq. (A10) with R_b in Eq. (A9) and using the definition of the characteristic length L as given in Eq. (A2) we get

$$h = C_8\, t^{0.25} \tag{A11}$$

If the advancing ice sheet is very strong or thick or if the ridge is not high enough, the weight of the pushed-up ice may not be sufficient to break it. Then the process of ridge building may take place on the side of the sail or the keel in a manner similar to ice piling as described by Kovacs and Sodhi (1980).

We have presented three failure scenarios that result in ridge height being a function of ice thickness (Eqs. A6, A8, A10) of the form

$$h = C\, t^n \tag{A11}$$

where $n = 0.5$ for buckling, $n = 0.25$ for flexural failure, and $n = 0$ for crushing. It is clear that height must go to zero with thickness (it can with crushing if we consider a discontinuity existing at zero thickness). Examination of Fig. 2 makes it obvious that the straight line obtained from the crushing analysis fits the observed data poorly. Clearly, buckling ($n = 0.5$) or flexural failure ($n = 0.25$) provide more representative trends. In thin ice it appears that the buckling curve is the best approximation; thus buckling failure may dominate in thin ice. For thicker ice, where the scatter of heights is quite large, it appears that either buckling or flexural failure could be occurring.

REFERENCES

Cox, G.F.N. and Weeks, W.F. (1974). *J. Glaciol.* *13*, 109.

Kerr, A.D. (1975). *USACRREL Research Report 333*, U.S. Army Cold Regions Res. and Eng. Lab., Hanover, NH.

Kovacs, A. (1972). *In* "Proceedings of an International Conference on Sea Ice" (T. Karlsson, ed.), p. 276. Icelandic National Research Council, Reykjavik.

Kovacs, A. and Mellor, M. (1974). *In* "The Coast and Shelf of the Beaufort Sea" (J.L. Reed and J.E. Sater, eds.), p. 113. Arctic Institute of North America, Arlington, VA.

Kovacs, A. and Morey, R.M. (1978). *J. geophys. Res.* *83*, 6037.

Kovacs, A. and Sodhi, D.S. (1980). *Cold Reg. Sci. and Tech.* *2*, 209.

Kovacs, A. and Sodhi, D.S. (1981). *In* "Proceedings of the Sixth International Conference on Port and Ocean Engineering Under Arctic Conditions," p. 985. Quebec.

Kovacs, A., Morey, R.M., Cundy, D.F., and Decoff, G. (1981). *In* "Proceedings of the Sixth International Conference on Port and Ocean Engineering Under Arctic Conditions," p. 912. Quebec.

Kry, P.R. (1980). *Canad. geotech. J. 17*, 97.

Parmerter, R.R. and Coon, M.D. (1973). *AIDJEX Bull. No. 19*, 59.

Reimnitz, E., Toimil, L., and Barnes, P. (1977). *AIDJEX Bull. 36*, 15.

Schwarz, J. and Weeks, W.F. (1977). *J. Glaciol. 1*, 499.

Sisodiya, R.G. and Vaudrey, K.D. (1981). *In* "Proceedings of the Sixth International Conference on Port and Ocean Engineering Under Arctic Conditions," p. 755. Quebec.

Sodhi, D.S. (1979). *In* "Proceedings of Fifth International Conference on Port and Ocean Engineering Under Arctic Conditions," p. 797. Norweg. Inst. Tech., Trondheim.

Sodhi, D.S. and Nevel, D.E. (1980). *In* "Working Group on Ice Forces on Structures, a State-of-the-Art Report" (T. Carstens, ed.), p. 131. Special Report 80-26, U.S. Army Cold Regions Res. and Eng. Lab., Hanover, NH.

Sodhi, D.S., Haynes, F.D., Kato, K., and Hirayama, K. (1982). *Ann. Glacid. 4*, 260.

Tucker, W.B. III, Weeks, W.F., and Frank, M. (1979). *J. geophys. Res. 84*, 4885.

Tucker, W.B., Weeks, W.F., Kovacs, A., and Gow, A.J. (1980). *In* "Sea Ice Processes and Models" (R.S. Pritchard, ed.), p. 261. Univ. Washington Press, Seattle.

Tucker, W.B., III and Govoni, J.W. (1981). *Cold Reg. Sci. and Tech. 5*, 1.

Vivatrat, V. and Kreider, J.R. (1981). *In* "Proceedings of the Offshore Technology Conference 1981," p. 417. Houston.

Weeks, W.F. and Assur, A. (1969). *USACRREL Research Report 269*. U.S. Army Cold Regions Res. and Eng. Lab., Hanover, NH.

Weeks, W.F. and Kovacs, A. (1970). *In* "Special Report to U.S. Coast Guard." U.S. Army Cold Regions Res. and Eng. Lab., Hanover, NH.

Weeks, W.F. and Gow, A.J. (1978). *J. geophys. Res. 83*, 5105.

Weeks, W.F. and Gow, A.J. (1980). *J. geophys. Res. 85*, 1137.

Wright, B.D., Hnatiuk, J. and Kovacs, A. (1978). *In* "Proceedings of the International Association for Hydraulic Research Symposium on Ice Problems," p. 249. Lulea, Sweden.

FAST ICE SHEET DEFORMATION
DURING ICE-PUSH AND SHORE
ICE RIDE-UP

Lewis H. Shapiro
Ronald C. Metzner
Arnold Hanson[1]
Jerome B. Johnson

Geophysical Institute
University of Alaska
Fairbanks, Alaska 99701

[1]Department of Atmospheric Sciences
University of Washington
Seattle, Washington 98195

I. INTRODUCTION

Studies of ice pile-up and ride-up along arctic beaches have tended to emphasize the geometry of cross sections and the mechanisms of ice failure. This is consistent with the objective of using the results to calculate the forces required to form the features. These forces, in turn, are taken as indicators of the far-field force levels which structures fixed to the seafloor in a fast ice sheet may have to contend with. In general, the areal characteristics of these features, other than length along the beach, have received little attention. However, if questions regarding environmental controls, mechanisms, and force transmission are to be answered, then more will have to be learned about the areal characteristics of the deformation of ice sheets involved in ice-push events.

This paper presents results of field studies of shore ice pile-ups and associated deformation of the fast ice which occurred during several ice-push events in the Point Barrow area in 1975-78. We will formulate a preliminary model of the areal characteristics of the deformation of a landfast ice sheet during the formation of ice-push ridges and ride-ups. While the data base is limited, such a model may be useful as a stimulus to further studies.

II. BACKGROUND

The problem of determining the force associated with the formation of shore ice pile-ups and grounded ridges is often approached in the manner suggested by Parmerter and Coon (1972) for free-floating pressure ridges. The force calculated is that required to do the work of storing potential energy in a ridge and of overcoming friction between the advancing ice sheet and the beach (see for example, Kovacs and Sodhi, 1979; Vivatrat and Kreider, 1981). Additional energy losses from work done in (1) bulldozing of the beach, (2) fracture of the ice during ridging, (3) opposing hydrodynamic forces, and (4) gouging of the seafloor by the drag of deep keels are not accounted for. While the contributions from these processes may not be large, calculations done without them are at best estimates of the lower limit of the force acting at the beach line. Still, such calculations give values of stress as high as 3.5×10^5 Pa (Kovacs and Sodhi, 1979), which are significantly higher than the estimated maximum stress which can be transmitted through pack ice in the open sea (approximately 10^5 Pa; Pritchard, 1977). This suggests that shore ice pile-up and ride-up features reflect a mechanism for stress concentration, through which stresses transmitted through the pack ice are amplified at the beach (Kovacs *et al.*, 1982 applied this concept to buckling of ice sheets against Fairway Rock in the Bering Sea). Such a mechanism probably involves a variety of factors.

The distribution of stress along the seaward edge of the fast ice is one aspect of the problem. It may be thought obvious that pack ice moving approximately parallel to the edge of the fast ice would transmit little stress to the fast ice; however, under some conditions this may not be the case. Shapiro (1975) demonstrated that drag effects due to shear at the boundary can propagate for some distance into the pack ice when the velocity vector of the moving ice is as close as 5^0 to the edge of the fast ice. A symmetric transmission of stress into the fast ice sheet is also implied, although it might be distorted around grounded ice features. If pack ice impacts normal to the fast ice edge there is, naturally, no shear component. In general, shear enters only to the extent that the effective coefficient of friction between the pack ice and fast ice is significant. The study of events and processes along this boundary is difficult and dangerous, and there are few data available, although W. D. Harrison (in Shapiro and Harrison, 1976) has presented an analysis of the problem.

Other questions involve the boundary conditions where ice meets the beach and how they affect ice pile-ups and ride-ups. On a large scale, the shape of the shoreline influences the geometry of the stress field in the fast ice sheet. On a smaller scale it might determine the locations of potential stress concentrations where fractures could be initiated. The strength of the bond between the ice and the beach is also important; the variables which influence this include the slope of the seabed near the beach, the depth to which the seabed is frozen, and the nature of the beach material.

Finally, the stress field within the advancing ice sheet is subject to boundary conditions which can vary in both space and time. For any given driving force, an ice sheet can advance as a unit or break into segments which move different distances up the beach depending upon factors such as the slope of the beach and whether the beach is frozen or thawed so that the advancing ice sheet can gouge or bulldoze the beach material. Similarly, fractures bounding the individual segments become the loci of shear or crushing, depending upon their orientation. Redistribution of stress due to these processes can make shore ice pile-ups, ridges within the ice sheet, and other features reflect local stress concentrations rather than the far-field stress.

The above discussion can be summarized into three related questions:

(1) What stress concentration mechanism operates during shoreward ice movements or, more generally, what are the properties of the stress field in the fast ice at various stages of such an event?

(2) What is the interaction at the boundary of pack ice and landfast ice?

(3) What variables at the fast ice-shoreline boundary influence the formation of shore ice pile-ups and ride-up features?

The remainder of this paper addresses the first of these questions.

III. DESCRIPTIONS OF EVENTS

The events discussed here (Table I) all occurred in the Point Barrow area. The features that formed have been described elsewhere (Shapiro et al., 1977; Hanson et al., 1979) and will be reviewed here only briefly.

The first two events we discuss occurred during spring breakup of the fast ice adjacent to the Naval Arctic Research Laboratory (NARL), 4 km north of Barrow on the Chukchi Sea coast. In both cases, the movements were sporadic, occurring in short episodes of up to about 1/2 hour each spaced over several days.

In early July 1975 an event occurred when the fast ice was melted back from the beach, leaving a shore lead up to 10 m wide, and the shoreward ice edge was free-floating. The ice at the time was about 1.0 m thick, and its temperature was at or very near the melting point. The part of the fast ice sheet which advanced shoreward included numerous multiyear ice floes and first-year ridges. The ridge heights, coupled with the bathymetry of the area, indicate that many ridges were grounded. Radar observations the previous autumn showed that the multiyear floes drifted into the area when the first-year ice sheet was just forming. The implication is that many drifted shoreward until they contacted the seafloor.

In 1976, a similar event occurred in late June. The characteristics of the fast ice sheet were similar to those described above; in fact, many of the multiyear ice floes and fragments had remained in place through the summer of 1975. The ice was about 1.5 m thick and its temperature was still below freezing. A shore lead was not well developed, and where the base of the ice sheet was exposed after the shoreward movement, the presence of a layer of gravel (about 30 cm thick at one locality) indicated that the freezing front of the shoreward edge of the ice sheet still

TABLE I.   *Ice Pile-Up and Ride-Up Events in the Point Barrow Area, 1975-1978*

Location	Date	Major features
Along beach at NARL	July 4-6, 1975	Offshore boundary of fast ice sheet moved shoreward about 250 m. Extensive ice pile-up and ride-up along about 2.5 km of beach.
Along beach at NARL	June 30, 1975	Fast ice sheet moved 15-25 m shoreward. Ice pile-up, ride-up and overthrusting along the same area as 1975 event.
City of Barrow	Dec. 30, 1977	Ice advanced up the beach on a 725-m front; maximum advance of 35 m. Ice pile-up along the entire front.
Point Barrow	Dec. 30, 1977(?)	Small ice pile-up on Chukchi Sea side of Point Barrow.
Tapkaluk Island	Late January 1978	Ice advanced along a 900-m front and overrode the island in several locations.
Martin Island	Late January 1978	Ice advanced onto island about 55 m along a 215-m front.
Igalik Island	Late January 1978	Ice advanced 105 m and overrode the island along a 400-m-wide front.

extended down into the seafloor. In some localities a thin ice foot, which had extended several meters toward the beach from the edge of the thicker fast ice sheet, was deformed during the movement (Fig. 1); gravel frozen into the base of the ice foot indicates the depth of freezing into the seafloor.

FIGURE 1. *Folded ice foot near NARL, June 1976. Note the gravel frozen into the lower surface of the ice sheet.*

Ice pile-ups and thrust features were distributed continuously along approximately the same 2.5 km of the beach in both years. Outside of this reach, the ice sheet moved shoreward far enough to close the shore lead for at least a few kilometers in each direction, but with only minor deformation along its shoreward edge. Within the zone of deformed ice at the beach, the zones of most intense deformation were almost identical for both events, suggesting control by local features of the beach or nearshore subsea topography. However, the style of deformation was different. In 1975 most of the shore-ice pile-ups were formed of broken rubble (Fig. 2). In 1976 the movement was largely taken up by thrusting of the advancing ice into superposed sheets (Fig. 3).

During the 1975 event, the ridge complex at the offshore edge of the fast ice was driven toward the beach through distances up to 250 m, about one-fifth of the initial width of the fast ice sheet at the time of the events. This motion was absorbed in forming the shore-ice pile-ups or thrusts at the beach and by pressure ridging at other locations in the fast ice. The boundary ridge and the ice sheet were also segmented by widely spaced transverse fractures; the segments moved through different distances as described below. In the 1976 event, while the extent of the

FIGURE 2. *Ice pile-up near NARL, July 1975. Pile-ups formed in this event consisted primarily of rubble as shown. In some localities, the rubble piles were partially overridden by the advancing ice sheet.*

fast ice was about the same as in 1975, the displacement of the offshore ridge was not more than 15 to 25 m, and the ridge did not break into segments. Instead, it was overridden or penetrated by the incoming pack ice along part of its length. In addition, no new ridges were formed within the fast ice sheet during the event.

The weather conditions at the times of the two events were similar. Local surface winds were essentially parallel to the coast from the southwest; weather maps denote the passage of a low-pressure system through the Chukchi Sea along an approximately northerly track, to the west of the coast. LANDSAT images show that there was little or no open water north of the pack ice edge in the Chukchi Sea prior to the time the events occurred, indicating that the pack ice had converged under the influence of southerly and southwesterly winds. This was verified in the local Barrow area during observational flights. In both years the pack ice offshore was continuous, with numerous new pressure ridges and no open water or thin ice.

During the 1975 event, the ice sheet broke into large plates along a series of cracks oriented almost normal to the shore. The cracks were spaced approximately 0.5 km apart and extended at least that distance offshore; some completely traversed the fast ice sheet, segmenting the bounding ridge as described above. The segment of the ice sheet defined by each pair of cracks moved independently and was displaced a different distance shoreward. In 1976 short fractures with similar orientation were

FIGURE 3. Overthrust ice sheet near NARL, June 1976. The leading edge
of the lower sheet is several meters inland from the water line.

present at a few locations, and differential shoreward motion was observed
in progress along one of them. However, long fractures through the entire
fast ice did not form in 1976, and it appeared that the fast ice sheet moved
almost as a unit.

As noted, in both years numerous first-year ridges and multiyear ice
floes appeared to be grounded within the fast ice sheet. Most of these
apparently moved smoothly with the ice sheet during the two events
(exceptions occurred only in shallow water where grounded multiyear floes
were overridden by the advancing ice sheet).   Field studies revealed no
evidence of rotation or fracturing around any of the grounded features,
which is taken to indicate that no differential motion occurred between the
features and the surrounding ice sheet during the movement. In addition,
side-scan sonar surveys of the seafloor after these events showed that the
dominant orientation of gouges paralleled the directions of ice motion of
the 1975 event as recorded by the University of Alaska shore-based radar
system (P. W. Barnes and L. H. Shapiro, unpublished data). Assuming that
the gouges were made by ice keels during the 1975 event then leads to the
conclusion that the seafloor sediment offered little resistance to the
movement.   This follows from the assumption because the keels were
strong enough to plow the seafloor, but the sediments did not resist them
enough to cause differential motion between the grounded features and the
remainder of the ice sheet.

The five remaining ice-push events to be discussed (Table I) occurred
during the winter of 1977–78 at the locations shown in Fig. 4.

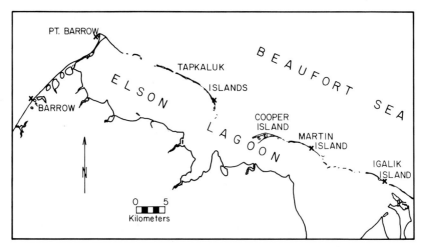

FIGURE 4. Locations (X's) of ice pile-ups and ice overrides in the Point Barrow area, December-January 1977-78.

The event at Barrow took place over a 12-hour period on December 30, 1977. At that time, a low-pressure system was moving northerly through the Chukchi Sea to the west (as was true during the spring events), and the local winds at Barrow were from the southwest. These winds persisted for about 3 weeks after the event and opened a large lead along the edge of the fast ice north of the barrier islands which bound Elson Lagoon. No shore ice ridges had formed on the islands by January 5, 1978 when an observational flight was made over the area. However, ice pile-ups, ride-ups, and overrides were observed on several of the islands on February 9, 1978. NOAA/VHRR satellite images for the intervening time showed that the lead remained open under the influence of southerly winds until January 18; then the wind reversed and closed it.

The area of the ice push at Barrow is shown in Fig. 5. The shoreward movement involved a single thrust sheet, which was pushed onto the beach slope along a front about 725 m wide. Additional minor movements occurred along the shoreline for 1 km to the southwest and 2 km to the northeast. The thrust sheet was bounded by fractures trending offshore, both of which appeared to terminate at the line of an older shear ridge about 700 m from the beach (details are shown in Fig. 5). This ridge was almost parallel to the shore; a segment (A in Fig. 5) which is slightly offset shoreward is visible on the thrust sheet. The fact that the fractures bounding the thrust sheet end at the shear ridge line suggests that this ridge was the boundary of the fast ice at the time of the event. The large ridge just south of the thrust sheet (B in Fig. 5) was examined in the field and found to be composed of blocks consisting of rounded fragments of ice of varying sizes in a matrix of finer grained ice. This is characteristic of the brash ice found in shear ridges and associated shear zones (Weeks *et al.*, 1971; Shapiro, 1975). It is probable that this ridge formed as a pressure ridge during the ice advance and incorporated ice from the pre-existing shear zone.

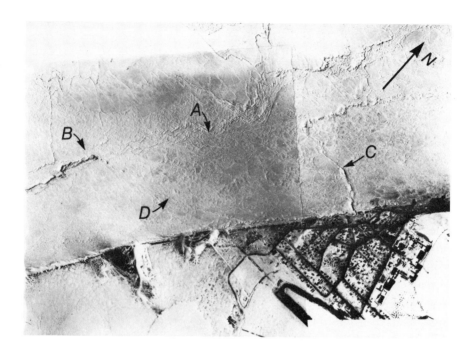

FIGURE 5. Ice push at the village of Barrow. The length of coastline shown is 3 km. The northeast boundary of the thrust sheet (C) is the line of ridges extending almost normal from the coast; the southern boundary (D) is a shear fracture shown as a sharp line curving toward shore from the end of the large pressure ridge B. 'A' indicates a segment of the shear ridge which marked the offshore boundary of the fast ice at the time of the event and which was translated shoreward during the movement.

The irregular fracture which bounds the northeast margin of the thrust sheet (C in Fig. 5) was marked by several small pressure ridges; some buckling of the thrust sheet into folds at a low to intermediate angle to the fracture also occurred. In contrast, the southwest boundary of the thrust sheet (D in Fig. 5) was a sharp, smoothly curving fracture with only minor deformation along its length, and trending toward tangency with the beach.   These observations suggest that, while the dominant dis-placement of the thrust sheet was shoreward, there may have been an additional rotation parallel to the smooth crack along the southwest boundary. This movement would result in a normal stress component across the northeast boundary that would account for the ridging and buckling along that line.

The ice, about 1.3 m thick at the time of the event, was pushed up the beach a maximum of 35 m at the easternmost corner of the thrust sheet, where it reached a beach elevation of 2.8 m above sea level. The heights of the shore ice ridges were generally between 4 and 6 m, with a maximum of 7 m at one location. Gravel berms, which are characteristically pushed up ahead of an advancing ice sheet, were generally absent (or small where present), reflecting the frozen nature of the beach surface at the time of the event.

Figure 4 shows the location of another ice pile-up at Point Barrow. Its time of formation is unknown, except that it was first observed and photographed on May 1, 1978 (it was not studied on the ground until summer). The photos show a high-angle fracture bounding the thrust sheet on the south. Because the northern part of the thrust sheet was not photographed, the shape of the fracture along that boundary is not known. The ice advanced only at the southeast corner of the thrust sheet, where it formed an ice-cored gravel mound alongshore about 1.6 m high, 14 m long, and 8 m from the shore. This feature, which was similar to those described by Hume and Schalk (1964), was destroyed by melting and wave action in late summer.

The remaining events occurred on the barrier islands north of Elson Lagoon (Fig. 4). The ice in the ice pile-ups ranged in thickness from 0.5 to 0.7 m; however, the unbroken ice sheet pushed onto Tapkaluk Island was 0.9 m thick. The difference in observed thickness between the ice sheet and the broken blocks is probably due to horizontal splitting of the ice sheet during piling, as well as the difficulty of locating samples of the full ice sheet thickness in the snow-covered ridges.

The ice was thrust onto Tapkaluk Island along a front nearly 900 m wide. Figure 6 shows the area as it appeared in late winter. Figure 7, taken in mid-June when the snow had melted off, shows the distribution of the ice pile-ups relative to the island and indicates that the thrust sheet included numerous cracks almost perpendicular to the beach and approximately parallel to the direction of motion of the advancing ice sheet (as shown by grooves in the beach surface). The contrast between these cracks and the surrounding ice is probably the result of clean water freezing into the cracks formed during the movement, adjacent to dirtier ice formed from muddy water during freeze-up. The cracks can, in some instances, be traced into the ridges at the front of the advancing ice sheet, and appear to define the boundaries between segments of the ice sheet which moved through different distances.

The island was completely overridden by three segments of the thrust. The largest of these was 120 m wide; there the ice advanced about 140 m. The island was 32 m wide at the override. The smaller overrides were 43 and 34 m wide and developed where the island was about 50 m wide. Ice pile-ups with elevations up to 10 m above sea level were distributed over the remainder of the width of the movement front. The maximum elevation of the island was about 1.5 m.

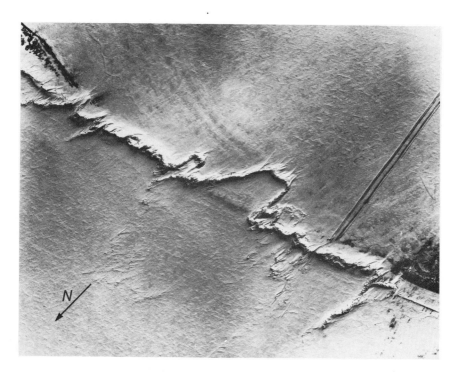

FIGURE 6. Override at Tapkaluk Island as it appeared in late winter. The length of the island included in the picture is about 1 km. The linear feature at right is the trail left by a seismic exploration party. The western boundary of the thrust sheet is in the lower right part of the picture.

Figure 8 is another view of the Tapkaluk Island override taken in late spring. The thrust sheet is bounded on the east by a curving fracture which trends to tangency with the beach, and on the west by a fracture zone at a high angle to the beach. The similarity to the geometry of the fractures bounding the thrust sheet at Barrow is apparent. Figure 8 also shows a distinct banding in the landfast ice offshore from the island. The banding probably resulted from the drift of slush ice into the area during freeze-up, and provides markers that can be used to determine the magnitude of the onshore motion through their displacement across the fractures which bound the thrust sheet. In vertical aerial photographs, this offset shows about 140 m of onshore movement, which agrees with the measurement of the maximum advance of the ice over the island as noted above. Differences between this value and the advance of the ice at any other point must be taken up by increasing the volume of ice in the ice pile-ups.

On Martin Island (Fig. 9), the ice advanced a maximum of 55 m, without completely overriding the island. The width of the overthrust

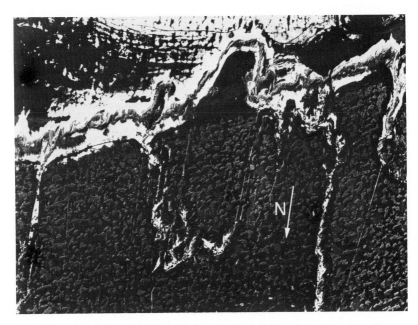

FIGURE 7.   Mid-June view of Tapkaluk Island showing the relationship between the ice pile-ups and overrides and the island.   The length of the island shown is about 800 m.   Note the fractures indicated by the light-toned lines nearly normal to the trend of the island, and their relationship to the distribution of remnants of the ice-push features.

sheet was 215 m, and the maximum elevaton on the island which the ice reached was 1.7 m.   Note the similarity between the geometry of the fractures bounding the thrust sheet with those of the thrusts on Tapkaluk Island discussed above.   However, photographs taken in late spring show no pattern of cracks internal to the thrust sheet and parallel to the direction of motion as formed at Tapkaluk Island.

The form of the thrust sheet at Igalik Island was more similar to that of the spring 1975 event at Barrow than to those of the winter events just described.   The maximum advance of the ice sheet was about 105 m along the front of a segment 400 m wide which completely overrode the island. The fractures bounding the segment were parallel and almost perpendicular to the shore.   The advancing ice sheet moved as a single unit with no segmentation or differential displacement at the leading edge (Fig. 10) and no apparent internal fractures parallel to the direction of motion.   Minor advance of the ice sheet occurred along an additional 1.4 km of the island beach, and motion of the ice into the pass east of the island was also indicated by the presence of pressure ridges in that area.

The December 1977 event at Barrow occurred just outside the 4 km field of view of the University of Alaska sea ice radar system in operation at NARL.   However, the radar data show that during the time of the event

FIGURE 8. Oblique view to the northwest showing Tapkaluk Island in the foreground and Point Barrow in the distance. Note the shape of the fracture along the east side of the thrust sheet and the offset of the banding in the ice sheet (foreground).

ridging occurred along a line about 3 km offshore from the radar site. This suggests that pressure from the pack ice was exerted along at least 8 km of the edge of the fast ice and possibly 13 km, if it is assumed that the ice advance at Point Barrow occurred at the same time. Within this distance, the only major advance of the ice up the beach was at the village of Barrow along a 725-m front. Minor movements of less than 3 m occurred along an additional 3 km of the beach, but in general the motion was taken up by ridging either within the landfast ice or at its outer boundary. Similarly, if it is assumed that all of the events along the barrier islands occurred at the same time, then the movements in that area occurred along a front about 30 km wide. However, the length of the beaches which were overridden is less than 3 km of the approximately 13 km of island beaches exposed to movement in that direction. No evidence of motion of the ice into the passes between the islands was found other than at the east end of Igalik Island. Thus, though the satellite imagery indicates that the pack ice closed against the edge of the landfast ice along a wide front, ice pile-up and ride-up occurred along a small fraction of the beach in this same area.

During the spring events at Barrow, the advancing ice sheets pushed up gravel berms at their leading edges. In the 1975 event these were important in initiating piling of the ice in areas of low beach slope; the leading edge of the ice sheet tended to ride up on the berm, then break as it dropped to the onshore side. This produced an irregular surface for the

FIGURE 9. Early summer view of the ice pile-up on Martin Island. The length of the island shown is about 850 m. Note the shape of the thrust sheet and the absence of fractures normal to the beach (compare with Fig. 8).

ice sheet to advance over, and additional breaking and piling resulted. No berms were noted at the sites of the winter events on the barrier islands, and their absence may have been partially responsible for the large advances of the ice sheets.

In sum, we can make six generalizations from the observational data:

(1) In ice ride-up events, the length of beach affected is probably significantly shorter than the length of the edge of the fast ice along which the driving forces are applied.

(2) The pack ice through which the driving force is transmitted is likely to be compact, indicating that ridging of thin, weak ice has already occurred at the time of an ice push event. Thus, the stresses transmitted to the fast ice edge and on toward the beach may be greater than those in normal pack ice where thin ice is present. This also implies that ice push or ride-up events are driven by wind stress imparted to the ice over large areas, and that local winds are not alone responsible.

(3) During the spring events the ice sheet was initially driven toward the beach as a unit, closing the shore lead. Then, fractures formed parallel to the direction of the motion (usually at high angles to the shore). During small advances, the cracks did not propagate far offshore but for large

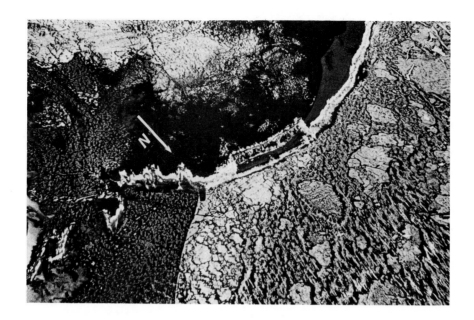

FIGURE 10. Early summer view of ice override on Igalik Island. The width of the override along the island is 400 m. North is to the bottom of the picture.

advances, some of the fractures crossed the entire 1.5-km width of the fast ice sheet. This indicates that the length of the cracks depends upon the extent of the advance. (A similar conclusion follows from comparing the winter event at Tapkaluk Island and those at Barrow and Martin Island.) The longer fractures separated the ice sheet into segments, each of which advanced through a different distance.

(4)   The three winter events at Barrow, Tapkaluk Island, and Martin Island show a similar pattern of fractures bounding the segment of the fast ice sheet which advanced. The pattern consists of two fractures; one is approximately perpendicular to the beach and may show evidence of active compressive stress (ridging or buckling) along its length. The second fracture extends from a point at or near the intersection between the first fracture and the beach. At its origin, it is approximately tangent to the beach but it curves offshore with increasing distance. No movement of the ice sheet occurs until these fractures form.

(5)   The Igalik Island event, which occurred in winter, was like the spring events at Barrow, in that fractures bounding the thrust sheet were at a high angle to the shore, and the thrust sheet moved as a plug with little internal fracturing.

(6)   The spring events caused extensive bulldozing of the beach gravel into ice-push ridges as much as 1 m high, which enhanced the probability of

formation of ice pile-ups. In contrast, bulldozing of the beach during the winter events was minor.

IV. DISCUSSION

The cracks bounding the winter thrust sheets at Barrow, Tapkaluk Island, and Martin Island share a geometric configuration which has an analogy in the theory of plasticity. The pattern of two cracks meeting at a right angle at the boundary of a field of deforming material, one curving and tangent to the boundary and the other normal, occurs frequently in slip-line field theory (see for example, Prager and Hodge, 1951). An example is shown in Fig. 11, which is part of the solution to Prandtl's problem for the flow field of a perfectly plastic material squeezed in plane strain between two rigid plates with rough (no-slip) surfaces. The lines indicate the trajectories of the maximum shear stresses and, according to von Mises' criterion, can be viewed as lines along which failure in shear can occur. This interpretation has been used to explain fracture patterns observed in rocks (Varnes, 1962; Cummings, 1976).

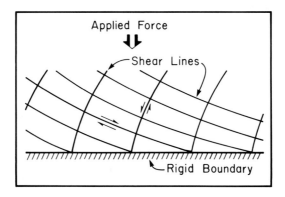

FIGURE 11. *Example of the relationships between shear lines and a rigid boundary from Prandtl's solution for an infinite slab of a perfectly plastic material compressed between rough, rigid plates (from Prager and Hodge, 1951).*

If we assume the analogy to apply at least qualitatively to the plane stress problem for an ice sheet in winter, an immediate implication is that the ice was firmly frozen to the beach when the cracks were initiated, in accordance with the requirement that the boundary be rough. There is also the suggestion that the direction of loading was normal to the beach. However, this does not require that the force arising from the interaction between the pack ice and the fast ice be directed normal to the shore. Some deviation can occur, depending upon the value of the effective coefficient of friction at the boundary. Clearly if this is low, then only the normal component of the applied force will be transmitted.

As noted, in the two spring events at Barrow the edge of the ice sheet was melted back from the shore when the first motion occurred. Thus, movement through several meters was possible before the ice contacted the beach. Subsequently, the advance was irregular with some sections of the leading edge stopped while others continued to move. A similar pattern of displacement apparently developed in the Tapkaluk Island event. Possible patterns of stresses and displacements in the fast ice sheet resulting from movements of this type were examined using finite-element methods.

The SAP IV structural analysis finite-element program (Bathe *et al.*, 1973) was used to model a fast ice sheet in irregular contact with the shore under a distributed load along its offshore boundary. A two-dimensional array of linearly elastic membrane elements was used to form a 300-element rectangular mesh to represent an ice sheet adjacent to a beach. The state of the ice sheet-beach interface was specified by the boundary conditions along one of the long edges of the rectangle. These determined the lengths of reaches of the boundary along which the ice edge was fixed or free to move in response to the applied loads. The configuration used in most of the calculations consisted of a fixed boundary with central free segments of various lengths.

A uniformly distributed load was applied on the seaward boundary of the ice sheet (the grid boundary opposite the ice-beach interface) for each calculation run. The loading direction was varied to simulate normal loading (load directed normal to the beach), simple shear loading (load directed parallel to the beach) and combined shear and normal loading (load directed at 45° to the beach).

The calculations did not include a mechanism by which fixed node points along the beach could be released and moved according to some criterion in the program. Further, the SAP IV program has no provision for the use of stress-strain laws other than linear elasticity. Therefore, the results of the computations could only be used as indicators of the geometries of the stress and displacement fields at the instant the load is applied. These fields were interpreted on the assumption that stress concentration points and zones of high stress or displacement gradients indicated the points of origin and directions of propagation of fractures in the ice sheet. While a more flexible model would have been useful, the results are adequate for our qualitative interpretations.

The most important results of the calculations are that, for cases of normal loading and combined normal and shear loading, the highest stress concentrations occur (as expected) at the points of transition from the fixed- to free-boundary condition along the ice-beach interface. In addition, the zones of high displacement gradient which trend offshore from these points suggest the formation of shear zones oriented approximately parallel to the line of action of the boundary stresses. Thus, if only part of the ice-beach boundary is free to be displaced, the ice sheet within this segment will tend to advance up the beach as a plug, as would be anticipated for a punch test. The same displacement pattern would also result if the ice sheet was in motion and then stopped along sections of its leading edge due to interaction with the beach or other obstacles. In that case, a crack at a high angle to shore and parallel to the direction of motion would be expected to form. An example from the 1975 Barrow event is shown in Fig. 12. The crack pattern within the thrust sheet at Tapkaluk Island also shows these characteristics with the fractures separating the segments of the ice sheet that moved through different distances (see Fig. 7). The implication that the length of the fractures is related to the magnitude of differential displacement was noted above, but the restriction of the model to linear elasticity prevented this possibility from being examined.

FIGURE 12. Shear offset in the fast ice sheet at NARL, July 1975.

V. PROPOSED MODEL

The results of the observations and analysis above can be combined to suggest a model for the fracture and displacement of a fast ice sheet during a typical winter ice-push or ice ride-up event (Fig. 13). The ice is assumed to be initially fixed to the beach, and the possible effects of grounded features are ignored. Figure 13A shows the geometry of the initial cracks which form according to von Mises' criterion. Some minor landward movement all along the front can occur at this time. In the second stage (Fig. 13B), the thrust sheet is pushed up the beach with a tendency to rotate along the curving crack as indicated by the arrows. This results in the formation of compressive features along the high angle crack. In addition, the normal force transmitted across the curving crack causes onshore movement of the wedge of ice between that crack and the shore. Figure 13C shows the final form of the process (represented by the December 1977 event at Barrow) if there is little or no differential motion

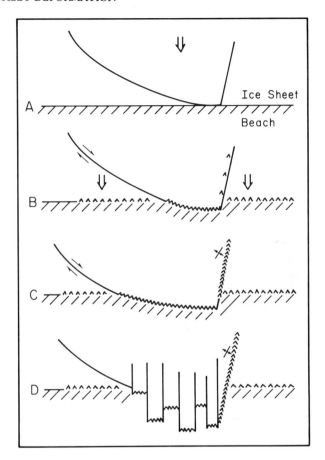

FIGURE 13. Proposed model for the fracture and displacement of a fast ice sheet during an ice-push event. See text for discussion.

at the leading edge of the thrust sheet as it moves up the beach, or if the total displacement is small. The alternative of segmentation of the thrust sheet as occurred at Tapkaluk Island is shown in Fig. 13D. In this case, compressive features at either side of the diagram may not be present if there is little resistance to movement along the leading edge of the thrust sheet. Instead, plug geometry (as described above) develops rapidly, and motion occurs along the high-angle cracks that bound the plugs. Additional cracks with similar orientation then develop as the movement progresses.

For an ice-push event during break-up, melting along the shore weakens the bond between the ice sheet and the beach. Thus, the first three steps in Fig. 13 do not apply. Instead, cracks at a high angle to the beach form in the early stages of the process and govern subsequent displacement of the ice sheet. A similar process might also be expected to occur for an event during freeze-up, when the ice sheet is not firmly bonded to the beach.

The geometry of the initial stress field at the time when the fractures of Fig. 13A are formed can be assumed to be similar to Prandtl's solution, modified to reflect the distribution of grounded features, the shape of the shoreline and of the pack ice-landfast ice boundary, and the distribution of forces along the latter. The fractures begin at some local stress concentration along the shoreline. If no suitable stress concentration exists, then the stresses are relieved by ridging at the edge of the fast ice, or elsewhere in the fast ice sheet, and no ice advance will occur. After the fractures form or the movement of the thrust sheet begins, the stress distribution will vary in accordance with changes in the boundary conditions and the distribution of new fractures within the thrust sheet. The final form of the thrust sheet (including ice pile-up and ride-up at the beach) will depend upon the sequence of changes and the magnitudes of the stress variations which occur throughout the process. As a result, the extent to which the local stresses deduced *ex post facto* are representative of the far-field stresses at the start will vary from event to event.

This model assembles the available information into a pattern; clearly the interpretation will require modification as more data become available. However, even a preliminary model can serve to emphasize questions that can usefully be raised regarding the processes of ice push and ice ride-up. Before proceeding to these, however, we should note that the January 1978 event at Igalik Island does not fit the model because its crack pattern is more similar to a spring event. There are several possible explanations, but none are supported by data. It is possible that the open lead was closer to (or at) the beach here than at other localities along the barrier islands. The resolution of the satellite imagery does not permit this to be evaluated. Alternatively, it is possible that the bonding between the ice sheet and the beach was uniformly weak; thus motion occurred all along the beach when the load was applied, as is postulated during spring events. Subsequently, the low, uniform slope of the island did not resist the motion enough to break the ice sheet up into smaller segments.

VI. CONCLUSIONS

The suggestion that ice pile-up and ride-up represent the result of a stress concentration mechanism is supported by three points: (1) The length of these features along the beach is generally less than that along which pack ice-fast ice interaction occurs; (2) For the case of winter events, the fractures bounding the thrust sheets tend to make the dimensions of the thrust sheet wider at the offshore edge than at the beach. (3) For an ice sheet advancing as a series of discrete segments, the resistance to the motion (and thus the force) can vary between segments, depending upon local conditions. This emphasizes the need for care in deducing far-field stresses from those required to form local features. It is likely that such values will be conservative in the sense that they will overestimate the far-field stresses, but the extent to which this is true depends on the contribution of other energy sinks that may not be accounted for in the calculations.

The discussion above addresses a possible sequence of events for breakup of an ice sheet during an ice-push event. Other aspects which require study include (1) possible seasonal variation in the efficiency of grounded features in anchoring the ice sheet and thus affecting shoreward ice movement, (2) the influence of the nature of the boundary between the fast ice and the beach, and (3) the nature of the interactions between pack ice and fast ice.

ACKNOWLEDGMENTS

We thank A. Kovacs, D. Sodhi, G. Weller, and G. Cox for their thoughtful reviews.

REFERENCES

Bathe, K-J., Wilson, E.W., and Peterson, F.E. (1973). *Earthquake Engineering Research Center Report 73-11.* Univ. of Calif., Berkeley.

Cummings, D. (1976). *Bull. geol. Soc. Amer. 87,* 720.

Hanson, A., Metzner, R., and Shapiro, L. (1979). *In* "Environmental Assessment of the Alaskan Continental Shelf," Ann. Rept. Vol. 9, p. 371. NOAA, Boulder.

Hume, J.D. and Schalk, M. (1964). *Amer. J. Sci. 262,* 267.

Kovacs, A. and Sodhi, D.S. (1979). *In* "Proceedings POAC '79," vol. 1, p. 127. Technical Univ. of Norway, Trondheim.

Kovacs, A., Sodhi, D.S., and Cox, G.F.N. (1982). *CRREL Report 82-31.* U.S. Army Cold Reg. Res. and Eng. Lab., Hanover, NH.

Parmerter, R.R. and Coon, M.D. (1972). *J. geophys. Res. 77,* 6565.

Prager, W. and Hodge, P.G., Jr. (1951). "Theory of Perfectly Plastic Solids." Wiley, New York.

Pritchard, R.S. (1977). *In* "Proceedings, POAC '77," vol. 1, p. 494. Memorial Univ. of Newfoundland, St. John's.

Shapiro, L.H. (1975). *In* Proceedings, POAC '75," vol. 1, p. 417. Univ. of Alaska, Fairbanks.

Shapiro, L.H., Bates, H.F., and Harrison, W.D. (1977). *In* "Environmental Assessment of the Alaskan Continental Shelf," vol. 15, p. 1. NOAA, Boulder.

Shapiro, L.H. and Harrison, W.D. (1976). *In* "Environmental Assessment of the Alaskan Continental Shelf," vol. 14, p. 117. NOAA, Boulder.

Varnes, D.J. (1962). *U.S. Geol. Survey Prof. Paper 378-B,* B-1.

Vivatrat, V. and Kreider, J.R. (1981). *In* "Proceedings of the Offshore Technology Conference," p. 471. Houston, TX.

Weeks, W.F., Kovacs, A., and Hibler, W.D., III (1971). *In* "Proceedings POAC '71," vol. 1, p. 152. Technical Univ. of Norway, Trondheim.

PACK ICE INTERACTION WITH STAMUKHI SHOAL
BEAUFORT SEA, ALASKA

Erk Reimnitz
E. W. Kempema

U.S. Geological Survey
Menlo Park, California

I. INTRODUCTION

The morphology of the Beaufort sea continental shelf is characterized by a series of linear shoals occurring slightly landward of the 20-m depth contour. These shoals interfere with the shifting ice pack and localize the formation of grounded ice ridges and hummocks, which in turn serve as "strong points" in the establishment of the seasonal ice zonation. Despite the scouring action of drifting ice, the crests of the shoals are not worn down, indicating rapid reconstruction by unknown processes. Shoreward migration of several shoals supports this notion (Reimnitz *et al.*, 1978a, b; Reimnitz and Maurer, 1978).

Piles of grounded ice on shoals, called stamukhi, protect the inner shelf from pack ice forces, allow the growth of relatively smooth, immobile fast ice, and thereby indirectly facilitate the development of oil resources. The shoals have more direct value to petroleum development in artificial island construction because, as a rule of thumb, each vertical foot of fill costs 2 to 3 million dollars. Thus, an understanding of processes affecting the stability of the shoals and the mechanisms of ice interaction has considerable importance.

Major ice piles seen repeatedly during the first several years of Landsat coverage in the same area of the Beaufort Sea shelf suggested the presence of a large topographic high where none was charted (Fig. 1). The USGS R/V KARLUK was used in 1977 to survey a 17-km-long linear shoal that rises as much as 10 m above the surrounding seafloor. The shoal, called Stamukhi Shoal, stands out on satellite images obtained in most summers and winters since then. Because Stamukhi Shoal, as a well-defined and dynamic feature, occupies a key position relative to ice

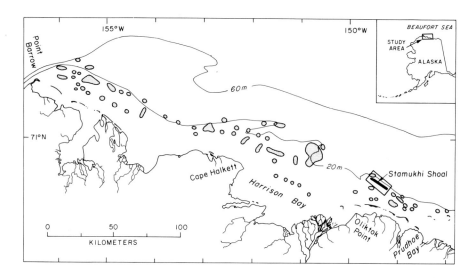

FIGURE 1. Map of study area showing all isolated shoals covered by less than 11 m (6 fm) of water (stippled areas), as taken from NOS chart no. 16004. Stamukhi Shoal lies where no shoal was charted. The two shoals indicated offshore and inshore of the west tip of Stamukhi Shoal, along with certain other shoals shown on chart 16004, do not exist where charted, but the indicated belt of bathymetric anomalies is characteristic of the stamukhi zone. The box around Stamukhi Shoal delineates Figs. 2, 7, 8, and 10.

zonation, we have extended our studies in recent years to make it the best known shoal on the Alaskan shelf.

In this report we use Landsat images to show the effects of Stamukhi Shoal on winter and summer ice regimes of the last 10 years. We also use side-scan sonar records, fathometer records, and direct diving observations to provide details of seafloor morphology and its changes resulting from ice keel interaction and currents affecting the shoals. Finally, other shelf surface anomalies along a line east and west of Stamukhi Shoal and forming the seaward boundary of the fast-ice zone will be discussed.

II. BACKGROUND INFORMATION

Soviet investigators and scientists were the first to note the role of shoals in the establishment of yearly ice zonation in the Laptev and East Siberian Seas. Zubov (1945) reported: "The importance of shallows also is manifested in the fact that ice heapings of various sorts, having considerable vertical measurements, ordinarily become grounded on shallower places like banks, rocks, and shoals. Later, these heapings, under the pressure of the ice from the sea, increase in size, become more durable, and play the role of offshore islands in the development of fast ice."

In recent years increasing numbers of observations and studies on the shelves of northern Alaska have shown the interaction between isolated shoals and pack ice. Reimnitz et al. (1972) and Reimnitz and Barnes (1974) observed the shadowing effect of topographically high regions on the seafloor that protect the seafloor from drifting ice keels, and they further noted increased ice concentrations and ice gouging (also called ice scouring; see Barnes et al., this volume) on shoals compared with deeper surrounding terrain. With the advent of repetitive coverage by Landsat-1 satellite imagery, certain continental shelf regions were conspicuous because of the recurrence of ice heapings and the subsequent stability of ice piles, suggesting ice interaction with the seafloor (Stringer, 1974a, b; Stringer and Barrett, 1975a, b; Kovacs, 1976; Toimil and Grantz, 1976; Stringer, 1978; Barry et al., 1979). Using a combination of satellite imagery and seafloor data, Reimnitz et al. (1978a, b) first noted the role that shoals play in establishing the annual sea-ice zonation in the Beaufort Sea. The Russian term stamukha (plural stamukhi) refers to large ice heaps that form on shoals along the outer margin of the smooth land-fast ice and commonly remain through much of the following summer. Following that usage, Reimnitz et al. (1978b) introduced the term "stamukhi zone" for the belt of major grounded-ice ridge systems seaward of the fast ice.

Large stamukhi act as fences and accumulate smaller ice floes. Thus ice commonly prevents access by the small survey vessels used for most studies in this region. The bathymetry, geology and sediment distribution in the stamukhi zone, which straddles the midshelf within the 18- to 35-m depth range, are therefore only poorly known. The belt of isolated shoals stretching along this zone between Point Barrow and Prudhoe Bay shown in Fig. 1 (from NOS chart no. 16004) only indicates where shoals are concentrated; their precise locations are highly uncertain (Reimnitz and Maurer, 1978). However, an unusual ice-free season in 1977 allowed a detailed survey of the Stamukhi Shoal area. Some of the results and background information on linear shoals inshore of Stamukhi Shoal were presented by Reimnitz and Maurer (1978). They concluded that the shoals are reshaped and moved under the influence of ice and are not relict barrier islands dating from times of lower sea level. Thus for an understanding of the shoals, a comparison with superficially similar features on the eastern seaboard of the United States is not useful (Reimnitz and Maurer, 1978).

For detailed descriptions of the regional setting of sea ice and marine processes, refer to Barnes and Reimnitz (1974), Kovacs and Mellor (1974), Reimnitz and Barnes (1974), and Reimnitz et al. (1978b). In general, the shelf is ice-covered most of the year, except for the period from mid-July to the end of September when open water, having variable concentrations of ice, exists. The ice motion on the shelf is predominantly from east to west, parallel to the isobaths. Several times during the last 10 years at the onset of winter, no multiyear ice or stamukhi existed in the Stamukhi Shoal area (see Reimnitz et al., 1978b), but big grounded ice ridges were found there several months later. In these years the ice piles were constructed entirely of thin ice that, until it was deformed, could not touch bottom. The relationship between the shoal and the formation of grounded ice ridges along the shoal crest is unknown.

III. METHODS OF STUDY

Two types of data form the basis for this study: satellite imagery and seafloor and ice observations. The best available Landsat images from 1972 through 1981 were studied for ice features related to the presence of Stamukhi Shoal. The results have been compiled for all years of record to distinguish winter and summer effects. Seafloor data in the Stamukhi Shoal area, gathered from the USGS R/V KARLUK in 1977, 1980, and 1981, include side-scan sonar, fathometer, and high-resolution seismic records. Most of the 1977 coverage had precise range-range electronic positioning control and was run at fast speed using the fathometer and 7-kHz subbottom profiler. Only two 1977 crossings of Stamukhi Shoal were surveyed using a Uniboom system* and side-scan sonar at the slow speed required for this work (Maurer et al., 1978). Two diving traverses supplement the 1977 data. All of the 1980 coverage was run at a boat speed of about 3.5 knots using side-scan sonar, a 7-kHz subbottom profiler, and a Uniboom. In 1980, however, only satellite navigation was available, providing intermittent fixes of only ± 1/2-km accuracy and with larger errors in the intervening dead-reckoning periods. Therefore, these records cannot be matched precisely to the contour chart of Stamukhi Shoal based on the 1977 survey. The navigation control nevertheless is adequate for locating each shoal crossing in its approximate place along the length of the shoal. Numerous grab samples were collected during the 1980 survey and analyzed for grain-size distribution. In 1981 we made two side-scan sonar mosaics using the EG&G SMS 960 seafloor mapping system and precision navigation, one over the northwest tip and one over the southeast tip of the shoal. During the 1981 field work, we also reran three of the accurately positioned 1977 fathometer lines to record changes in shoal morphology; in addition, we towed divers on a sled along one of these traverses.

IV. RESULTS

A. Bathymetry

Stamukhi Shoal is a 10-m-high ridge oriented northwest-southeast parallel to regional isobaths at a depth of 18-20 m (Fig. 2). Except for two small knolls seaward of the northwest tip, the surrounding shelf is smooth with a gentle seaward slope. The shoal terminates abruptly at both ends, with the highest point being near the northwest end. Ten representative

*Any use of trade names is for descriptive purposes only and does not imply endorsement by the USGS.

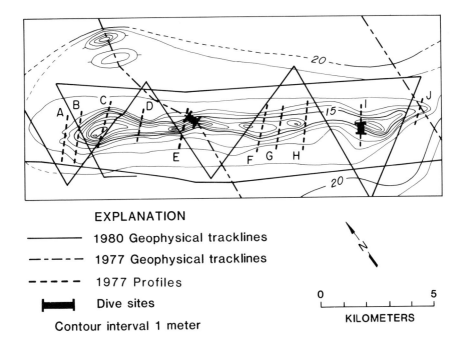

EXPLANATION

———————— 1980 Geophysical tracklines

— · — · — · 1977 Geophysical tracklines

— — — — — 1977 Profiles

▬▬▬ Dive sites

Contour interval 1 meter

0   5

KILOMETERS

FIGURE 2. *Bathymetric map of Stamukhi Shoal based on 1977 surveys, also showing 1977 and 1980 geophysical tracklines, cross sections of Fig. 3, and two diving traverses. Area covered by this figure is indicated in Fig. 1, and is the same area as Figs. 7, 8, and 10.*

shoal cross sections, labeled A through J in Fig. 2, are shown in Fig. 3. No cross section can be called "typical," and all have considerable microrelief from ice scouring. Jagged local relief occurs on the seaward toe of the shoal. Profiles I and J were smoothed visually to eliminate false relief resulting from rough seas during the survey.

B. Ice Patterns from Satellite Images

From the first Landsat image of the study area in July 1972 and extending through the winter of 1981, the presence of Stamukhi Shoal is revealed either by characteristic ice types or as a boundary for sea-ice distribution. Winter ice patterns are best exhibited in satellite images taken just prior to sea-ice breakup. At that time, high features, from which meltwater has drained, are white. These high areas contrast sharply with smooth, low-lying ice that collects meltwater and appears dark. Figure 4 is a Landsat image covering extensive regions east and west of Stamukhi Shoal. The crest of the shoal is marked by strong lineations from ice ridging and separates smoother ice on the seaward side from a

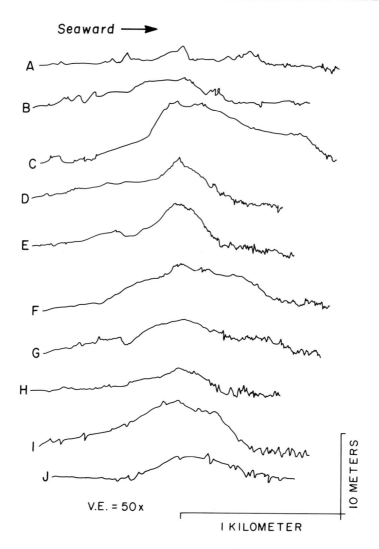

FIGURE 3. *Representative cross sections of Stamukhi Shoal in 1977, including ice-gouged microrelief. Locations are shown in Fig. 2.*

triangular hummock field reported by Stringer (1974b). The ice landward of the shoal is immobile at this time, held by grounded ridges on the shoal, while some of the adjacent ice is beginning to move. Reimnitz and Barnes (1974) used this image to delineate the outer edge of the fast ice following the crest of Stamukhi Shoal without knowledge of its existence. A lack of

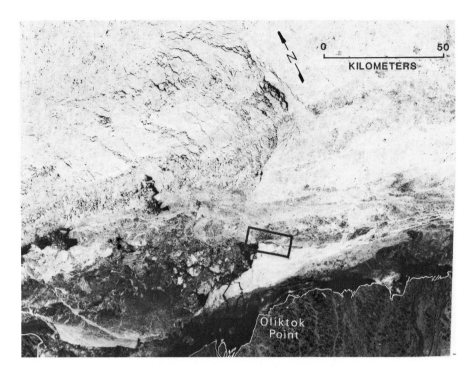

FIGURE 4.   Landsat image of July 2, 1973 showing the two most
characteristic effects of Stamukhi Shoal on the winter ice regime: (1) The
shoal is marked by grounded ridges (E in Fig. 6), and (2) these ridges
separate two distinct ice types (F in Fig. 6).   The box shows the area of
Figs. 2, 7, 8, and 10.   For a more detailed analysis of this scene see
Reimnitz et al. (1978b).

multiyear ice on the Beaufort Sea shelf during the previous freeze-up
indicated that the grounded ridges on Stamukhi Shoal (Fig. 4) are
constructed of thin, first-year ice, which in the undeformed state could not
touch bottom (Reimnitz et al., 1978b). This lack of multiyear ice during
freeze-up and through the winter was repeated for at least two more
seasons (1977-78 and 1978-79) during the study period.

Two typical winter ice effects that resulted from Stamukhi Shoal
serving as a strong point are shown in Fig. 4: (1) the linear shoal is a focal
point for ice ridging, and (2) the shoal is a boundary between two distinct
ice fields.

Figure 5 shows the effects of Stamukhi Shoal and other shoals in the
region on drifting ice in a recurring summer pattern. These two Landsat
images were taken in 1974 and 1977. The strikingly similar pattern develops
under dominant northeasterly wind and westerly current, causing ice to
drift onto shoals, and suggests that shoal attrition from ice impacts is
occurring. The ice lineation that continues east of Stamukhi Shoal, having

FIGURE 5.   Landsat images of Sept. 6, 1974 (top) and Aug. 12, 1977, showing a typical recurring summer scene of shoals collecting westward-drifting ice.   The crest of Stamukhi Shoal, which obviously suffers a very large number of ice impacts per year, marks the inner edge of the ice fields north of Oliktok Point.   The shore-parallel, en-echelon ice pattern east of Stamukhi Shoal is controlled by a slight break in bottom slope, which is dotted by 3- to 4-m-high shoals (Rearic and Barnes, 1980).   The box northeast of Oliktok Point shows the area of Figs. 2, 7, 8, and 10.

a slight seaward en-echelon offset, marks a set of poorly charted shoals 3-4 m high (Rearic and Barnes, 1980).

Besides the two winter manifestations of ice dynamics and the summer situation described above, other patterns can be recognized. We grouped ice patterns or types that are related to, or caused by, Stamukhi Shoal into four summer (Fig. 6A,B,C,D) and three winter categories (Fig 6E,F,G). The seasons in which these seven categories are recognized on available Landsat images are listed in Table I with scene identification numbers. A brief discussion of the categories follows.

(A) Stamukhi Shoal and stamukhi act as barriers to drifting sea ice, generally providing shelter to the inner shelf. The situation depicted in this image followed the indistinct pattern G seen slightly earlier that same season.

(B) Stamukhi Shoal crest is marked by line of stamukhi, with relatively open water on either side.

(C) Stamukhi Shoal corresponds to one margin of a lead. The adjacent ice can move either landward or seaward away from stamukhi on the shoal.

(D)   Stamukhi Shoal is marked by a chain of grounded ice-island fragments, tabular massive glacial ice derived from the Ward-Hunt ice shelf (Breslau et al., 1971; Skinner, 1971; Brooks, 1973). In 1972, a large drifting ice island broke up in the Beaufort sea and scattered over 400 fragments, commonly having diameters of 40-100 m but ranging up to 3000 m, along the coast of northern Alaska (Kovacs and Mellor, 1974). Brooks (1973) reported more than 40 fragments were aligned along the 18-m isobath over a distance of 32 km in the area of Stamukhi Shoal.

(E) Ice-ridge lineation is coincident with Stamukhi Shoal, as discussed above.

(F)   Ice boundary coincides with Stamukhi Shoal and separates two distinct ice types, as discussed above.

(G)   Large, indistinct ice piles accumulate along general trend of Stamukhi Shoal. In two winters (1978 and 1979) of such poorly defined ice piles, we flew low-level reconnaissance flights along the shoal and landed in several spots, confirming the existence of massive grounded ice piles. In both of these seasons, no multiyear ice existed in the area, and in May 1978 pressure ridges of new ice contained sand, pebbles, and shells, demonstrating interaction of thin first-year ice with the crest of the shoal.

The summary of available satellite observations made from 1972 through 1981 (Table I) shows that Stamukhi Shoal interacted with pack ice in all but two summers (1973 and 1979). Stamukhi Shoal probably interacted with the pack during these two summers as well, but during periods undocumented by Landsat images.

C. Bedforms

The bedforms delineated in the Stamukhi Shoal region by geophysical surveys using side-scan sonar and fathometer include (1) ice gouges, (2) ripple fields, (3) sand waves, and (4) jagged outcrops. Figures 7, 8, and 9 present compilations of bedform data from 1980 and 1981 surveys. These compilations would be quite different if they were made using the data collected in 1977, as shown later. The 1980 tracklines were not shifted to fit the accurate 1977 bathymetry, and the sinuous shoal crest plotted for reference on Figs. 7, 8, and 10 is a result of position inaccuracies. However, this is of no consequence to the following discussion.

1. Ice Gouges

Visual comparisons of all 1980 segments of side-scan sonar records against counted segments of sonar records allowed us to group ice gouge densities per kilometer of trackline into four classes (Fig. 7): (1) areas with high gouge densities, estimated 100 or more gouges per kilometer of trackline, an example of which is shown on the left in Fig. 8C; (2) areas with medium gouge densities, estimated 30 to 100 km^{-1}, (3) areas with low

SUMMER   WINTER

TABLE I. *Ice Patterns A Through G (see Fig. 6), Seasons Observed, and Representative Landsat Images Identified by Number*

Year	Winter		Summer	
	Type	ID number	Type	ID number
1972	No data		A	1002-21300
			D	1020-21281
1973	E	1326-21284	No ice anywhere on shelf	
	F	1344-21283		1397-21220
1974	F	1723-21260	A	1775-21124
1975	F	1812-21172	B	2233-21213
			C	2178-21165
1976	F	2538-21095	A	2592-21082
			B	2556-21092
1977	G	2896-20434	A	2915-20483
1978	G	30095-21281	A	30164-21115
			B	30182-21121
1979	G	21635-21044	No data	
1980	G	21980-21265	A	30866-21021
			B	RBV Scene – Aug. 23
1981	F	22339-21193	B	22372-21013

gouge densities, estimated 10 to 30 km^{-1}, (as in Fig. 8B); and (4) areas with very low gouge densities, an occasional scratch on the seafloor, or no gouges at all, as exemplified in Figs. 8A and 8C on the right side.

FIGURE 6. Seven types of summer and winter ice patterns controlled by Stamukhi Shoal, but not necessarily persisting through entire seasons: A, drift-ice barrier (July 25, 1977); B, stamukhi lineation (Sept. 3, 1978); C, lead boundary (July 19, 1975); D, ice-island lineation (Aug. 12, 1972); E, ice-ridge lineation (July 2, 1973); F, ice-type boundary (October 13, 1974); and G, indistinct ice piles (July 7, 1977). Map at bottom right shows Stamuki Shoal and other major shoals in a stippled pattern.

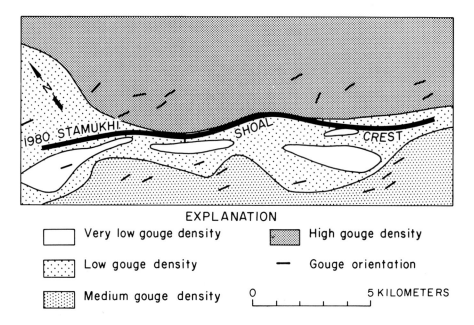

EXPLANATION

☐ Very low gouge density   ▦ High gouge density

▦ Low gouge density   — Gouge orientation

▦ Medium gouge density   0 ⌞⌞⌞⌞⌞⌞⌞⌟ 5 KILOMETERS

FIGURE 7. Map of ice-gouge densities and dominant gouge trends on Stamukhi Shoal. The single trend indicator on, and at right angles to, the shoal crest represents only a few gouges and is therefore of little significance. Mapped area matches that of Figs. 2, 8, and 10, and is keyed to Fig. 1.

High gouge densities occur on the seaward flank of Stamukhi Shoal at depths greater than about 17 m (Fig. 7). At the western end of the shoal the boundary of the intensely gouged terrain swings seaward around small topographic highs, one cresting at 15 m (Fig. 2). Downdrift (south and west) of this small rise is a tongue of low gouge counts, or a "shadow," that is the result of the high ground protecting the deeper water behind it from the scouring action of ice. Medium gouge densities are found in flat terrain landward of the shoal. The landward slope of the shoal, its crest, and the tongue extending seaward off the western end are marked by low gouge densities. Large patches, elongated parallel to the shoal crest on the landward side, have very low gouge densities. The dominant trend of gouges is about east to west, oblique to the shoal. The average gouge incision depth in the intensely scoured terrain is 0.3 to 0.5 m.

The detailed bathymetry of the 1981 eastern Stamukhi Shoal is shown on Fig. 9 with ice gouge sets and texture traced from complete side-scan sonar coverage. In tracing ice gouge patterns we have eliminated the striped quality of the digital records by enhancing faint gouges to an average level. The dominant lines in this product represent individual gouges traced, and they show the distance over which each gouge can be

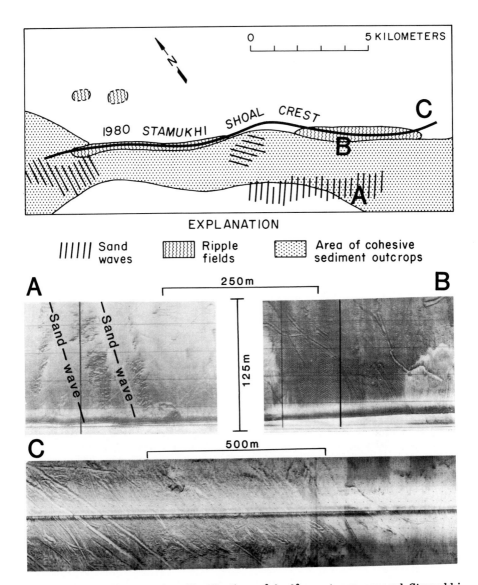

FIGURE 8.   Map showing distribution of bedform types around Stamukhi
Shoal in 1980 and 1981 plus sample sonographs showing: (A) sand waves
that have outcrops of firm cohesive sediments in the troughs and very few
gouges on the crests, (B) irregular patch of rippled gravel (dark area) and
adjacent, overlapping patches of sand (light), all only lightly scoured by ice,
and (C) swath from seaward area of high-gouge density through pock-
marked transition zone into crestal area that has few gouges. Mapped area
matches Figs. 2, 7, and 10, and is keyed to Fig. 1.

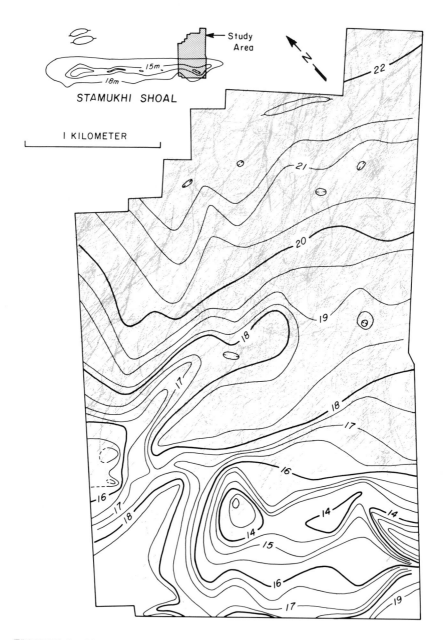

FIGURE 9. Line drawing of all ice gouges in an area in the eastern part of Stamukhi Shoal (taken from complete and overlapping side-scan coverage in 1981) superimposed on detailed bathymetry. This demonstrates that individual gouges continue through several meters of relief and that the shoal does not steer the ice.

followed with certainty. Figure 9 reveals a strong dominant trend of ice gouges from east to west, undeflected by the shoal serving as an obstacle to ice motion. Some individual gouges are over 2 km long, even on relatively steep slopes. For example, a large gouge that cuts east-west across the upper left corner of the mosaic persists through a depth range of at least 3 m. Smaller gouges in the center of the mosaic (for example, below the 18-m contour) rise obliquely across a 1.5-m-high shoal and descend the lee slope. There is no obvious widening of individual gouges with decreasing depth and increasing weight of ice masses ascending the stoss side of Stamukhi Shoal. A rather sharp boundary in the vicinity of the 17-m isobath separates the intensely disrupted seafloor region on the seaward side of the shoal from the only slightly gouged crestal region. The transition from high to low gouge density occurs over a distance of 200 to 300 m (Fig. 8C). In 1981 this transition zone was commonly marked by numerous small, isolated, irregular depressions (center of Fig. 8C), changing upslope into fields of small ripples. Bathymetric details resulting from close line spacing in this mosaic area show that the crest of Stamukhi Shoal is not a long sinuous and continuous feature as depicted in Fig. 2.

2. Ripple Fields

Extensive fields of rippled bottom were observed along the crest of Stamukhi Shoal in 1980 (Fig. 8, top). Figure 8B shows their slightly sinuous pattern that is cut by several ice gouges. The wavelength of these ripples is 1-1.5 m. The height is estimated at 8-10 cm, below the resolution of the fathometer. The trends of the ripples range from azimuth 130° to 175°, but most fall in the narrow range from 150° to 170°. Figure 8B suggests that the ripple field boundary corresponds with a boundary separating coarser sediments from finer (sandy?) sediments outside of the field. The detailed surveys of the eastern Stamukhi Shoal in the 1981 mosaic revealed smaller ripple fields just landward of the heavily gouged terrain. These ripples have the same orientation and spacing as in the previous year, but are absent along the shoal crest. As in the previous year, ripple fields commonly are associated with patchy background textures suggestive of alternating sandy and gravelly deposits.

3. Sand Waves

Irregularly spaced sand waves with wavelengths from 50 to over 200 m, heights from 0.5 to 1 m, and crest length from 125 m to more than 250 m, occur in several patches in the lee of Stamukhi Shoal (Fig. 8, top). The surfaces of the sand waves are smooth, while the troughs commonly are marked by jagged relief, described under the section "Outcrops." The trends of the sand-wave crests are uniform within each area, but differ widely between the areas of occurrence (10°, 43°, and 140°, see Fig. 8, top). In the westernmost patch of sand waves, the waves are asymmetrical to the east, suggesting an influence of easterly currents.

4. Outcrops

Irregular, jagged outcrops are widely distributed on the lee side of the shoal crest, and they extend seaward of the crest off the northwestern tip (Fig. 8, top). These outcrops appear similar to those occurring over wide regions of the shelf (Reimnitz et al., 1980). The outcrops consist of overconsolidated, cohesive silty clay, which apparently forms under modern processes in arctic shelf regions. Diving observations in numerous areas reveal that the silty clay outcrops range in appearance from highly irregular, jagged outcrops recently disrupted by ice keels to outcrops totally rounded and polished by swift currents. The occurrence of such outcrops in the troughs between hydraulically shaped bodies of granular sediments, as here in the area of sand waves (Fig. 8A), is typical for the inner shelf of the Beaufort Sea.

D. Bottom Sediments

Fourteen surface sediment samples were collected during the 1980 survey on and around Stamukhi Shoal. The locations of the stations and the mud/sand/gravel percentages, determined from textural analyses, are given in Fig. 10. Mud characterizes the low, flat terrain around Stamukhi Shoal, where gouge densities are medium to high (Fig. 7). Coarse granular sediments, that range from clean sand to gravel, make up the body of the shoal, and gravel marks the very crest. As with gouge density, the very northwestern tip of Stamukhi Shoal is different in that sand extends seaward to include the pair of small shoals shown in Fig. 2. Patches of jagged outcrops, representing cohesive deposits on the lee slope of the shoal (Fig. 8, top), indicate that the body of coarse granular material also contains lenses of mud.

The surface samples generally contain numerous clamshell fragments and some very small clams. All samples in the regions of sand and gravel show pronounced iron-oxide staining. Two mud samples seaward of the shoal have a 1- to 2-mm reddish-brown mud layer between firm materials below and a soft ooze layer above. A few pebbles are found even at the sites here labeled as mud, and one of the samples seaward of the shoal included a pebble 5.5 cm in diameter. Pebbles are subrounded to rounded. Areas with gravelly sediments generally are recognized on the sonographs by a dark background. This is a result of a multitude of echoes originating from individual clasts. Clean sand on the sonographs is characterized by a light-toned and even background (Fig. 8B), commonly separated from gravel by a sharp boundary. Cohesive mud in the regions of high ice-gouge density on the seaward side of the shoal also produces a dark background on

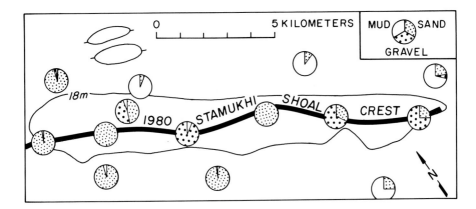

FIGURE 10. Sediment sample stations with pie diagrams showing the percentages of mud, sand, and gravel in relation to the shoal crest in 1980. A tongue of sand off the west end of the shoal extends seaward and includes the two small shoals shown. Mapped area matches Figs. 2, 7, and 8, and is keyed to Fig. 1.

sonographs and therefore cannot be distinguished from gravel by shades of darkness of the sonographs alone. In these regions the dark background is the result of reflections from rough surfaces that were generated by the churning action of ice and preserved in cohesive materials.

E. Diving Observations

Direct observations of bedforms, sediments, and organisms, made by scuba diving along two 300- to 400-m traverses in 1977, serve to support survey trackline data and spot samples discussed above. The western traverse is located about 6 km east of the western tip of the shoal, and the eastern traverse is approximately 3 km west of the east tip of the shoal (Fig. 2).

The western diving transect extends from near the crest down the seaward slope of the shoal. Slightly sandy gravel, commonly 2–3 cm in diameter but as large as 6 cm, occurs near the crest. This material, highly disrupted by intense ice action, forms crisscrossing gouges that have 1–1.2 m relief. All recent relief forms are sloping at the angle of repose. Medium-grained sand overlies the gravel in several patches 5–10 m wide. The sand patches are marked by oscillation ripples that have wavelengths of 10 to 15 cm and heights of 3 to 5 cm and are oriented roughly parallel to the shoal. Sparse brown filamentous algae, worm tubes, and a few hydroids are seen in local depressions. The sediments gradually become finer with increasing depth seaward and range from gravelly sand to cohesive mud. Sharp vertical relief forms lacking any signs of bioturbation are seen in this mud.

Along the lee slope of the shoal on the eastern transect, medium-grained sand predominates and has ripples similar to those seen on the western dive. The ripples seem recent, as they are crisscrossed by only a few tracks and trails of organisms. In several places gravelly material underlies the sand. Small sharp ledges underlain by mud layers occur in some gouge flanks. A few snails, small clams, coelenterates, and pectens were seen, but there were no attached organisms.

F. 1977, 1980, and 1981 Bedform Comparison

Lack of precise navigation in 1980 precludes direct comparison of sonographs from 1977 and 1980. The patterns observed in both years, however, are consistent and show that the fairly heavily gouged crestal region of 1977 was replaced by ripple fields, which have crests spaced about 1.5 m apart and oriented at 150°, in 1980. In other regions of the shoal, gouge patterns in those two years show no noticeable difference.

Three of the 1977 bathymetric profiles (B, C, and D of Fig. 2) were rerun in 1981 for a comparison (Fig. 11). Although the vessel cannot be

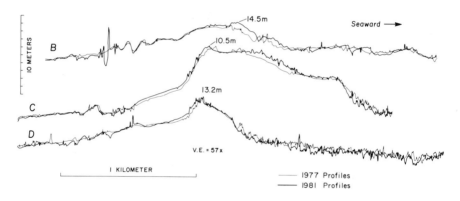

FIGURE 11. Comparison of profiles B, C, and D (Fig. 2) in 1977 and 1981. The 1981 cross section lines deviate horizontally up to 10 m from the 1977 lines and cross the dominant gouge trend at an oblique angle, so it is impossible to match individual gouges. However, this comparison does show that major changes have occurred on the shoal.

steered precisely enough to duplicate traverses exactly, major changes in gouge pattern are evident. In particular, numerous newly cut large gouges in the 1981 records did not exist in 1977. Large-scale changes in profiles B and C evidently also occurred, but further monitoring is required before any discussion of these is possible. Additional changes observed between 1980 and 1981 are the disappearance of ripples in the crestal region and the apparent development of ripple fields, which have a similar wavelength and orientation, on the seaward flank in the transition zone between heavily gouged and lightly gouged terrain.

IV. DISCUSSION

Hartmann (1891) strongly emphasized the effects of drifting ice in grinding, leveling, and polishing the shelf surface and shoreline in polar regions. He summarized observations from numerous explorers and ship captains and pointed out that the continuous motion of extensive and heavy ice masses in arctic regions may prevent the formation of sandy shoals that other processes tend to create under local conditions in the marine environment. The dredging and leveling action of ice may even help to explain the characteristically wide, smooth, and very gently sloping arctic shelves, where water depths of only 10 to 15 m are not uncommon several tens of kilometers from shore and where there is very little relief over wide regions. However, Stamukhi Shoal and other arctic shoals are made predominantly of sand accumulated since the last transgression (Reimnitz and Maurer, 1978); this suggests that ice processes may actually contribute to the formation and maintenance of sandy shoals in certain areas.

The satellite data presented here document both the interference that Stamukhi Shoal causes to drifting floes and ice islands and the deformation of extensive winter ice sheets focused on the shoal year after year. Stamukhi Shoal is able to resist the motion of tabular ice islands up to 200 m across (Brooks, 1973) and produces ice jams extending 100 km or more updrift. The precise matching of ice features seen in satellite images to the crest of the shoal indicates the scouring action of ice is most frequent and intense along the crest. Like a harrow dragged over a field of furrows, the keels of the drifting ice pack displace materials from the crests of shoals toward the sloping flanks. Here additional downslope movement of particles is aided by gravity. Only the scale of the processes on a farmer's field and the arctic shelf surface differs.

The ice drag marks on Stamukhi Shoal do not deviate from the regional trends in the years of record, indicating little topographic steering effect by the shoal. Ice either bulldozes across the shoal or plows into the side and stops. How have Stamukhi Shoal and other similar shoals survived?

An interplay between ice scouring and hydraulic shaping of shoals must be considered to explain their maintenance. From 1970 through 1977 we were impressed by the intensity of scouring on the shoal crests in the Beaufort Sea compared to that on the low-lying surrounding terrain, as seen both in bottom observations and in the distribution of grounded ice (Reimnitz et al., 1972, 1977, 1978b; Reimnitz and Barnes, 1974; Barnes and Reimnitz, 1977; Barnes et al., 1978). A sample sonograph and fathogram recorded on Loon Shoal (Fig. 12A) in 1977 demonstrates the intensity of scouring. In contrast, Fig. 12B also shows a shoal profile representing the other extreme, a profile recently shaped by waves or currents. This example was recorded in 1981 over a sand shoal in the stamukhi zone 250 km east of the present study area. A similar smoothing of the crest of Stamukhi Shoal was documented from 1977 to 1980. The falls of 1977, 1978, and 1979 may well have provided the conditions for reworking the shoal by waves and currents: the fall of 1977 was marked by a shelf entirely clear of

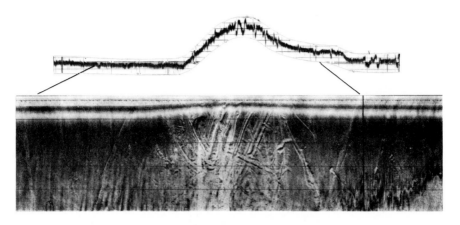

A. PRE-1977

B. POST-1977

FIGURE 12. Comparison of a shoal reflecting the high scour intensity on the crest (typically seen prior to 1977) to another shoal typical of those seen in the years since 1977. A, sonograph and fathogram recorded on Loon Shoal (approximately 10 km southeast of Stamukhi Shoal) in 1977. B, sonograph of a sand shoal in 1981 in the stamukhi zone 250 km east of study area. Both shoals rise approximately 4 m above the surrounding seafloor.

ice and resulting long fetches for wave generation, and the falls of 1978 and 1979 were marked by unusually strong northeasterly winds. The ripple patterns on top of Stamukhi Shoal, characterized by their long, continuous, slightly sinuous crests and their apparent symmetry, suggest orbital flow in a wave train as the most likely ripple generating process. The orientation of the ripples (150° to 170°) is aligned for easterly to northeasterly storm waves.

Using the most severe wave conditions on record, one may estimate the maximum particle sizes that could be moved on the crest of Stamukhi Shoal. Short (1973) recorded a 2- to 2.5-m-high swell with a 9- to 10-s period in early September 1972 inshore of Stamukhi Shoal. Following procedures outlined by Komar (1976), orbital velocities of up to 86 cm s^{-1} are estimated for these conditions at 15-m depths. This velocity is near the threshold of motion for particles of 7-mm diameter, or small pebbles. In September 1977, when most of the ice had disappeared from the shelf, we measured 2-m-high waves with a period of 6 s in water 15 m deep during a northeasterly wind of approximately 20 knots. Based on the above procedures, these waves could result in 40 cm s^{-1} orbital velocities, capable of moving medium-grained sand at a depth of 15 m. According to preliminary analysis, orbital velocities of about 100 cm s^{-1} (approximately 2 knots) could have produced the ripple fields of 1- to 1.5-m wavelength in gravel 20 mm in diameter.

Knowledge of the distribution and concentration of ice during major storms is required for wave hindcasting, but such information is essentially nonexistent for the periods during which the gravel ripples formed on Stamukhi Shoal. We therefore refrain from such analysis. However, two days of easterly winds having daily average velocities of 16 m s^{-1} (35 mph) and higher, recorded by the National Weather Service at Barter Island during a freeze-up storm in 1978, was an unusual event that produced large amounts of sediment-laden slush ice in the Beaufort Sea. We suspect that little drift ice was present to calm the seas and that the gouges on the crest of the shoal were transformed into ripple fields during the event.

The large sand waves that have steep faces to the east along the lower south flank of Stamukhi Shoal (Fig. 8) must have formed from continuous currents flowing toward the east. Differing orientations of other sand waves in the area suggest that strong currents may have been funneled and deflected by large grounded pressure ridges.

Stamukhi Shoal, a deposit of noncohesive sand and gravel, stands as an anomaly above the surrounding shelf surface that is covered with cohesive Holocene mud. On the basis of shallow seismic stratigraphy, Reimnitz and Maurer (1978) interpreted the shoal to be a Holocene constructional feature. They rejected the possibility that Stamukhi and other linear shoals in the area are drowned barrier islands and argued that shoals are built and maintained by modern ice-related processes from surrounding shelf deposits. Grounding ice, churning and softening bottom deposits and at the same time producing rough relief, makes materials readily available for removal by waves and currents, thereby aiding the winnowing of fine materials. The surrounding shelf deposits not only provide the range of coarse particle sizes that make up the shoal, but winnowing by the combined action of ice and currents in the vicinity also maintains the shoal as a coarse deposit.

Frequent ice scouring in a localized area, repeated over long time intervals, results in local coarsening of existing bottom deposits. However, we cannot envision an ice-related process that would result in slow and systematic construction of a major topographic high like Stamukhi Shoal. The repeated action of ice could only have the opposite result—that of leveling. We know of no evidence that a major topographic high in the Arctic was built by a single catastrophic event, and we believe that this is unlikely. Once a shoal is constructed to an elevation that exceeds the depth range through which grounded ice can be pushed upward on a steep slope in the natural environment, some ice would get stuck on the stoss side. Each event terminating on the stoss side would move material toward the crest. However, the amount of ice that continues to move over the shoal, instead of stopping on the stoss side and adding material to the shoal, is much larger and more efficient in lowering and reducing a shoal.

If the boundary of heavily gouged terrain on the stoss side of the shoal marks the area of long-term grounding after floes are shoved up the slope, we should see such characteristic signs such as increasing gouge size and terminal ridges. If, on the other hand, that boundary is controlled by the water depth to which wave reworking is active, it should show signs of sand patches inundating areas of ice-gouge relief along the crest of the shoal. The area mosaicked using total side-scan sonar coverage shows no signs of either process (Fig. 9). The boundary typically is a 300-m-wide mottled or pitted transition zone. We speculate that this characteristic bottom type may be the footprint of a pressure ridge formed in place, where relatively small ice slabs are shoved into the bottom and subsequently melt.

Pebbles provide an ideal base for biologic growth in certain ice-sheltered areas on the shelf. However, the continuously repeated grinding by ice and reworking by waves and currents makes Stamukhi Shoal a hostile environment for fauna and flora. Diving traverses reveal desert-like conditions where attached or burrowing organisms are almost totally absent. Iron-staining on all coarse clasts suggests frequent turnover of the clasts, so that all faces of pebbles are exposed to oxidizing conditions.

Stamukhi Shoal plays a key role in establishing the regional ice regime and provides considerable shelter locally against drifting floes and ice islands. It is the best known of the shoals that mark the edge of the stamukhi zone, but reconnaissance studies have been made on several othershoals. Weller Bank, marked by the large ice accumulations west of Stamukhi Shoal in the two summer satellite photos of Fig. 5, was compared by Barnes and Reiss (in press) to Jaws Mound, another shoal north of Prudhoe Bay. They reported that both shoals are elongated parallel to regional isobaths and consist of sand and iron-oxide-stained gravel, partly inundated by sand blankets on their east ends. In 1980 both of these shoals had only short irregular gouges on the crest, none on the flanks, and many large gouges seaward of the shoal. Ripples with wavelengths of 1.5 to 2 m, oriented at 130° to 150°, were also observed. These two shoals are, at least superficially, similar to Stamukhi Shoal, which indicates that it may serve as a model for all shoals in the stamukhi zone. However, there are also important differences. Stamukhi Shoal is the most linear shoal and has the greatest relief of any of the shoals yet studied in the stamukhi zone. Also, Stamukhi Shoal is more consistently marked by grounded ice than other shoals of the region. Until more studies of the entire stamukhi zone are made, it would be unwise to apply the findings of this study to all shoals in the zone.

The landward edge of the stamukhi zone east of the study area is marked for at least 150 km by a line of morphologic features much more subtle than Stamukhi Shoal (Rearic and Barnes, 1980; Barnes and Reiss, in press). This boundary generally follows the 18-m isobath and shows up as an anomaly in numerous ice-gouge parameters (Barnes *et al.*, this volume). For the first 35 km east of Stamukhi Shoal, the boundary is characterized by 3- to 4-m-high shoals that gradually decrease in size eastward to form a 2- to 4-m-high bench that has a sharp seaward edge (Barnes and Reimnitz, 1974; Reimnitz and Barnes, 1974; Barnes *et al.*, 1980; Rearic and Barnes,

1980). Very small shoals are present along the sharp seaward edge of the bench in some areas, perhaps marking the initial stages of major future shoals. Eastward of longitude 146° W, the boundary is again marked by subtle shoals (Rearic and Barnes, 1980).

Soviet navigators apparently have long taken advantage of large grounded ice piles present in the stamukhi zone of the East Siberian Sea, as reported in H.O. Publication No. 705 (1957): "In summer ice-free water is found between the stamukhi and the coast, providing a fine shelter for ships from the drift ice still present in the northern part of the sea. This area also may be used as an anchorage by a ship forced to winter over." In the Alaskan Beaufort Sea, petroleum development could also use the presence of shoals in the stamukhi zone to advantage (Reimnitz *et al.*, 1978a). The shoals may become sites for artificial islands used for exploration and production because the shallower seafloor would greatly reduce construction costs. We need, however, a better understanding of how ice interacts with these shoals, especially in the fall when the ice is thin. Conditions on the shoal, which is a focal point for ice dynamics, must be considered extremely hazardous. Lastly, the sand and gravel that compose Stamukhi Shoal are valuable as construction materials, and mining the shoal will be considered. But removal of the shoal could change the ice regime over wide regions of the shelf to the west and southwest and thus should be avoided.

IV. SUMMARY

Since the first Landsat images were taken in 1972, anomalies in the ice cover observed in the study area have suggested the presence of an uncharted topographic feature. A bathymetric survey over the area in 1977 revealed Stamukhi Shoal, a 17-km-long linear shoal with up to 10 m relief. In eight out of ten summers, satellite images have shown at least one of four characteristic ice features coinciding with the shoal: (a) sharp boundary separating open inner shelf waters from dense pack ice offshore, (b) the edge of a major lead, (c) an isolated belt of stamukhi, and (d) an isolated belt of grounded ice islands. A lack of imagery is probably the reason no characteristic ice pattern was recorded in two summers. In all ten winters of satellite data, the shoal has coincided with at least one of three characteristic ice features: (e) major pressure and shear ridges, (f) a boundary between extensive fields of different ice types, and (g) an indistinct line of ice piles.

All of the observed ice patterns require grounding on the crest of the shoal. The shoal has no apparent topographic steering effect on pack ice; thus a large amount of energy is consumed there by ice scouring. Before 1977, the crests of Stamukhi Shoal and other shoals in the grounded-ridge zone were marked by high numbers of ice gouges. In recent years, active hydraulic reworking of material on the crests of the shoals has smoothed the gouges, and left wave-generated ripples in gravel patches. Physical disruption of the shoals by ice keels alternating with scouring by currents results in removal of fine materials and concentration of coarse materials in the shoal. Stamukhi Shoal is thus maintained against an energy gradient of presumably destructive forces that are focused on the crest. On the other hand, the shoal is a constructive feature postdating the last transgression. This dichotomy exposes a major gap in our understanding of processes on the shoal. Because shoals of the stamukhi zone may play roles in the offshore petroleum developments, further research is highly desirable.

ACKNOWLEDGMENTS

We thank Douglas K. Maurer for his work on the first bathymetric and geophysical surveys of Stamukhi Shoal. Numerous able assistants before and since conributed to this study by their work aboard the KARLUK, and by participation in numerous discussions about the shoals. We owe special thanks to Peter Barnes for his thoughts on the subject. Jeanne A. Blank drafted the illustrations.

REFERENCES

Barnes, P.W., McDowell, D.M., and Reimnitz, Erk (1978). *Open File Rept. 78-730.* U.S. Geol. Surv.

Barnes, P.W., and Reimnitz, Erk (1974). *In* "The Coast and Shelf of the Beaufort Sea," (J.C. Reed and J.E. Sater, eds.), p. 439. Arctic Institute of North America, Arlington.

Barnes, P.W., and Reimnitz, Erk (1977). Geol. Soc. Canada, Ann. Meeting, Program with abstracts, Vol. 2, p. 6. Vancouver, B.C.

Barnes, P.W., and Reiss, Thomas (1981). *In* "Environmental Assessment of the Alaskan Continental Shelf," Ann. Repts. NOAA. (In press).

Barnes, P.W., Ross, Robin, and Reimnitz, Erk (1980). *In* "Environmental Assessment of the Alaskan Continental Shelf," Ann. Repts., Vol. 4, p. 333. NOAA.

Barry, R.G., Moritz, R.E., and Rogers, J.C. (1979). *Cold Reg. Sci. and Technol.,*[1] 129.

Breslau, L.R., James, J.E., and Trammell, M.D. (1971). *In* "International Conf. on Port and Ocean Engineering under Arctic Conditions," Vol. 1, p. 119. Tech. Univ. of Norway.

Brooks, L.D. (1973). *Rept. RDCGA-36.* U.S. Coast Guard Academy, New London, Conn.

Hartmann, Georg (1891). *Leipzig, Verlag von Duncker and Humbolt, Vol. 1, Part 3,* 175.

H.O. *Publication No. 705* (1957). "Oceanographic Atlas of the Polar Seas, Part II, Arctic," p. 47. Washington, D. C.

Komar, P.E. (1976). "Beach Processes and Sedimentation." Prentice-Hall, Inc., Englewood Cliffs, NJ.

Kovacs, Austin (1976). "Grounded Ice in the Fast Ice Zone Along the Beaufort Sea Coast of Alaska." U.S. Army Corps of Engineers, Hanover, N.H.

Kovacs, Austin, and Mellor, M. (1974). *In* "The Coast and Shelf of the Beaufort Sea," (J.C. Reed and J.E. Sater, eds.), p. 113. Arctic Institute of North America, Arlington.

Maurer, D.K., Barnes, P.W., and Reimnitz, Erk (1978). *Open File Rept. 78-1066.* U.S. Geol. Surv.

Rearic, D.M., and Barnes, P.W. (1980). *In* "Environmental Assessment of the Alaskan Continental Shelf," Ann. Rept., Vol. 4, p. 318. NOAA.

Reimnitz, Erk, Barnes, P.W., Forgatsch, T.C., and Rodeick, C.A. (1972). *Mar. Geol. 13,* 323.

Reimnitz, Erk, and Barnes, P.W. (1974). *In* "The Coast and Shelf of the Beaufort Sea," (J.C. Reed and J.E. Sater, eds.), p. 301. Arctic Institute of North America, Arlington.   •

Reimnitz, Erk, Barnes, P.W., Toimil, L.J., and Melchoir, John (1977). *Geology,* 5, 405.

Reimnitz, Erk, and Maurer, D.K. (1978). *Open File Rept. 78-666.* U.S. Geol. Surv.

Reimnitz, Erk, Toimil, L.J., and Barnes, P.W. (1978a). *J. Petrol. Technol., July,* 982.

Reimnitz, Erk, Toimil, L.J., and Barnes, P.W. (1978b). *Mar. Geol.,* 28, 179.

Reimnitz, Erk, Kempema, E.W., Ross, C.R., and Minkler, P. W. (1980). *Open File Rept, 80-2010.* U.S. Geol. Surv.

Short, A.D. (1973). Ph.D. Thesis. Louisiana State Univ., Baton Rouge.

Skinner, B.C. (1971). *Rept. RDCGA-23.* U.S. Coast Guard Academy, New London, Conn.

Stringer, W.J. (1974a). *In* "The Coast and Shelf of the Beaufort Sea," (J.C. Reed and J.E. Sater, eds.), p. 165. Arctic Institute of North America, Arlington.

Stringer, W.J. (1974b). *Northern Engineer,* 5, [No. 4,] 36.

Stringer, W.J. (1979). *In* "Environmental Assessment of the Alaskan Continental Shelf," Research Unit no. 257, Final Rept., Vol. 2, p. 367. NOAA.

Stringer, W.J., and Barrett, S.A. (1975a). *Northern Engineer,* 7, [No. 1,] 2.

Stringer, W.J., and Barrett, S.A. (1975b). *In* "Proceedings of the Third Conference on Portland Ocean Engineering under Arctic Conditions," p. 527. Trondheim, Norway.

Toimil, L.J., and Grantz, A. (1976). *In* "AIDJEX Bull.," No. 34, p. 11.

Zubov, N.N. (1945). "Arctic Sea Ice," Translated by Naval Oceanographic Office and American Meteorological Society under contract to Air Force Cambridge Research Center, 1963. U.S. Naval Electronics Laboratory, San Diego, CA.

ICE GOUGING CHARACTERISTICS AND PROCESSES

Peter W. Barnes
Douglas M. Rearic
Erk Reimnitz

U. S. Geological Survey
Menlo Park, California

I. INTRODUCTION

Over much of the Arctic Shelf, scouring of the seafloor by ice disrupts and modifies the seabed, affecting seabed sediments, ice zonation, and petroleum development activities. Scouring occurs where sea ice comes into contact with the seafloor to form ice gouges. As sediments are disrupted, atmospheric and oceanic energy is absorbed, ice movement is arrested, the ice canopy on the shelf is stabilized, and an areal ice zonation results. Development activities that place pipelines and subsea structures on the seafloor are affected by the plowing forces involved in ice scouring (Grantz *et al.*, 1980).

Since 1972, we have recorded morphologic data of the ice-scoured continental shelf of the Alaskan Beaufort Sea using side-scan sonar and fathometers. The primary objective has been to assemble quantitative data on ice-gouge characteristics and processes and to analyze these data for trends. Initial comparison of seabed morphology and shelf-ice zonation suggested a relationship between ice gouging and sea ice ridges on the inner Beaufort Sea shelf (Reimnitz and Barnes, 1974). In this report we update earlier work, summarize new data regarding the character and variability of ice gouges on the Beaufort Sea shelf (Fig. 1), and discuss the gouging process, suggesting relationships to seabed morphology, sediments, and ice dynamics.

II. TERMINOLOGY

Terminology for features produced by ice interaction with the seafloor has not been standardized. Researchers have used one term to describe both a process and the resulting feature. Terms such as "ice plow mark" (Belderson and Wilson, 1973), "ice score" (Kovacs, 1972; Pilkington and

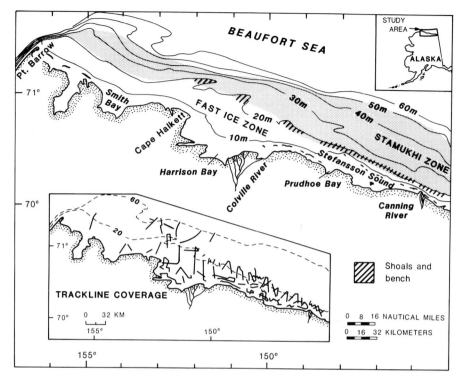

FIGURE 1.   The shelf and coast of northern Alaska showing the bathymetry, ice regime, tracklines, and locations used in the text.

Marcellus, 1981), "ice scour" (Pelletier and Shearer, 1972; Brooks, 1974; Lewis, 1977a,b), "ice-scour track" (Wahlgren, 1979; McLaren, 1982), and "ice gouge" (Reimnitz and Barnes, 1974; Thor and Nelson, 1981) have been used to describe a single feature.   Accordingly, the processes were plowing, scoring, scouring, and gouging.   A 1982 National Research Council of Canada workshop elected to use the term *ice scouring* for the processes of ice interaction with the seafloor.   We use the term *ice gouging* interchangeably in this paper for the same processes to clearly separate ice scouring from hydraulic scouring. But only the term *ice gouge* is used here for the characteristic seafloor furrow and associated morphology caused by ice gouging. Each furrow is considered a separate gouge even when many gouges result from the same ice scouring event.   We consider each gouge separately, as we are primarily interested in seafloor processes and secondarily in the events that caused them.   The following terminology is used for the quantitative enumeration of an ice-gouged seafloor.

   Gouge density – the density of all ice-produced sublinear features preserved on the seafloor.   The measurement expresses the number of preserved gouges per square kilometer of seafloor by the normalizing of trackline data (Barnes *et al.*, 1978).   Scour density or frequency as used by

Lewis (1977a) and McLaren (1982) identifies and enumerates scouring events, each of which may have resulted in one or more gouges.

Gouge depth - the depth of a gouge measured vertically from the average level of the surrounding seafloor to the deepest point in the gouge (Fig. 2). Due to sedimentation and slumping, this depth is usually not equivalent to the original incision depth made by the ice. This value is similar to Lewis's (1977a) scour depth. Gouge depth is not to be confused with depth below sea level.

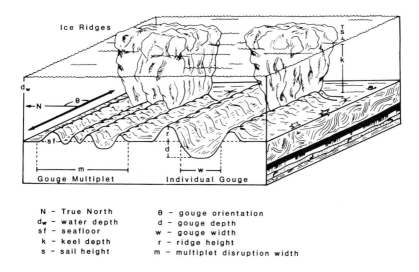

N – True North
d_w – water depth
sf – seafloor
k – keel depth
s – sail height

θ – gouge orientation
d – gouge depth
w – gouge width
r – ridge height
m – multiplet disruption width

FIGURE 2. An idealized ice gouge and gouge multiplet, showing terms used to quantify the character of ice gouges.

Gouge width - the width of a gouge measured horizontally at the average level of the surrounding seafloor (Fig. 2). This measurement does not include sediment ridges which commonly bound the gouges. Gouge width is equivalent to Lewis's (1977a) scour width.

Ridge height - the height of the ridge of sediments bounding a gouge, measured vertically from the averaged seafloor depth to the highest point on the ridge (Fig. 2). Lewis (1977a) used the term lateral embankment for the ridges bounding a "scour."

Gouge relief – the sum of gouge depth and ridge height.

Gouge orientation – the orientation of an ice gouge relative to true north (T). We report orientation as a vector between 180° and 360°. Using this convention, we imply a sense of motion, but recognize that gouging may occur in either of two directions (Reimnitz and Barnes, 1974). Considerable variation in the gouge orientations commonly made these observations subjective.

Gouge intensity – a quantitative estimate of visible sediment disruption calculated as the product of gouge density, maximum gouge depth, and maximum gouge width. No units are assigned to this measure.

Gouge multiplet – A gouge multiplet is defined as two or more gouges, closely paralleling or overlapping one another, suggesting formation by a single multiple-keeled ice mass (Fig. 3). Lewis (1977b) called such features "multiple scour tracks" but did not clearly distinguish them from "ice scours," which are features that also may have multiple tracks. We consider each individual gouge within a gouge multiplet as a separate geologic feature created by a single ice event (Fig. 2).

Gouges per multiplet – the number of individual gouges making up a single gouge multiplet.

Multiplet disruption width – the width of seabed disrupted by a scouring event, measured normal to a multiplet incision and including the ridges on either side (Fig. 2). Disruption widths of individual gouges were not measured but are approximately 25% greater than the gouge width.

Multiplet orientation – the orientation of a gouge multiplet relative to true north.

III. BACKGROUND

The Beaufort Sea shelf can be characterized as a narrow, shallow shelf, whose prominent features are broad shallows off major rivers (Dupre' and Thompson, 1979), sand and gravel island chains trending in echelon parallel to the coast, and a series of sand and gravel shoals in water 10 to 20 m deep (Fig. 1). The surficial sediments are characterized by textural variability over short lateral and vertical distances (Naidu and Mowatt, 1975; Barnes *et al.*, 1980a). In nearshore areas (water depths to 15 m), surficial sediments may be reworked to depths of tens of centimeters by episodic storm waves and currents (Barnes and Reimnitz, 1979). In water depths of 0 to 40 m or more, the seafloor is episodically reworked by ice. Thus, the seafloor is exposed to an interplay between hydrodynamic and ice-related processes (Barnes and Reimnitz, 1974).

A. Ice Regime

Temporal and spatial studies of ice zonation and the distribution of ice ridges and keels are critical to an understanding of the correlation between sea ice and the scouring events it causes. Regional ice ridge distributions and discussions of the ice regime have been presented by Reimnitz *et al.* (1978) and by Stringer (1978). The relation of ice ridge sail height to ice keel depth, primarily in the central part of the arctic ice pack, has been

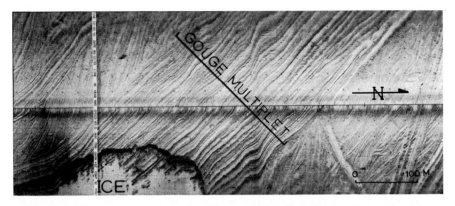

FIGURE 3. *Sonograph record of a gouge multiplet from 25 m water depth east of Barter Island. Note grounded ice floe along the margin of the record. This floe scoured the gouge multiplet in a southeastward direction.*

studied by Weeks *et al.* (1971), Hibler *et al.* (1972), Kovacs and Mellor (1974), and Wadhams (1975, 1980). However, ridging intensities and energy expenditures in ridge building are greatest on the edge of the polar pack, where it rubs against the coast (Hibler *et al.*, 1974; Reimnitz *et al.* 1978; Stringer, 1978; Pritchard, 1980). As Wadhams (1975, p. 44) notes: "the coastal areas of the Arctic, such as the Beaufort Sea, are probably the site of the deepest keels in the Arctic Ocean, since they have a combination of high ridge frequency and a preponderance of first year ridges of dense ice which results in deeper keels for the same ridge height."

The seasonal ice patterns change in the following general manner. As winter progresses, ice motion inside the barrier islands and in shallow water are small, while at the seaward boundary of the fast ice, repeated incursions of the polar pack cause ice ridging. Along this boundary, grounded first-year and multi-year ridges form a stamukhi zone (zone of grounded ice ridges) (Fig. 1). This zone forms in water depths of about 15-45 m, strung from promontory to promontory or from shoal to shoal along the inner shelf (Kovacs, 1978; Reimnitz *et al.*, 1978). In Harrison Bay, two stamukhi zones form (Reimnitz *et al.*, 1978; Stringer, 1978). An inshore zone occurs near the 8 to 12 m isobaths. Further offshore, the major stamukhi zone is located along the 15 to 20 m isobaths and appears to be limited in shoreward extent by shoals in the northeast part of Harrison Bay and farther east. Additional ridges are commonly added to the stamukhi zone throughout winter, expanding the zone to 35 to 45 m water depths (Reimnitz *et al.*, 1978).

In spring (May and June), Arctic rivers flood the nearshore ice, hastening the onset of melting and deterioration of the fast-ice canopy, which is finally broken and dispersed by the wind. Grounded remnants of the stamukhi zone may persist through the summer open-water period. As sea ice melts and pack ice retreats during summer, the nearshore wave and current regimes intensify as more water surface is exposed to wind stress.

Maximum open water generally occurs in September and early October and corresponds to the period of most intense storms (Reimnitz and Maurer, 1979).

B. Ice Scouring

Studies by Carsola (1954) and Rex (1955) were the first directed at seabed relief features related to ice scouring, although reports indicate that early arctic explorers had known scouring to occur (Kindle, 1924; Wahlgren, 1979). Studies during the early 1970's culminated in a series of descriptive papers on these features (Pelletier and Shearer, 1972; Kovacs and Mellor, 1974; Reimnitz and Barnes, 1974; Lewis, 1977a; and McLaren, 1981). Subsequent studies have concentrated on quantifying the processes and, in particular, have attempted to ascertain the annual rate of gouging (Lewis, 1977b; Reimnitz *et al.*, 1977; Barnes *et al.*, 1978; Toimil, 1978; Barnes and Reimnitz, 1979; Wahlgren, 1979; Thor and Nelson, 1981; Pilkington and Marcellus, 1981; Weeks *et al.*, this volume).

In a paper describing ice characteristics in relation to seabed gouging Kovacs and Mellor (1974) examined ice keel structure and the forces required and forces available from wind and momentum for gouging. They found that virtually all ice keels have enough strength for scouring. Enough wind energy was accumulated by the ice pack to easily cause gouging by an ice keel protruding from the pack. They found that when energy would be in the form of momentum of individual drifting floes driven by winds and currents, only short (tens of meters) and shallow (maximum about 60 cm) scour tracks would be created. Chari and Guha (1978) considered the gouging forces available from the movement of the massive icebergs of the east coast of Canada. When their data are extrapolated to the smaller ice masses of the Beaufort Sea, only shallow (less than 1 m deep) or short gouges would result from ice momentum alone. Thus the most intense gouging should be associated with ice keels driven by forces amassed from an encompassing ice pack.

In studies by Reimnitz and Barnes (1974) and Barnes and Reimnitz (1974), ice–gouge character was related to ice and sediment type. These authors noted that the bulk of the gouges were less than 1 m deep with a maximum depth of 5.5 m. Dominant gouge orientations were parallel to isobaths. They indicated that in water less than 20 m, lower gouge densities could reflect sediment reworked by waves and currents filling gouges rapidly. Other areas of low gouge density included shoals, lagoons and the lee of islands. Gouges in water deeper than 50 m were thought by Kovacs (1972) and Pelletier and Shearer (1972) to be relict since present ice keels are not that deep. However, Reimnitz and Barnes (1974) thought deep water gouges were possibly modern. They reasoned that ridge keels on the shelf may be deeper than in the deep sea, because here the highest concentrations of ridges occur. They also pointed out that average sedimentation rates are not applicable to gouge troughs, which serve as traps.

Lewis (1977b), in his landmark paper on Canadian ice scouring, indicated that the floor of the Canadian Beaufort Sea is saturated with gouges between 15 and 40 m water depths and that gouges are best preserved in cohesive silt and clay sediments. The less cohesive sand usually found inshore is seasonally reworked by waves and currents. Scouring also diminished in deeper water. Gouge depths averaged less than 1 m but ranged up to 7.6 m below the seafloor. Lewis was the first to note that the numbers of shallow and deep gouge depths followed an exponential distribution. He also suggested that the maximum water depth for modern gouging was the 50-m isobath as the deepest reported ice ridge keels are 47 m deep.

IV. METHODS

A. Data Collection

Data were gathered using a 105-kHz side-scanning sonar system and 12- and 200-kHz fathometers recording at 3 to 5 knots ship speed. Seafloor profile data were obtained almost exclusively with the 200-kHz recording fathometer, which has a resolution of approximately 10 cm in calm seas. The side-scan sonar was operated at slant ranges of 100 to 125 m, covering a swath of the seafloor up to 250 m wide. Many features were visible on the sonar that were not resolved by the fathometer, indicating that this system could resolve seabed features less than 10 cm high. Navigational accuracy varied according to the methods employed, which ranged from dead reckoning to the use of precision range-range systems. Estimated location errors range from a maximum of 1 km at distances greater than 20 km offshore to a few meters in nearshore surveys. A more complete discussion of equipment and techniques is given in Rearic et al. (1981).

The data presented in this report result from examination of more than 2000 km of trackline records and the observation and measurement of more than 100,000 ice gouges. Tracklines were selected to give continuous coverage of the Alaskan shelf from near shore to the shelf break at approximately 60 to 90 m depths and from Smith Bay to Camden Bay (Rearic et al., 1981, and Fig. 1).

B. Data Analysis

The trackline spacing on the inner shelf is approximately 10 km and the spacing on the outer shelf is approximately 25 km. The survey tracklines, sonographs and fathograms were divided into 1-km segments for analysis. Sonographs were used to measure gouge density, gouge width, orientation, and gouge multiplet characteristics. Gouge depths and ridge heights were measured from the fathograms. In each kilometer segment, the total number of gouges were counted and the dominant orientation estimated. This allowed us to normalize the gouge numbers to arrive at a gouge density by accounting for the angle at which the gouges were crossed (Barnes et al., 1978). A distribution was prepared from the fathograms of gouge depths in 20-cm increments for each kilometer segment. Gouges less than 20 cm deep were entered as the difference between the number counted on the fathogram in the depth distribution and the number counted on the sonograph in determining gouge density. The maximum gouge depth, maximum width, and maximum ridge height were determined in each segment, as were the number and dominant orientation of multiplets and the maximum number of gouges per multiplet. Maximum gouge relief was computed from maximum gouge depth and maximum ridge height which are not normally found on the same gouge in the segment.

Subjective judgment was required in interpretating the data because equipment malfunctions, weather, or natural randomness in gouge occurrence and orientation made the quality of the data variable. To keep this judgment factor consistent, one of us (Rearic) examined and interpreted all records.

V. RESULTS

A. *Typical and Maximum Gouges*

1. *Individual Gouges*

The "typical" gouge from our data, the one embodying the mean values of all parameters, occurs in water about 18 m deep, forms a furrow 56 cm deep with flanking ridges 47 cm high, and has a width of 7.8 m; it has a total relief of more than 1 m (Table I). In the vicinity of this gouge, the bottom is scoured to a density of 70 gouges per square kilometer with a dominant orientation of 273°. These gouge data represent an average of maximum values from 1-km-long trackline segments. The wide scatter and variability of the data are shown by the standard deviations which, in many cases, are as large as the mean values (Table I).

The maximum values show that gouge densities reach almost 500 km^{-2}. Gouges up to 67 m wide and up to 4 m deep[1] and flanking ridges as much as 5 m high have been measured. Maximum relief of a single gouge has been measured at 8 m.

2. *Gouge Multiplets*

Gouge multiplets occur an average of 1.6 times per kilometer of trackline, contain an average of almost 5 gouges per multiplet, and disrupt the seabed over a width of about 30 m (Table I). The average orientation of gouge multiplets is nearly east-west (266°), less than 10° from the mean orientation of all gouges.

The maximal values for gouge multiplets from our data show as many as 15 multiplets per kilometer of trackline. These multiplets contain up to 27 gouges with a seabed disruption width up to 150 m (Table I). Records taken in 1981 contain an even larger multiplet 275 m wide composed of 64 gouges (Reimnitz *et al.*, 1982, and Fig. 3).

---

[1] *A single gouge 5.5 m deep was measured in water 39 m deep northwest of Cape Halkett. Poor fathometer records due to rough weather precluded enumeration of gouges on this 1-km segment except for this large one; therefore this gouge does not show in our routine statistics.*

TABLE I. *Means and Extremes of Data on Gouges and Gouge Multiplets (1972 to 1980 Data).*

Parameter	Mean	Standard deviation	Range	Number of observations
Water depth (m)	18.0	14.4	1.2 - 125	2400
Gouge density (no.km^{-2})	70	71.8	0 - 490	2191
Gouge orientation (^{o}T)	273°	30.1		1917
Individual gouges				
Incision width (m)	7.78	7.96	0.5 - 67	2184
Incision depth (m) (A)	0.56	0.65	0.2 - 4	2179
Ridge height (m) (B)	0.47	0.49	0.2 - 5	2176
Gouge relief (m) (A+B)	1.02	1.09	0.2 - 8	2176
Gouge multiplets				
Density (multiplets km^{-1})	1.6	2.3	0 - 15	1842
No. of gouges multiplet^{-1}	4.8	3.7	2 - 27	884
Disruption width (m)	28.4	21.4	2 - 150	884
Orientation (^{o}T)	266°	40.6		884

The volume of sediment excavated by gouging can be impressive (Fig. 4). A gouge couplet noted in outer Harrison Bay had a width of 78 m, total relief of 6.3 m, and a cross-sectional area of the incision estimated at 234 m^2.

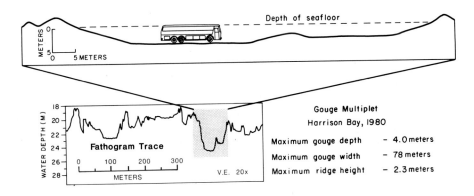

FIGURE 4. *Major gouge multiplet observed in outer Harrison Bay in 21-m water depth redrawn to remove vertical exaggeration. The total cross-sectional area of the incision cut by the double ice keels is approximately 234 m^2 (3 m by 78 m).*

Multiplets can be divided into two distinct classes. Multiplets with more than 4 or 5 gouges rarely contain any deep ones and are commonly composed of gouges of nearly equal depth. These depths are usually less than 20 cm (Fig. 3). Multiplets made up of fewer than 4 or 5 gouges may be shallow, but usually are more deeply and unevenly incised (Fig. 4).

B. *Distribution of Data With Water Depth*

1. *Individual Gouges*

Gouge parameters plotted as means against water depth create bell-shaped curves, with highest mean values of the parameters in 20 to 50 m water depths (Fig. 5).

Highest gouge densities are in water between 20 and 40 m deep, with mean values of more than 100 km^{-2}; low gouge densities there are almost nonexistent. Trackline segments in these water depths always contained significant scouring. Lowest density values occur in water less than 5 m deep or more than 45 m deep (Fig. 5A). The maximum depths of gouges (Fig. 5B), maximum width (Fig. 5C), and maximum ridge height (Fig. 5D) follow a pattern similar to gouge density except that the deepest gouges occur in water 30-40 m deep. The peak maximum widths (Fig. 5C) occur in even deeper water (40-50 m). The figures show that the frequency of features associated with ice gouges diminishes abruptly in water deeper than about 40 m (Fig. 5). Another feature of the curves is a persistent nickpoint in the data at about 18 meters, and another at 30 to 40 m.

Gouge depths were enumerated in 20-cm increments. As not all gouges observed on the sonographs (areal observations) were crossed by the fathometer (linear observations), the number of gouges reported as less than 20 cm deep should be anomalously high. The plot of these depth values is an exponential distribution from the shallowest gouges to gouges 2.5 m deep (Fig. 6) and suggests that our approximation of the less-than-20-cm gouges is reasonable.

The relationship between gouge depth and ridge height was examined. In taking the measurements, we noted that the maximum height and maximum depth in each segment were from different gouges and that ridges were normally asymmetric. Using the maximum ridge height and the maximum gouge depth in each 1-km segment, the mean maximum gouge relief (Fig. 7A) displayed the same bell-shaped curve as density, depth, width, and ridge height (Fig. 5). In an idealized gouge the ridges might be expected to be approximately half as high as the gouge is deep, with half of the debris piled on either side, or a ratio of about 1:2. The data, plotting maximum heights versus maximum depths (Fig. 7B), show that gouges up to 1 m deep are associated with ridges of equal height; a 1:1 ratio. The ratio of the mean values becomes closer to 1:3 for deeper gouges; that is, ridges are not as high as gouges are deep. This suggests that the material from incisions deeper than 1 m is distributed over a larger flanking area or compressed.

Dominant orientations of gouges plotted against water depth (Fig. 7C) show that in water more than 10 m deep the gouge trends are generally within 20° of being parallel to the coastline orientation. In waters less

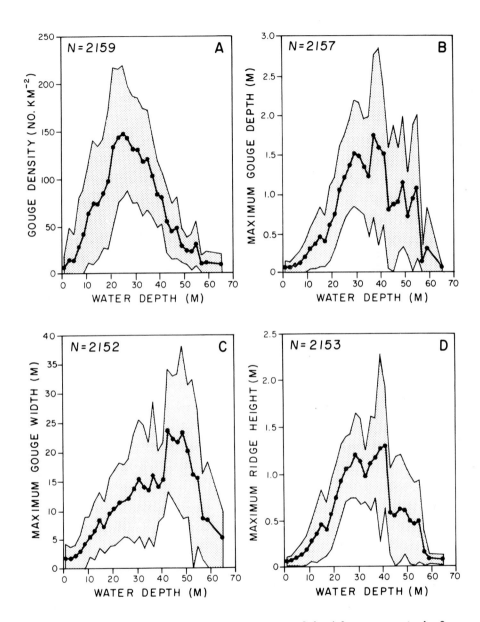

FIGURE 5. Mean gouge parameters measured in 1-km segments in 2-m depth increments. Standard deviation (shaded areas) is shown about the mean (connected dots). N refers to the number of observations in the distribution. Note the bell-shaped curves and the nickpoints in the data at 15-20-m depth. A, gouge density, B, maximum depth, C, maximum width, and D, maximum ridge height.

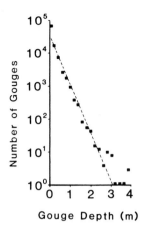

FIGURE 6. *Total number of gouges observed versus their depth.*

than 10 m and more than 50 m deep the deviation from coast-parallel scouring increases to almost $50°$ onshore.

Multiplying three gouge parameters - maximum depth, maximum width, and density - approximates the volume of the sediments involved in scouring and may be the best measure of gouge intensity. The derivative graph of mean intensity versus water depth (Fig. 7D) emphasizes the similar bell-shaped character as seen in the individual components (Figs. 5A, B, and C). Gouge intensity increases with depth very slowly to water depths of 17 to 19 m, then increases rapidly to peak values in water depths of 30 to 40 m before decreasing to very low values in depths over 55 m (Fig. 7D). The scatter of values about the mean, expressed as the standard deviation, is commonly greater than the mean value (Fig. 7D). This may be due in part to the fact that the data composing this plot are maximum values and not mean values for each segment. Mean values for each segment could show less variation.

2. Gouge Multiplets

Multiplets are most abundant in water 25 to 35 m deep and are relatively uncommon in shallow water and in deeper parts of the shelf (Fig. 8A). The number of gouges per multiplet and the disruption widths increase to water depths of 25 to 35 m deep, then decrease as water depth continues increasing (Fig. 8B and C). Disruption widths triple from 10 m in water less than 10 m deep to more than 35 m in water depths greater than 25 m (Fig. 8C). Wide multiplets are prevalent from 35 m to the seaward limit of the data set.

Gouge multiplets are oriented slightly onshore from the trend of the coastline and isobaths (Fig. 8D and Table I). Multiplets do not show the increasing onshore trend that was observed in the distribution of all gouge orientations inshore of the 20-m isobath (Figs. 7C and 8D).

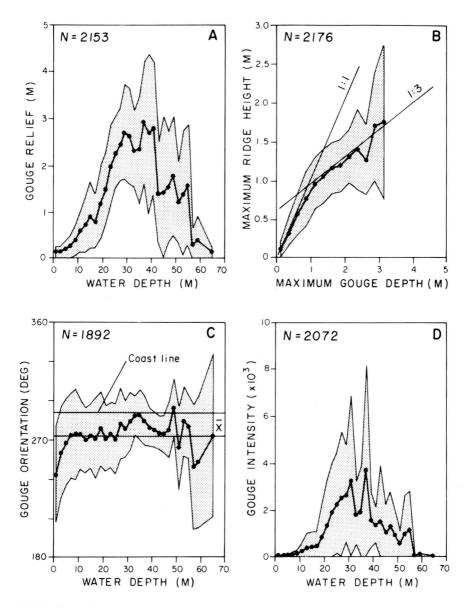

FIGURE 7. Mean parameters (connected dots) measured in 1-km segments (solid line) and standard deviation (shaded area) are shown. N refers to the number of observations in the distribution. A, gouge relief versus water depth, B, ridge height versus water depth, C, gouge orientation versus water depth, and D, gouge intensity versus water depth.

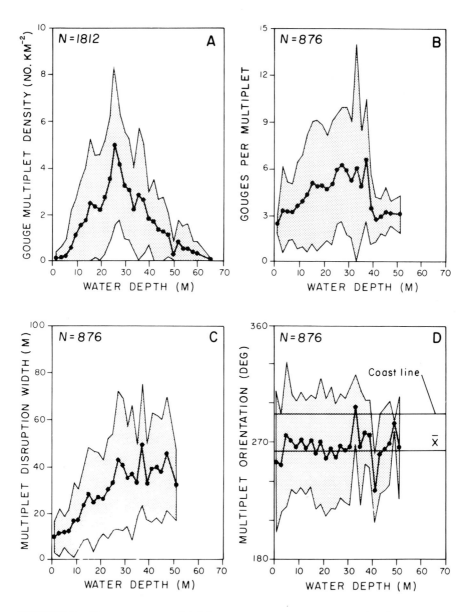

FIGURE 8. Mean gouge multiplet parameters versus water depth measured in 1-km sonograph segments and divided into 2-m depth increments. Mean value is shown by solid line, standard deviation by shaded area. N refers to the number of observations in the distribution. A, gouge multiplet density, B, gouges per multiplet, C, multiplet disruption width, and D, gouge multiplet dominant orientation.

3. Parameter Correlations

Although the measured gouge parameters share similar bell–shaped curves, correlation coefficients (Table II) show generally poor correlation between them. The low correlation value may be due to the slight positive and negative skewedness exhibited in the graphs of these parameters or to the fact that hydraulic reworking has reshaped many of the gouges since their inception. The low correlation could also indicate that the parameters are unrelated. Exceptions are the positive correlation between ridge height and gouge depth (0.84), between gouge density and the number of gouge multiplets (0.72), and between the number of gouges per multiplet and the total disruption width of that event (0.70). These correlations suggest that (1) higher ridge heights are found in segments with deeper gouges, (2) high gouge densities are associated with areas of numerous gouge multiplets, and (3) the widest sediment disruptions are from multiplets containing many gouges.

Table II. Pearson Correlation Coefficients[a].

	Gouge depth	Gouge width	Ridge height	Gouge multiplets	Gouges per multiplet	Multiplet disruption width
Gouge density	0.54	0.35	0.58	0.72	0.32	0.14
	Gouge depth	0.56	0.84	0.57	0.14	0.27
		Gouge width	0.51	0.33	0.02	0.24
			Ridge height	0.59	0.14	0.25
				Gouge mutiplets	0.31	0.26
					Gouges per multiplet	0.70

[a]Values are statistically significant at the 0.05 level.

C. Regional Distribution of Data

To provide an understanding of the regional distribution of ice keels contacting the shelf surface gouge densities, maximum gouge depths, gouge relief, gouge intensities, and gouge multiplets were contoured. In this effort the data had to be treated as if all records were of equal quality and that data were synoptic. However, where tracklines from different years crossed each other, there were commonly disparities in the data because the records were of uneven quality and the non–synoptic data represent various stages of scouring and reworking by waves and currents. As a result subjective compromises were made to accomplish the contouring.

Highest densities of gouges are found in the stamukhi zone, in water 20 to 30 m deep. Gouge densities are lowest inshore and at the seaward edges of our data in zones paralleling the general trend of the coast (Fig. 9). Low densities also appear in the lee of the islands and to the southwest of the offshore shoals. The central portion of Harrison Bay also has relatively low gouge densities.

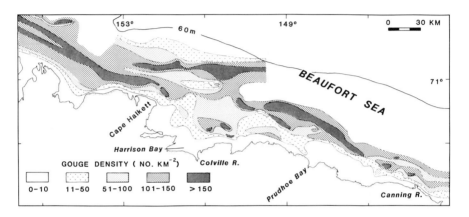

FIGURE 9. Regional distribution of gouge densities observed in 1-km segments. See Fig. 1 for distribution of tracklines used.

Gouge depths are greatest in a zone parallel to the isobaths in water depths between 20 and 40 m (Fig. 10), in deeper water than the corresponding values of high gouge densities. Lower gouge depths are associated with central Harrison Bay east of Cape Halkett and in the vicinity of shoals.

Gouge relief in excess of 2 m is common in a band of varying width that extends across the central part of the shelf (Fig. 11). Gouge relief is generally less than 1 m in the coastal embayments and inside the coastal island chains. Other areas of low gouge relief occur in the central part of Harrison Bay and at the seaward limit of our data.

Gouge intensities are greatest in a band of varying width on the central shelf and in an inshore area off the Colville River (Fig. 12). Low values occur within the coastal embayments, inside the coastal island chains, and at isolated locations in the central portion of Harrison Bay, as well as at the seaward limit of the area studied.

The regional distribution of gouge multiplets is patchy. Multiplet densities are highest in the vicinity of the 20-m isobath, particularly off the Prudhoe Bay area (Fig. 13). Low multiplet densities are present in the central part of Harrison Bay. Occasional multiplets occur inside the islands or in the shallow portions of the coastal embayments.

In the analysis of regional gouge orientation variability the shelf was divided into 26 regions. The boundaries of each of these regions encompass

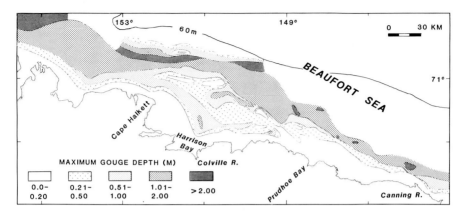

FIGURE 10. Regional distribution of maximum gouge depths observed in 1-km segments on the shelf. See Fig. 1 for distribution of tracklines used.

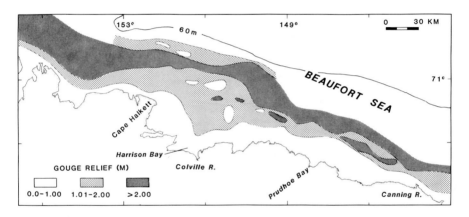

FIGURE 11. Regional distribution of gouge relief (the sum of maximum gouge depth and maximum ridge height) in each 1-km fathogram segment. See Fig. 1 for tracklines used.

what we judge to be uniform settings in terms of bathymetry and ice zonation. The dominant orientations within these regions were plotted as rose diagrams (Fig. 14). The orientation of gouges between the 20- and 40-m contours is essentially coast-parallel but slightly onshore.   The dominance of isobath-parallel orientations also holds in the shallow water of Stefansson Sound and the shallow area off the Colville River delta. A slight counterclockwise rotation of orientations is observed nearshore. This rotation is most pronounced just seaward of the islands, and along sections of the open coast, southeast of Cape Halkett.

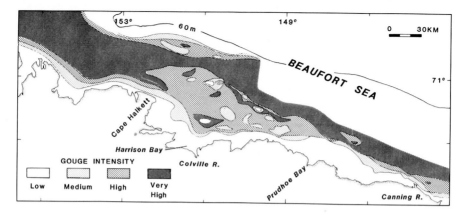

FIGURE 12. *Regional distribution of gouge intensity (product of maximum gouge depth, maximum gouge width, and gouge density) in each 1-km segment. Compare this figure with Figure 16 (ice zonation and ice ridging). See Fig. 1 for tracklines used.*

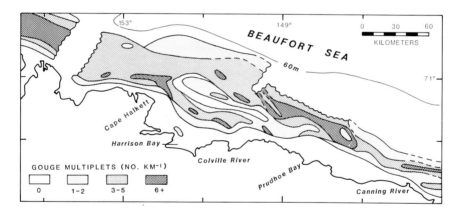

FIGURE 13. *Regional distribution of number of multiplets in each 1-km segment. See Fig. 1 for location of tracklines used.*

D. Distribution of Ice Ridges

Compressional and shearing forces in the ice pack commonly cause failure of the ice sheet and piling of ice blocks. The result is an ice ridge, composed of a submerged keel which isostatically supports a subaerial sail (Fig. 2). As it is difficult to measure keel depth, efforts have been made to determine the relationship between depth geometry and the more readily measured ridge sail height (Weeks *et al.*, 1971; Kovacs and Mellor, 1974;

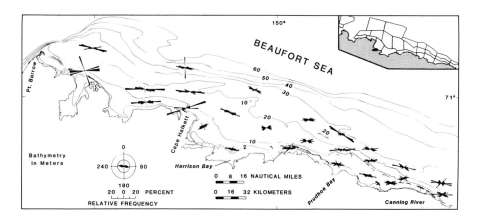

FIGURE 14. Dominant gouge orientations from 1-km sonograph segments shown as rose diagrams for 26 areas outlined in the upper righthand corner.

Kovacs and Sodhi, 1980; Wadhams, 1980; Tucker and Govoni, 1981). This work suggests a sail-to-keel ratio of about 1:4.5 for first-year ice ridges (formed during the most recent winter) and 1:3.3 for multiyear ridges.

Laser profile studies reveal considerable annual variation in ice ridges on the Beaufort Sea shelf (Tucker and Govoni, this volume). Tucker *et al.* (1979) analyzed the distribution of ridge sails on three profiles across the shelf (Fig. 15), off Barter Island, off Prudhoe Bay, and west of Cape Halkett. Ice 20 to 80 km from the coast over the central and outer shelf contained the most ridges. These authors suggested that grounded ice floes (stamukhi) stabilize ice inshore of about 20 km and limit ridging, and thus the development of sails. Further offshore, where no grounding occurs, weak first-year ice is subject to increased ridging (Fig. 15). The 1978 Prudhoe Bay profile reflects the fact that no multiyear ice was encountered on the inner 150 km of trackline; thus no stamukhi zone formed to protect the inner shelf and ridging extended to the coast (Tucker and Govoni, 1981). This suggests that year-to-year variability in ice ridging depends upon when the stamukhi zone develops.

Stringer (1978) examined satellite imagery for the period from 1973 to 1977 to assess the distribution of ice sails. The 5-year composite map produced by this study smooths the considerable seasonal and year-to-year variability. We have overlaid a 5 km by 5 km grid onto the 5-year composite ridge map to quantify the density of ice ridges on the shelf. The result (Fig. 16) shows that densities are highest (more than 6 ridges per 25 km^2) between the 20 and 50 m isobaths in the area between Prudhoe Bay and eastern Harrison Bay. Low ridge densities occur inshore, in central Harrison Bay, and in isolated areas on the outer shelf. The regional ice ridge abundance observed on satellite imagery also illustrates the increased occurrence of ridges in the stamukhi zone (Fig. 17).

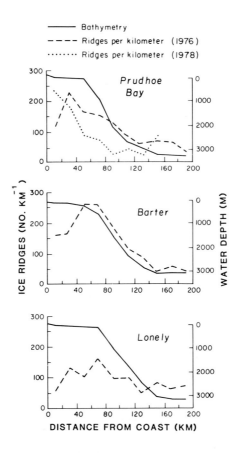

FIGURE 15. Number of ice ridges (sails) per kilometer on three transects perpendicular to the coast compared to shelf depths on the same transect (modified after Tucker et al., 1979).

VI. DISCUSSION

Those familiar with the literature recognize that this study of ice gouging on the Beaufort Sea Shelf and the relationship to ice regime are a quantification and reinforcement of earlier work presented with much sketchier data and older techniques (Reimnitz and Barnes, 1974; Lewis, 1977a). Several aspects deserve additional discussion. A useful tool for the study of sediment dynamics would be a measure of the severity of modern gouging. The trend of gouges are indications of the direction of ice motion during the plowing actions which has implications for the direction of

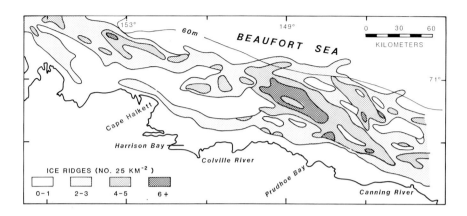

FIGURE 16. Density of ice ridges on the Beaufort Sea shelf (modified from Stringer, 1978). Compare with Figs. 9, 10, 11, 12, and 14.

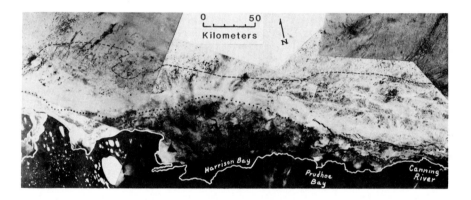

FIGURE 17. LANDSAT photomosaic of ice-covered Beaufort Sea, showing areas of winter ice ridging. Dashed line is the 60-m contour and dotted line is the inner edge of the stamukhi zone. Imagery taken June 1977.

sediment transport on the shelf. The break in gouge character at 15-20 m suggest a relationship between seafloor geologic character and ice zonation. Gouge multiplets form a unique set of gouges which may be indicative of only certain ice conditions, which would indicate the character and location of these ice conditions on the shelf now and in the past tens to hundreds of years.

A. *Severity of Gouging*

The severity of ice gouging is a result of the recurrence rate and intensity of ice-seabed interaction. High gouge density values do not always indicate severe gouging but may reflect predominantly shallow, narrow, and infrequent scouring in an area with relatively little sediment movement. Conversely, areas with relatively low gouge densities may experience many large gouge events whose record in the form of gouges has been partially or completely erased by sedimentation or hydraulic reworking (Barnes and Reimnitz, 1979; Reimnitz and Kempema, this volume).

For determining actual gouge severity, either the spatial and temporal distribution of ice keels or temporal occurrence of new gouges is needed. Few data on gouge recurrence rates and the character of new gouges exist. There are no public data on the temporal distribution of keels, only a qualitative knowledge of ice sail distribution, and an even sketchier knowledge of the quantitative relationship between sails and keels (Reimnitz *et al.*, 1978; Kovacs and Mellor; 1974, Tucker *et al.*, 1979; Wadhams, 1975). The rate of seabed reworking by ice as determined from repetitive surveys is limited to only a small part of the shelf, primarily inshore of the stamukhi zone, or to statistical considerations of gouge distribution (Lewis, 1977a,b; Barnes *et al.*, 1978; Wahlgren, 1979; Pilkington and Marcellus, 1981; Weeks *et al.*, this volume).

The most severe ice scouring should result with deeper, wider, and longer gouges and by this definition is approximated where gouge intensities are highest (Fig. 12). Our implications about gouge severity are therefore limited to a discussion of the general physical characteristics of gouge features and the overlying ice canopy.

The stamukhi zone is an area in which ice forces from the polar pack are expended, in part by building ice ridges (Thomas and Pritchard, 1980), but also on the seabed by disrupting sediments to form gouges. Reimnitz *et al.* (1978) showed that the most severe ice ridging occurs on the shelf. Sail height data (Tucker *et al.*, 1979) support this earlier concept (Fig. 15). The ice data also suggest that ridging and presumably grounding occur in this zone on a yearly basis (Stringer, 1978; Reimnitz and Kempema, this volume; Tucker *et al.*, 1979). Sediment cores from the stamukhi zone are turbated and lack horizontal laminations, while seaward and landward of the zone current-related laminations are common (Barnes and Reimnitz, 1974). This suggests frequent bottom reworking by ice in the stamukhi zone and also implies that all gouges could be modern features. Thus, we believe that seabed disruption is most severe where the stamukhi zone develops.

In Harrison Bay the relationship between ice regime and sea-floor processes is espcially clear. The two zones of ice ridging near 10-m and 20-m water depths (Fig. 16) correlate well with the highest gouge densities, maximum gouge depths, and highest gouge intensities as contoured in Figs. 9, 10, and 12.

In waters shallower than 10-15 m, the values of gouge intensity (Figs. 6D and 12) may not be true indicators of the rate of ice-seabed interaction. Here, hydraulic reworking of the seabed by waves and currents is frequent and the gouges represent fewer years of ice action (Barnes and Reimnitz, 1974, 1979). This interplay of ice and current is pronounced on shoal crests. The shoals are composed primarily of sand and gravel (Reimnitz and Maurer, 1979; Reimnitz and Kempema, this volume) on which gouges may readily fill through failure of the gouge ridges or through hydraulic reworking of sediments, either by storms or by intensified flow in the vicinity of grounded, or nearly grounded, ice keels.

Considering the ice regime alone, we would expect the number of ice gouge events in shallow water to increase while the depth and width of these events would decrease. Hibler *et al.* (1972) and Wadhams (1975, 1980) showed that the distribution of ice ridges and keels is exponential; thus, many more shallow keels are available to scour in shallow water than there are deep keels available in deep water. The depth and width of gouges in shallow water should reflect the smaller size of the keels, resulting in shallow, narrow gouges. Furthermore, shallow-water sediment may be able to resist gouging to a higher degree being coarser and more consolidated (Barnes and Reimnitz, 1974; Reimnitz *et al.*, 1980).

B. Ice Motion During Gouging

Inshore of the stamukhi zone (Fig. 1), ice motion in winter (and hence scouring) is restricted to tens of meters by the coast and the grounded ridges of the stamukhi zone. During the summer open-water period, this zone is often ice-free (Barry, 1979; Stringer, 1978). The most likely time for scouring within the fast ice zone is during spring breakup (June–July) and during fall freeze-up (October–November), when considerable ice may be present and in motion.

During formation of the stamukhi zone in winter, grounding and thus scouring occurs (Kovacs, 1976; Reimnitz *et al.*, 1978; Reimnitz and Kempema, this volume; Stringer, 1978). Once grounding has stabilized the zone (Reimnitz *et al.*, 1978; Kovacs 1976; Kovacs and Gow, 1976), the possibility of further scour to occur is limited. In waters beyond the stamukhi zone, ice ridges of sufficient draft are more rare although ice is present and in motion throughout most of the year (Hibler *et al.*, 1974; Kovacs and Mellor, 1974).

C. Direction of Ice Motion

The dominant ice motions along the Beaufort Sea coast in winter, when most scouring occurs, are from east to west (Campbell, 1965; Hibler *et al.*, 1974; Kovacs and Mellor, 1974; Reimnitz *et al.*, 1978). Thus, the dominant gouge orientation slightly oblique to isobaths indicates slightly onshore components of ice motion. This southwestward scouring action results in scour shadows in the lee of shoals inshore of the stamukhi zone.

When orientations are analyzed by water depth (Fig. 7C), the shallow inshore regions show orientations that are directed more onshore than in regions farther seaward. This onshore-turning also is characteristic for gouges and for ice movement in the Point Barrow area (Barnes, Shapiro, unpublished data) and for Harrison Bay (Rearic, unpublished data). We suggest that the long-term ice motion related to boundary stresses of the polar pack on the ice of the inner shelf may produce this pattern with shear (shore-parallel) motion more prevalent offshore and compressional (onshore) motion more prevalent inshore.

D. The 15-20 m Boundary

Brooks (1974) was the first to note that a change occurs in ice gouge character in water 18 m deep. He reported that gouge density, width, and length decrease inshore of 18 m and held the opinion that the 18-m isobath marks the limit of the onshore motion of the deep draft ice-island fragments. These fragments were presumably responsible for the larger gouges seaward of 18 m.

The inshore edge of the stamukhi zone in many areas is associated with a change in geologic character near the 20-m isobath. This change is particularly pronounced from Prudhoe Bay to the Canning River. Cohesive but unconsolidated unstructured muddy gravel offshore abuts against overconsolidated layered muddy gravel inshore (Barnes and Reimnitz, 1974; Reimnitz and Barnes, 1974). Gouge depths are greater in the area of unconsolidated sediment due to its lower shear strength (Reimnitz *et al.*, 1980b). The sediment boundary is also associated with a bench or a shoal 2 to 4 m high (Reimnitz and Barnes, 1974; Barnes *et al.*, 1980b; Rearic and Barnes, 1980).

The boundary is also seen as a jog on graphs of mean values of ice-gouge characteristics (Fig. 5), including gouge multiplets (Figs. 8A and 8C), in water 15 to 20 m deep. Gouge characteristics show increasing means with increasing water depth to depths of 35 to 45 m. This general trend in means is broken consistently in water depths of 15 to 20 m with one or two decreasing values before the continued increases toward deeper water.

Lower than expected values at 15 to 20 m depth may be due to resistance to gouging by the overconsolidated sediments that are common shoreward of this depth zone (Reimnitz *et al.*, 1980). Alternatively, hydraulic reworking of unconsolidated sediments on the numerous shoals associated with this depth zone (Fig. 1) may be responsible for reducing the mean values. The small bench or shoal-like features (Barnes *et al.*, 1980b; Rearic and Barnes, 1980) and the large shoals, do provide shelter on the "down-drift side," where less scouring occurs. This sheltering, shown by a detailed study of Stamukhi Shoal (Reimnitz and Kempema, this volume) and discussed further below, is partially responsible for the anomaly in ice gouge parameters at the inner boundary of the stamukhi zone.

We are uncertain as to the origin of this geologic boundary and corresponding change in gouge character. However, either the inner edge of the stamukhi zone is controlled by this boundary or the seasonally reforming stamukhi zone somehow is responsible for the geologic boundary. The overconsolidated sediments may be the result of freeze-thaw processes (Chamberlain *et al.*, 1978) during the Holocene transgression when sea level was lower or they may be caused by dynamic vertical, and perhaps more important, horizontal forces (Chari and Guha, 1978) associated with the intense ice–seabed interaction at the inner edge of the stamukhi zone.   McLaren (1982) documented higher sediment shear strengths in gouge troughs which he attributed to compaction during gouging.

E. Gouge Multiplets and First-Year Ice Ridges

As stated above, gouge multiplets are divided into two types. The first type has commmonly two, and always less than five gouges, and scours into the seabed to a depth of 50 cm or more (Fig. 4). The second type creates many incisions and is almost always unresolvable on the fathograms, which indicates that the gouge depths are less than 20 cm (Fig. 3).

The parallel tracks of multiplets indicate single scour events. The uniformly shallow gouges cut into a horizontal shelf surface indicate scouring by adjoining ice keels that extend to the same depth below the sea surface. The formation of an ice ridge with multiple keels aligned as tines on a rake extending tens of meters laterally, all of about the same depth beneath the surface and creating gouge depths within as little as 20 cm of one another, is a highly improbable event. Yet we commonly observe gouge multiplets that suggest this characteristic (Fig. 3).

We propose that gouge multiplets are formed by ridge keels composed of first-year ice. This ice crumbled into piles of loose blocks, is shoved downward to conform to the seafloor over extensive areas. In order to gouge the bottom, the initially loose aggregate must be at least partially fused when its movement to another site takes place, otherwise short, interrupted, or irregular tracks would result, as blocks are rolled or dislodged from the keel. Instead, the tracks commonly are continuous, for hundreds of meters, as if made by a rake. A partial welding of the ice aggregate may occur during, or soon after ridge formation, because a heat sink from surface exposure to very cold temperatures is brought to the keels during ridge formation (Kovacs and Mellor, 1974). Seawater close to freezing point driven by oceanic circulation through such porous piles should result in rapid ice growth between the blocks. If such loosely bonded ridges were shoved into shallower water, its strength would be further increased by the resulting uplift (Kovacs and Mellor, 1974). The ability of first-year pressure ridges to gouge the bottom was observed in a study in Lake Erie (Bruce Graham, personal comm.).   If the multiplets under discussion really are formed by first year pressure ridges in the manner outlined, then they formed from ice tools made at the site. This means that multiplets can form in depressions that seem to be protected from ice scouring by surrounding shallow sills such as lagoons.

Single gouges and multiplets with few incisions are the deepest and widest gouges observed (Fig. 4). We believe that these features result from ice gouging by keels of multiyear ice ridges formed in deeper water. The multiple freezing seasons available for the welding of ice keels in a multiyear ridge makes for ice scouring tools more capable of forming deep gouges than newly formed first-year ridges (Kovacs and Mellor, 1974). The gouging of these deep features by multiyear ice ridges also implies that the keels of multiyear ridges are uneven in depth and gouge the bottom with only a few of their deepest keels.

VII. CONCLUSIONS

The most intense gouging on the Alaskan Beaufort Sea shelf is associated with the major ice ridging in the stamukhi zone. Gouge intensity is greatest in water between 15 and 45 m deep. The resultant gouges may be incised 4 m or more into the seafloor, have relief of 7 m or more, and saturate the seafloor with densities of more than 200 km^{-2}. Gouge orientations indicate an uphill scouring motion from east to west, principally parallel to shore. Gouging tends to decrease in intensity both inshore and seaward of the stamukhi zone. Gouge intensity inshore is less, even though ice-seabed impacts may be more frequent, because ice motion is less and the ice masses available to scour are small. The intensity of gouging is modified by non-ice-related factors such as shoals and seabed sedimentologic character, the increased rate of seabed impacts inshore, and increased rate of hydrodynamic reworking of the seabed in shallow water.

The inner edge of the stamukhi zone at 15 to 20 m is a geologic boundary marked by shoals, an abrupt decrease in the intensity of scouring, the presence of overconsolidated surficial sediments, and a change from offshore turbated to inshore bedded Holocene sediment.

Gouge multiplets consisting of many shallow gouges are believed to be caused by the formation of first-year ice ridges whose keels are forced to conform to the seafloor over a wide swath during formation and indicate that first-year ridges can scour the seafloor.

ACKNOWLEDGMENTS

Our thanks to many colleagues who helped in gathering and interpreting the data, and to E. Phifer, S. Clarke, P. McLaren, and N. Summer, who reviewed and greatly improved the clarity of thought and expression in our text.

REFERENCES

Barnes, P.W., McDowell, D.M., and Reimnitz, E. (1978). *Open-File Rept. 78-730.* U.S. Geol. Surv.

Barnes, P.W., and Reimnitz, E. (1974). *In* "The Coast and Shelf of the Beaufort Sea," (J.C. Reed and J.E. Sater, eds.), p. 439. Arctic Institute of North America, Arlington, Virginia.

Barnes, P.W., and Reimnitz, E. (1979). *Open-File Rept. 79-848.* U.S. Geol. Surv.

Barnes, P.W., Reimnitz, E., and Ross, C.R. (1980a). *Open-File Report 80-196.* U.S. Geol. Surv.

Barnes, P.W., Ross, C.R., and Reimnitz, E. (1980b). *In* "Environmental Assessment of Alaskan Continental Shelf," Ann. Rept. Vol. 4, p. 333. NOAA, Boulder.

Barry, R.G. (1979). *In* "Environmental Assessment of the Alaskan Continental Shelf," Ann. Rept. Vol. 7, p. 272. NOAA, Boulder.

Belderson, R.H., and Wilson, J.B. (1973). *Norsk Geol. Tidsc. 53*, 323.

Brooks, L.D. (1974). *in* "The Coast and Shelf of the Beaufort Sea" (J.C. Reed and J.E. Sater, eds.), p. 355. Arctic Institute of North America, Arlington, Virginia.

Campbell, W.J. (1965). *J. geophys. Res. 70*, 3279.

Carsola, A.J. (1954). *Bull. Am. Assoc. petrol. Geol. 38*, 1587.

Chamberlain, E.J., and Blouin, S.E. (1978). *In* "Third International Conference on Permafrost," Vol. 1, p. 629. National Research Council of Canada, Ottawa.

Chari, T. R., and Guha, S. N. (1978). *In* "Proceedings 10th Offshore Technology Conference," Vol. 4, OTC-3316. Houston.

Dupre', W. R. and Thompson, R. (1979). *In* "Proceedings 11th Offshore Technology Conference," Vol. 1, p. 657. Houston.

Grantz, Arthur, Barnes, P.W., Dinter, D.A., Lynch, M.B., Reimnitz, E., and Scott, E.W. (1980). *Open-File Rept. 80-94.* U.S. Geol. Surv.

Hibler, W.D. III, Weeks, W.F., and Mock, S.J. (1972). *J. geophys. Res. 77*, 5954.

Hibler, W.D. III, Ackley, S.F., Crowder, W.K., McKim, H.L., and Anderson, D.M. (1974). *In* "The Coast and Shelf of the Beaufort Sea" (J.C. Reed and J.E. Sater, eds.), p. 285. Arctic Institute of North America, Arlington, Virginia.

Kindle, E.M. (1924). *Am. J. Sci. 7*, 251.

Kovacs, A. (1972). *Oil and Gas J. 70, (43),* 92.

Kovacs, A. (1976). *CRREL Rept. 76-32,* U.S. Army Cold Reg. Res. and Eng. Lab., Hanover, NH.

Kovacs, A. (1978). *North. Eng. 10*, 7.

Kovacs, A., and Gow, A.J. (1976). *Arctic 29*, 169.

Kovacs, A., and Mellor, M. (1974). *In* "The Coast and Shelf of the Beaufort Sea" (J.C. Reed and J.E. Sater, eds.), p. 113. Arctic Institute of North America, Arlington.

Kovacs, A., and Sodhi, D.S. (1980). *Cold Reg. Sci. Technol. 2*, 209.

Lewis, C.F.M. (1977a). "Beaufort Sea Tech. Rept. 23." Department of the Environment, Victoria.

Lewis, C.F.M. (1977b). *In* "Proceedings, POAC 77," Vol.1, p. 568. Memorial University of Newfoundland, St. Johns.

McLaren, P. (1982). *Bull. 333* Geol. Surv. Canada, Ottawa.

Naidu, A.S., and Mowatt, T.C. (1975). *In* "Delta Models for Exploration" (M.L.S. Broussard, ed.), p. 283. Houston Geological Society, Houston.

Pelletier, B.R., and Shearer, J.M. (1972). *In* "Proceedings 24th International Geological Congress," Sec. 8, p. 251. Montreal.

Pilkington, G.R. and Marcellus, R.W. (1981). *In* "Proceedings, POAC 81," Vol.2, p. 684. Universite Laval, Quebec.

Pritchard, R.S. (1980). *In* "Sea Ice Processes and Models" (R.S. Pritchard, ed.), p. 49. University of Washington Press, Seattle.

Rearic, D.M., and Barnes, P.W. (1980). *In* "Environmental Assessment of Alaskan Continental Shelf," Ann. Rept. Vol. 4, p. 318. NOAA, Boulder.

Rearic, D.M., Barnes, P.W., and Reimnitz, E. (1981). *Open-File Rept. 81-950.* U.S. Geol. Surv.

Reimnitz, E., and Barnes, P.W. (1974). *In* "The Coast and Shelf of the Beaufort Sea" (J.C. Reed and J.E. Sater eds.), p. 301. Arctic Institute of North America, Arlington, Virginia.

Reimnitz, E., Barnes, P.W., Rearic, D.M., Minkler, P.W., Kempema, E.W., and Reiss, T.E. (1982). *Open-File Rept. 82-974.* U.S. Geol. Surv.

Reimnitz, E., Barnes, P.W., Toimil, L.J., and Melchior, J. (1977). *Geology 5,* 405.

Reimnitz, E., Kempema, E.W., Ross, C.R., and Minkler, P.W. (1980). *Open-File Rept. 80-2010.* U.S. Geol. Surv.

Reimnitz, E., and Maurer, D.K. (1979). *Arctic 32,* 329.

Reimnitz, E., Toimil, L.J., and Barnes, P.W. (1978). *Mar. Geol. 28,* 179.

Rex, R.W. (1955). *Arctic 8,* 177.

Stringer, W.J. (1978). "Morphology of Beaufort, Chukchi, and Bering Seas nearshore ice conditions by means of satellite and aerial remote sensing," Vol. 1. Geophys. Inst., Univ. of Alaska, Fairbanks.

Thor, D.R., and Nelson, C.H. (1981). *In* "The Eastern Bering Sea Shelf: Oceanography and Resources" (D.W. Hood and J.A. Calder, eds.), Vol. 1, p. 279. National Oceanic and Atmospheric Administration, University of Washington Press, Seattle.

Toimil, L.J. (1978), *Open-File Rept. 78-693.* U.S. Geol. Surv.

Tucker, W.B. III, and Govoni, J.W. (1981). *Cold Reg. Sci. Technol. 5,* 1.

Tucker, W.B. III, Weeks, W.F., and Frank, M.D. (1979). *J. geophys. Res. 84,* 4885.

Wadhams, P. (1975). Beaufort Sea Tech. Rept. 36. Department of the Environment, Victoria.

Wadhams, P. (1980). *In* "Sea Ice Processes and Models" (R.S. Pritchard, ed.), p. 283. University of Washington Press, Seattle.

Wahlgren, R.J. (1979). M. A. Thesis, Dept. Geog. Carleton University, Ottawa.

Weeks, W.F., Kovacs, A., and Hibler, W.D. III (1971). *In* "Proceedings, POAC 71," Vol. 1, p. 152. Technical University of Norway, Trondheim.

SOME PROBABILISTIC ASPECTS OF ICE GOUGING ON THE ALASKAN SHELF OF THE BEAUFORT SEA

W. F. Weeks

U.S. Army Cold Regions Research and Engineering Laboratory
Hanover, New Hampshire

Peter W. Barnes
Douglas M. Rearic
Erk Reimnitz

U.S. Geological Survey
Menlo Park, California

I. INTRODUCTION

A survey of the bathymetry of the Beaufort Sea shows that large areas of this marginal sea of the Arctic Ocean have water depths of less than 60 m. It is now known that in this region ungrounded pressure–ridge keels may protrude downward for nearly 50 m and that ice floes containing such keels drift in a general pattern from east to west along the Beaufort coast. Therefore it is reasonable to presume that these ice masses could interact with the seafloor. Indeed, ice–related disturbances of the seafloor have been inferred for some decades from observations of seafloor sediments entrained in obviously grounded ice masses (Kindle, 1924). As such processes were largely of academic interest at the time, there was little motivation to systematically explore them further.

With the discovery of oil and gas along the margins of the Beaufort Sea at Prudhoe Bay and off the Mackenzie Delta, processes modifying the floor of the Beaufort Sea became of interest due to their possible effect on offshore design and operations. Examination of early side–scan sonar and precision fathometry data coupled with diving observations (Shearer et al., 1971; Pelletier and Shearer, 1972; Kovacs, 1972) showed clearly that much of the seafloor was heavily marked by long linear depressions, which we will refer to as gouges, produced by the plowing action of ice. The depths and widths of gouge incisions in the seafloor reached several meters and several tens of meters, respectively, with gouges occurring both as individual isolated events and as multiple events, presumably produced by

213

many projections on a pressure-ridge keel gouging the seafloor (Kovacs and Mellor, 1974; Reimnitz and Barnes, 1974; Barnes *et al.*, this volume).

In the present report we will discuss some random-appearing aspects of the ice-produced gouges that occur along a 190-km stretch of the coast of the Alaskan Beaufort Sea between Smith Bay and Camden Bay. We will also include a brief discussion of the statistical concepts and techniques that are utilized. We believe that our results hold immediate interest to the engineering community involved in offshore design for the Beaufort Sea as well as long-term interest to the scientific community interested in near-shore processes in shallow, ice-covered seas. The present paper is a portion of a much longer study published as a USACRREL Report. This report can be obtained from the authors.

II. BACKGROUND

Because of their importance to offshore design in arctic areas, ice-produced gouges have been the subject of a number of investigations, especially since the time when they were recognized as a recurring seafloor feature in the shallow parts of ice-covered seas. Rather than review this literature here, we will simply mention publications of general interest that can be used to find more exhaustive reference lists. Reviews of early work can be found in Kovacs (1972) and Kovacs and Mellor (1974). Early studies off the Mackenzie Delta were described by Shearer *et al.* (1971); Kovacs and Mellor (1974); and by Pelletier and Shearer (1972). Early work off the Alaskan coast was reported by Skinner (1971), Reimnitz *et al.* (1972, 1973), Barnes and Reimnitz (1974), and Reimnitz and Barnes (1974). More recent work was discussed by Shearer and Blasco (1975), Reimnitz *et al.* (1977 a, b; 1978), Barnes *et al.* (1978), Hnatiuk and Brown (1977), Barnes and Reimnitz (1979), and Barnes *et al.* (this volume). In most studies little attention was paid to ways that the observed gouge parameters varied or to methods for estimating infrequent gouging events, such as the formation of deeper gouges. Lewis (1977a, b) and Wahlgren (1979a, b), however, examined the statistical aspects of the gouges located in the general area of the Mackenzie Delta.

III. DATA COLLECTION AND TERMINOLOGY

Seven years of data amounting to some 1500 km of tracklines, obtained between 1972 and 1979 (excluding 1974), were used in the present study. Data were collected from the Research Vessels *Loon* and *Karluk*, using a side-scan sonar and a precision fathometer (200 kHz). Both systems were capable of resolving bottom relief of less than 10 cm. The side-scan records covered either 200-m or 250-m swaths (depending on scale selection) of seafloor beneath the ship. The tracks were spaced to provide fairly evenly distributed sampling along the coast between Smith Bay and Camden Bay. Data were obtained both inside and seaward of the barrier islands to the 38-m isobath. Figure 1 shows the locations of the different

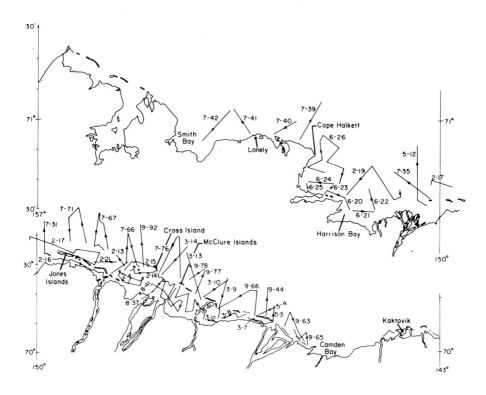

FIGURE 1. Map showing the location of the sampling lines. The arrows indicate the direction of ship movement.

sampling lines. The trackline navigation was plotted in 1-km segments. The sonographs and fathograms were also divided into 1-km segments and tied directly to the navigation. Some aspects of the data interpretation are subjective. To minimize variations due to this factor, all the counting and measuring was performed by one individual (D.R.). A complete ice-gouge data record sheet showing all measurements was given by Rearic et al. (1981). A description of the techniques used in analyzing the sonographs and fathograms can be found in Barnes et al. (1978) and in Barnes et al. (this volume). A few important points affecting the parameters used in the present study should be mentioned, however:

Dominant gouge orientation (θ) - Templates were used to remove horizontal exaggeration from the sonographs and to obtain all measurements of the estimated dominant orientation to within 5°. (It should be noted that the gouge orientations within each line segment are variable.)

Spatial gouge frequency (N_n) - In determining the number of gouges per kilometer of sampled track (N), every feature on the sonograph presumed to result from ice contact with the bottom was counted, including individual gouges produced by different segments of what was probably the same pressure ridge keel (our interest is in the number of gouges in the bottom, not the number of events); these N values were then corrected in order to estimate N_n, the expected number of gouges that would have been seen on a 1-km sampling line if the ship's track was oriented normal to the dominant gouge trend. This correction was made by using $N_n = N/\sin \alpha$, where α is the acute angle between the ship track and the gouge orientation. As most gouges are oriented parallel to the coast and the majority of the sampling lines were roughly normal to the coast, these corrections were usually small. Gouges with depths of less than 0.2 m were not counted as it commonly was difficult to positively identify all of them on the fathogram.

Gouge depth (d) - The gouge depth was measured on the fathogram vertically from the level of the (presumably undisturbed) adjacent seafloor to the lowest point in the gouge; each individual gouge was measured. Values were grouped in 20-cm class intervals; in some cases, because of factors such as swell and wind chop, it was only possible to determine the number of gouges that have depths greater than a specified value; because of these problems gouges having depths of less than 0.2 m were not considered.

IV. GOUGE DEPTHS

To examine the distribution of gouge depths we prepared histograms of gouge depths for different regions. The general nature of these graphs appeared to be a decreasing exponential with a rapid fall-off in the frequency of occurrence of larger gouges. A similar tendency was noted by Lewis (1977a, b) and Wahlgren (1979a, b) for gouges occurring north of the Mackenzie Delta. However, an examination of their data (Lewis, 1977a) shows that the number of small gouges is significantly less than would occur in an exponential model, suggesting that some other type of distribution might also be a possibility. As will be seen, this "non-exponential" decrease in the number of small gouges is not apparent in our data set. The assumption of an exponential distribution of initial gouge depths is, as a first approximation, reasonable in that the depths of pressure ridge keels measured to the north of the study area by submarine sonar can also be well described by the use of an exponential distribution (Wadhams and Horne, 1980).

Figure 2 shows a semi-log plot of the number of gouges with different gouge depths for four representative areas of the study region: (1) from the lagoons (41 data points), (2) from Harrison Bay (842 data points), (3) from off of Lonely (2869 data points), and (4) from the profiles seaward of the barrier islands and east of Harrison Bay (16,620 data points). Other groupings of the data and data from other areas gave similar plots. The four curves are well separated as the result of the coincidence that the numbers of gouges observed in the four regions are quite different. This is the result of differing lengths of sampling line and of differing spatial gouge frequencies associated with differences in water depth. If the same sets of data are plotted as relative frequency (the proportion of the total number of observations from that region that falls in each of the 0.2 m depth classes), the shapes of the curves are identical but there is

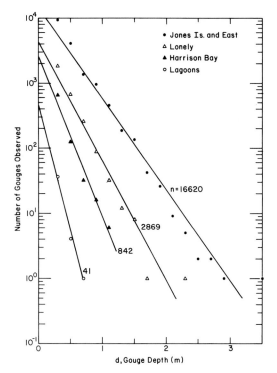

FIGURE 2. Semi-log plot of the number of gouges observed versus gouge depth for four regions along the Alaskan coast of the Beaufort Sea.

considerable overlap. All plots are reasonably linear over the complete range of four decades (r^2 values vary from 0.94 to 0.98). Thus the use of an exponential distribution in the Mackenzie studies as suggested by Lewis (1977a) appears reasonable as an initial approximation. However, even though the correlation coefficients are high, chi-squared tests for goodness of fit are commonly failed. In future studies of gouging, the possibility of either finding a more satisfactory distribution function or of better rationalizing the deviations from exponentiality should be explored.

The exponential distribution is a convenient, well-studied distribution. If the simple frequency distribution is a negative exponential, then the probability density function (PDF) of X will also be of a similar form

$$f_X(x) = ke^{-\lambda x} \qquad\qquad x \geq 0 \qquad\qquad\qquad (1)$$

(Here x represents the values that the random variable X may acquire.) Because the integral of $f_X(x)$ from 0 to ∞ must equal 1, as it contains all the sample points with non zero probabilities,

$$\int_0^\infty k e^{-\lambda x}\, dx = \frac{-k}{\lambda} e^{-\lambda x} \Big|_0^\infty = \frac{k}{\lambda} = 1$$

or $k = \lambda$

This gives the following PDF:

$$f_X(x) = \lambda e^{-\lambda x} \qquad\qquad x \geq 0$$

Here the maximum likelihood estimate of the free parameter λ is simply the reciprocal of the sample mean $\bar{x}$:

$$\hat{\lambda} = 1 / \bar{x}$$

The probability that a random variable will assume a value in the interval (x_1, x_2) is then

$$P[x_1 \leq X \leq x_2] = \int_{x_1}^{x_2} f_X(x)dx = \lambda \int_{x_1}^{x_2} e^{-\lambda x}dx$$

The cumulative distribution function (CDF) is, in turn, found by integration:

$$F_X(x) = P[X \leq x] = \int_0^x f_X(u)du = 1 - e^{-\lambda x} \qquad x \geq 0$$

Finally, because we are interested in the probability of occurrence of gouges that have depths greater than or equal to some specified value, we are largely concerned with the value of the exceedance probability, given by the complementary distribution function $G_X(x)$:

$$G_X(x) = P[X \geq x] = 1 - F_X(x) = e^{-\lambda x} \qquad\qquad (2)$$

$G_X(x)$ is a particularly simple function to graph as it is a straight line on semi-log paper and has a value of 1 at $x = 0$. Therefore the simple relation

$$P[D \geq d] = \frac{n[D \geq d]}{N} = e^{-\hat{\lambda} d} \qquad\qquad (3)$$

can be used to estimate n $D \geq d$ (the expected number of gouges having depths greater or equal to d given that N gouges are present). In determining $\hat{\lambda}$, the fact that the 0-0.2-m gouge depth class was excluded was handled by letting $d' = (d - c)$ where $c = 0.2$ m, the cutoff value. Note that in Figs. 3 and 5 the nominal $d = 0$ location is, in fact, $d = 0.2$ m. Note also that when the number of gouges is given, only gouges having depths equal to or greater than 0.2 m are counted.

It is, however, possible to sharpen the above analysis by noting that, at least off the Mackenzie Delta, the nature of the gouge depth distribution is known to change with water depth (Lewis, 1977a). This is hardly surprising in that the deep draft of the large ice masses that produce the deep gouges observed in deeper water prevents them from entering shallow water. We will now examine the effect of such a variation within our study area. From Fig. 2 it can be surmised that similar changes occur in lagoons and in Harrison Bay, which show no deep gouges. When the $\hat{\lambda}$ values corresponding to various 5-m water depth classes in the different regions are plotted against water depth z, there is clearly a general decrease in $\hat{\lambda}$ with increasing z within the range of the data set. For a discussion of the area in general, we have combined all the data for "offshore" areas unprotected by barrier islands (Lonely, Harrison Bay, Jones Islands, and East) into one data set. Figure 3 gives three representative plots of data from this combined set for three different 5-m water depth intervals and also shows the fitted curves based on equation (1). Figure 4 shows the seven $\hat{\lambda}$ values for this combined set plotted versus z. We have chosen to fit the $\hat{\lambda}$ versus z data with a negative exponential (r^2 = 0.95) purely as a matter of convenience. This curve should not be extrapolated beyond the range of the data. For instance, off the Mackenzie Delta, the peak in mean gouge density occurs at a water depth of 23 m, and no gouges were found at greater water depth than 80 m (Lewis, 1977a). Therefore, one might expect that in the present study area $\hat{\lambda}$ values may increase again at water depths greater than 35 m.

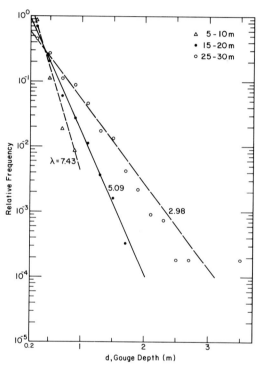

FIGURE 3. *Relative frequency of occurrence of gouges of differing depths based on all data from offshore areas unprotected by barrier islands.*

To obtain the exceedance probability for the occurrence of gouges of different depths given that gouging has occurred, the relation in Fig. 4 can provide an estimate of λ applicable to the water depth of interest. The exceedance probability is then obtained from Eq. (2). For instance for a water depth of 5 m, $\hat{\lambda}$ = 8.16 and

$$P \ [D \geq 1] \ = \ \exp \ [-8.16(1-0.2)] \ = 1.46 \ \text{x} \ 10^{-3}$$

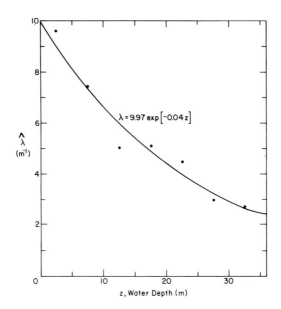

FIGURE 4. $\hat{\lambda}$ *values versus water depth based on the data set from offshore areas unprotected by barrier islands.*

gives the probability of a gouge exceeding 1 m in depth. Therefore using Eq. (3), one gouge in 685 would be expected to be at least 1 m deep. The 0.2-m correction in the above calculation is caused by the fact that the 0–0.2 m depth class was deleted in the estimation of λ. At the same water depth, one gouge in 2.39 million would be expected to be at least 2 m deep. For 35 m of water ($\hat{\lambda}$ = 2.46), things are very different: one gouge in 7 exceeds 1 m and one in 980 exceeds 3 m. A graphical display of the variations in the exceedance probability as a function of water depth for the offshore region is given in Fig. 5.

The $\hat{\lambda}$ values determined for lagoons appear to be in the 7 to 9 m^{-1} range, in short in general agreement with the $\hat{\lambda}$ values obtained from similar water depths in the offshore data set.

Clearly water depth is a most important parameter in studies of gouging.

V. GOUGE ORIENTATIONS

Determining the absolute cartographic orientation of every gouge would be very time-consuming. To provide some information on gouge orientations we visually estimated the dominant trend along each kilometer of sample track. These orientation values do not provide information on the actual direction of the ice movement (for instance, the direction 90° indicates only that the gouge runs along the 90°-270° line). Figure 6 shows

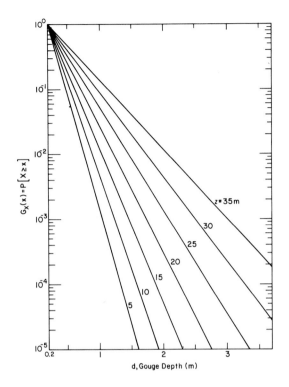

FIGURE 5. Plot of the exceedance probability $((G_X(x))$ versus gouge depth for different water depths (z) in the offshore region unprotected by barrier islands. In calculating the $G_X(x)$ values, the $\hat{x}$ values are obtained from the relation shown in Fig. 4.

linear histograms of the probability of the occurrence of different orientations. The data are displayed between $0°$ and $180°$. This proved to be convenient as there was a natural break in the observations at this orientation (that is, very few gouges were aligned north-south).

Figure 6 shows several obvious things. First, the dominant gouge orientations have a unimodal distribution that is reasonably clustered. Second, gouge orientations show more variability in lagoons and in other shallow (0-10 m) areas. Farther off the coast in deeper water, these variations generally decrease. The average orientation in water more than 20 m deep is $97°$-$99°$, which is just a few degrees less than parallel to the coast ($110°$). In shallow areas the gouges generally show a higher angle ($71°$-$83°$) to the coast, although this tendency is not evident in the measurements made off the Jones Islands. It is reasonable to expect a floe that is in the process of grounding to rotate and move toward the coast (this effect has been observed in radar imagery at Barrow by Shapiro (pers. communication)). However, it is not clear to us why this phenomenon should be more pronounced in shallow water. The mean gouge orientation

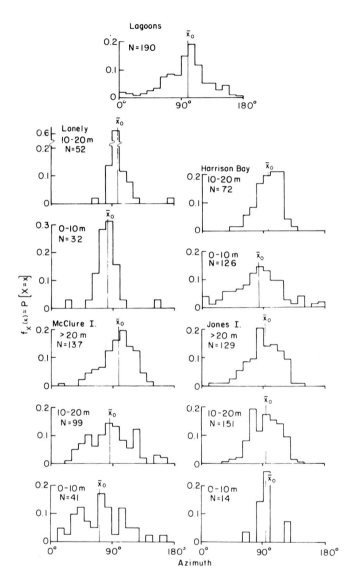

FIGURE 6. *Linear histograms of the observed probability of different dominant gouge orientations in different geographical areas and depth ranges within them.*

in the lagoons is 99°, which is similar to the gouge orientations in water deeper than 20 m.

There are three factors which presumably control gouge orientation. The first factor is the wind direction, which at Kaktovik is predominant in two directions: from the east–northeast to east (55°-100°) 35% of the time and from the west–southwest to west (235°-280°) 23% of the time. The mean wind speed is the same (6.7 m s^{-1}) in both directions. Therefore, the ice drift, which is roughly 45° to the right of the surface wind, would be expected to be between 100° and 145°, a value range just above that observed. Secondly, because the fast ice edge generally parallels the isobaths, ice-ice interactions tend to force the nearshore ice to drift parallel to the coast even when the free–drift direction is not exactly parallel with the coast.

Finally, the higher resistance commonly encountered on the nearshore side of a grounding floe will cause gouges to form at angles less than expected from a free–drift situation. The end result is, therefore, an average gouge direction in the range of 80° to 125°, as observed.

VI. GOUGE FREQUENCY

We now have a reasonable description of the probability of gouges having different gouge depths given that gouges have occurred. Next we need to determine how many gouges have occurred so that we can estimate N in Eq. (2). The number of gouges that is of primary interest is the temporal gouge frequency (the number of gouges that intersect a unit length of line per unit of time (e.g., gouges km^{-1} year^{-1}). As will be seen, data leading to such estimates are extremely sparse. What is available are measurements of the spatial gouge frequency (e.g. gouges km^{-1}) as seen at a given location at essentially a fixed instant in time. We will now discuss these two parameters.

A. Spatial Gouge Frequency

To study variations in the spatial gouge frequency the number of gouges deeper than 0.2 m was determined for each kilometer of sampling track. These values were then converted to N_n, the number of gouges per kilometer that would have been encountered if the sampling track were oriented perpendicular to the trend of the gouges. The values were then separated into five different groups (lagoons, Lonely, Harrison Bay, Jones Islands, and McClure Islands and East) and plots were made of N_n versus water depth. Examination of these plots showed that lagoons were different from the other four areas in that gouges were rare (92% of the 298 km sampled contained no gouges and the largest N_n value was 12 gouges km^{-1}). The four other regions showed differences, but these appeared to be largely caused by changes in the water depths sampled in the different areas. Therefore all four regions were combined and considered as one. Figure 7 shows the N_n versus z plot for the combined data. As was the case in the lagoons, in shallow water N_n values of zero (N_0 values) are common and N_n values greater than 50 are rare. In water 15 to 20 m deep, zero values become less common and larger N_n values are seen. As water depths increase above 22 m, all samples show 20 or more gouges km^{-1}. These changes can be shown (Fig. 8) by taking 10-m-wide vertical slices

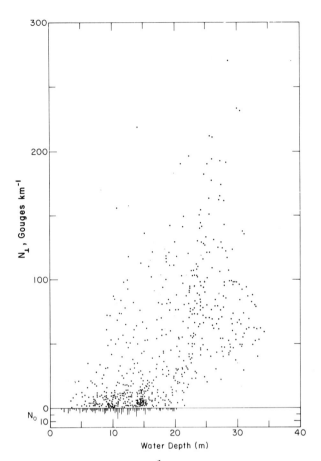

FIGURE 7. *Number of gouges* km^{-1} *measured normal to the trend of the gouges* (N_n) *versus water depth.*

through Fig. 7 and displaying the results as histograms giving relative frequency versus $N_n/10$. As can be seen, in the lagoons there is a rapid exponential drop-off in frequency as the $N_n/10$ value increases. In shallow water (< 10 m) outside of the barrier islands the trend is similar, although null values are not as frequent (42%). At depths of 10-20 m the null values compose only 24% of the total sample and $(N_n/10)$ values in excess of 10 are not rare. The distributions from 20-30 m and 30-38 m had nearly identical means and forms and were therefore combined. The histogram is here more nearly Gaussian with a slight positive skew. Again the nature of the distribution clearly is a function of the water depth. One additional piece of information should be added here. At one location (off of Lonely) a study was made of the distribution of the spacings between gouges (as measured along the sampling line). Again the distribution resembled a negative exponential.

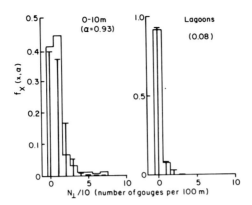

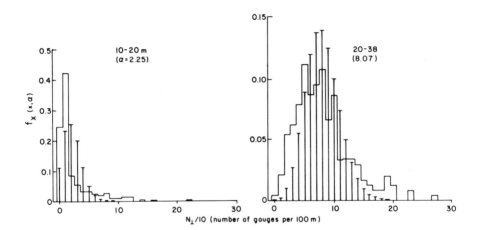

FIGURE 8. Relative frequency of different values of $N_n/10$ for lagoons and 3 different water depths offshore of the barrier islands.

It would be convenient to have one distribution function that describes all the histograms shown in Fig. 8. If possible this distribution should have the following characteristics:

1. It should be discrete in that we are describing a counting process (either a gouge is present or it is not).

2. It should be capable of dealing with the positive frequency of zero values.

3. It should have a shape which varies from a negative exponential to normal as the mean value of N_n increases.

4. The distribution of spacings between occurrences should be given by the exponential distribution.

The Poisson distribution has, in fact, all these characteristics and is given by

$$f_X(x,\alpha) = \frac{\alpha^x e^{-\alpha}}{x!}, \qquad x = 0, 1, 2, 3 \ldots; \alpha > 0 \tag{4}$$

where the parameter α is the sample mean, which in our case varies from 0.08 for lagoons to 8.07 for depths in excess of 20 m. As we have plotted $N_n/10$, these sample means correspond to N_n values of 0.8 and 80.7 gouges km^{-1}. The use of $N_n/10$ was necessitated by the fact that N_n values as large as 270 gouges km^{-1} occur. The Poisson distribution, on the other hand, is not convenient for values much in excess of 20. When $N_n/10$ is used, the Poisson probability for an integer such as 3 is used to represent the probability of N_n occurring in the interval $25 \leq N_n \leq 35$ gouges km^{-1}. Examination of Fig. 8 shows that the Poisson distribution does, in fact, give a reasonable representation of the frequency plots of the N_n values, although it does drop off too rapidly at large $N_n/10$ values. The Poisson distribution also possesses the additive property that the sum of two independent Poisson random variables with parameters α_1 and α_2 is also a Poisson random variable with parameter $\alpha = \alpha_1 + \alpha_2$.

The use of the Poisson distribution brings to mind its association with the Poisson process describing the occurrence of random events occurring at a constant rate along a continuous space (or time) scale. To be a Poisson process the underlying physical mechanism generating the events must satisfy the following three assumptions:

(1) Stationarity - the probability of at least one event in any short interval is proportional to the length of the interval.

(2) Nonmultiplicity - the probability of two or more events in a short interval Δx is negligible in comparison to $\alpha \Delta x$.

(3) Independence - the number of events in any interval is independent of the number of events in any non-overlapping interval.

The probability distribution of the number of events N in distance x for a Poisson process in given by

$$f_N(n; \nu x) = \frac{(\nu x)^n e^{-\nu x}}{n!}, \qquad n = 0, 1, 2, 3, \ldots; \nu x > 0 \tag{5}$$

where νx has replaced α in Eq. (4) and the parameter ν is the average spatial rate of occurrence of the event.

When gouging is looked on as an annual event, it apparently satisfies the requirements for a Poisson process reasonably well as a first approximation. However, when the spatial distribution of gouging is examined in more detail we find locations where gouges occur in groups (on the seaward sides of shoals). Also, gouges presumably are more common in areas where the surface sediments are poorly bonded than they are in regions where the surface sediments show a high strength. In addition, if gouging is examined on a time scale finer than yearly, the assumption of stationarity is clearly not satisfied as in many locations no gouging occurs during the summer months. However, these problems are probably no worse than in many other phenomena such as customer arrivals and number of telephone calls per unit time, where the Poisson process has been found to be a very useful model.

B. Temporal Gouge Frequency

In investigating problems concerning ice-induced gouging of the sea-floor it is highly desirable to have independent information on the rates at which new gouges form. Unfortunately such data are rather limited, and for our study area are largely contained in Barnes *et al.* (1978). This work describes replicate observations made on sample line 35 (see Fig. 1 for location) during the summers of 1973, 1975, 1976, and 1977 and on line 31 during the summers of 1975, 1976, and 1977. We have reanalyzed the data set from line 31 for the 1976-77 year and on line 35 for the 1976-77 and 1977-78 intervals so that the counts of new gouges are based on 1-km sampling lines. We also analyzed replicate runs on line 39 (north of Cape Halkett) for 1977-78. The results of this analysis plus that of Barnes *et al.* (1978) are combined giving a $\bar{g}$ value of 5.2 gouges km^{-1} $year^{-1}$ with values for individual years varying from 2.4 (1975-76) to 3.5 (1976-77) to 7.9 (1977-78). These are appreciably larger values than those off the Mackenzie Delta in 15 to 20 m of water ($0.19 + 0.06$ gouge km^{-1} $year^{-1}$) reported by Lewis (1977a), giving a return period per kilometer of 0.2 year as compared to 5.3 years. Figure 9 shows a plot of observed g values versus water depth. There is no strong trend. In addition, there is a large scatter, and zero values (1-km segments without new gouges) are rather evenly distributed at all water depths. Because of this we have treated all the observations as a single group.

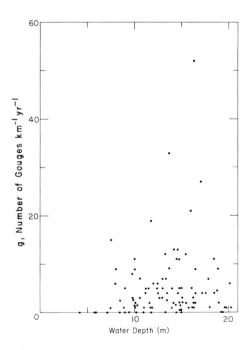

FIGURE 9. *Number of gouges km^{-1} yr^{-1} (g) versus water depth.*

An examination of whether new gouge depth values follow an exponential distribution similar to that obtained by sampling all the gouges on the seafloor is of interest, because the latter include many old gouges that presumably have been partially filled with sediment. The observations used (n = 76) were from both test lines 31 and 35 and were made between 1976 and 1977. The results (Fig. 10) again appear to show an exponential dropoff with a value of 4.52 m^{-1}. This value is close to but somewhat lower than the values obtained from the samples of all the gouges (taking 15 m as a mean water depth along the replicate sampling lines, we obtain a value of 5.5 m^{-1} from Fig. 10 as contrasted with 4.5 m^{-1} from the new gouges). New gouges should have a lower value than a corresponding distribution of old and new gouges (E. Phifer, pers. comm.) because deep gouges receive more fill per year than do shallow gouges (Fredsoe, 1979). There is no strong reason to doubt that the distribution of new gouge depths is exponential or that the values that will be obtained are greatly different (presumably slightly less) than values obtained from our earlier analysis of all the gouges.

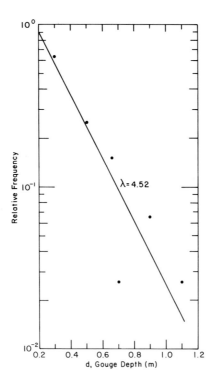

FIGURE 10. Semi-log plot of the relative frequency of occurrence of new gouges of differing depths.

VIII. APPLICATIONS TO OFFSHORE DESIGN

In the preceding sections we have attempted to systematize and clar-
ify some of the essential characteristics of a large set of measurements on
the geometry of ice gouges in the Alaskan Beaufort Sea.   These observations are, of course, extremely valuable in themselves.   They tell us for
instance that between the barrier islands and the 38-m isobath the deepest
gouge among 20,313 gouges measured along 1500 km of sampling track was
3.6 m.   In protected lagoons, on the other hand, the deepest gouge (0.7 m)
was much shallower (from a sample of 41 gouges obtained from 298 km of
sampling track) and a large percentage of the 1-km segments examined
(92%) contained no gouges at all.   In the remainder of this section we
attempt to use the data analysis performed earlier to determine pipeline
burial depths in various kinds of ice-gouged terrain.
First we will consider the problem of burying a pipeline at a depth so
that it is all covered (assuming an acceptably low probability of
encountering a gouge deeper than the burial depth that would leave the line
uncovered).   In this case we are dealing with existing gouge depths at a
given instant of time.   In this example we will consider a 20-km-long
pipeline routed (1) through a lagoon with water depths of 5 to 10 m and (2)
across the open shelf outside the barrier islands at 25 to 30 m water
depth.   We will also consider a pipeline route at 20° off the direction of the
gouges as well as one normal to the direction of the gouges.   For instance,
at a water depth of 25 to 30 m we would expect to encounter an average of
80 gouges km^{-1} if the line is normal to the gouges and 80 sin 20° = 27
gouges km^{-1} if the angle between the gouges and the line is 20°.   For the
20-km pipeline this corresponds to 1600 and 540 gouges, respectively.   Next
one must decide how many gouges can be tolerated deeper than the depth
of burial.   We will take two cases: 1 exceedance per 20 km and 1
exceedance per 100 km.   Burial depths x can then be calculated from Eq.
(3), which when rearranged and modified to treat the above cases becomes

$$\frac{n\ [D > d]}{N} = \frac{n\ [D \geq d\ ]}{N_n\ (\sin\theta\ )L} = e^{-\hat{\lambda}(x-\ 0.2)} \tag{6}$$

or, rearranging,

$$x = 0.2 - \frac{1}{\lambda}\ \ln(\frac{n\ [D \geq d\ ]}{N\ (\sin\theta\ )L}) \tag{7}$$

As stated, at a water depth of 5 to 10 m, $\hat{\lambda}$ = 7.3, N_n = 10, θ = 20 or 90°,
L = 20 or 100 km, and $n(D > d)$ = 1 inasmuch as we only wish to allow one
exceedance.   The results of several such calculations are given in Table I.
More meaningful, and also more difficult, is to estimate as a function
of burial depth how often a buried pipeline can be expected to be impacted
by a pressure ridge keel.   This problem requires knowledge of the rates of
occurrence of new gouges.
As described earlier, knowledge of this subject is too sketchy to attempt more than a preliminary estimate of burial depths.   The number of
gouges formed per kilometer per year $\bar{g}$ was found to range from 2.4 to 7.9
with a mean of 5.2, and apparently without correlation to water depth.   The

TABLE I. *Estimated Burial Depths Assuming that One Existing Gouge Will Exceed the Burial Depth along the Length of the Pipeline.*

			Line normal to gouges			Line at 20^0 to gouges		
Location	λ (m^{-1})	N_n (gouges (km^{-1})	Length of line (km)	Burial depth (m)	N_n $(sin20^0)$ (gouges km^{-1})	Length of line (km)	Burial depth (m)
Lagoons	7.7	0.8	20	0.56	0.27	20	0.42
			100	0.77		100	0.63
Seaward of barrier islands (water depth = 5 to 10 m)	7.3	10.0	20	0.93	3.42	20	0.78
			100	1.15		100	1.00
Seaward of barrier islands (water depth = 25 to 30 m)	3.2	80.0	20	2.51	27.36	20	2.17
			100	3.01		100	2.67

PDF for recent gouges was exponential with a λ value of 4.5 m^{-1}, or about 1 m^{-1} less than comparable λ values from all existing gouges.

With this information preliminary estimates can be made of the burial depths for a given length of pipeline to be hit once during a specified period of time. To do this, first estimate N, the total number of gouges that will occur during the proposed lifetime of the pipeline, by

$$N = \bar{g} \; T \; L \; \sin \theta \tag{8}$$

where $\bar{g}$ is the average number of gouges km^{-1} yr^{-1} occurring along the pipeline route, T is the proposed lifetime in years, L is the length of the line in kilometers, and θ is the angle between the route and the trend of the gouges. As we only consider 1 contact in T, $n \, [D \geq d]$ in Eq. (3) equals 1 and we obtain

$$e^{-\lambda(x-0.2)} = \frac{1}{\bar{g} \; T \; L \; \sin \theta} \tag{9}$$

or

$$x = 0.2 - \frac{1}{\lambda} \ln \frac{1}{\bar{g} \; T \; L \; \sin \theta} \tag{10}$$

In Table II we show a series of burial depth estimates made using Eq. (10). In these calculations we have used both the observed $\hat{\lambda}$ value for the existing gouge set from Fig. 10 and also λ-1 as an estimate of the corresponding parameter for new gouges. In using the table note that a 20-year lifetime

TABLE II. Estimated Burial Depths Required for an Average of One Ice Impact during a 100 yr Period. Calculations Made Using Eq. (10).

Location	$\bar{g}$ (gouges km^{-1} yr^{-1})	λ or (m^{-1})	Length of line (km)	Line normal to gouges		Line at 20° to gouges	
				Gouges crossing line during 100 yr lifetime	Burial depth (m)	Gouges crossing line during 100 yr lifetime	Burial depth (m)
Lagoons	5	7.7	20	10,000	1.40	3,420	1.26
		7.7	100	50,000	1.61	17,101	1.61
		6.7	20	10,000	1.57	3,420	1.41
		6.7	100	50,000	1.81	17,101	1.81
Seaward of barrier islands (water depth = 5 to 10 m)	5	7.3	20	10,000	1.46	3,420	1.31
		7.3	100	50,000	1.68	17,101	1.54
		6.3	20	10,000	1.66	3,420	1.49
		6.3	100	50,000	1.92	17,101	1.75
Seaward of barrier islands (water depth = 25 to 30 m)	5	3.2	20	10,000	3.08	3,420	2.74
		3.2	100	50,000	3.58	17,101	3.25
		2.2	20	10,000	4.39	3,420	3.90
		2.2	100	50,000	5.12	17,101	4.63

for a 100-km line is identical with a 100-year lifetime for a 20-km line. Obtaining better data for improved estimates of λ and $\bar{g}$ for new gouges is very important. As a general rule a few tens of centimeters increase in the burial depth results in appreciable increases in the safety of the line, particularly in shallow water where λ is large.

Table III compares our estimates of burial depths with those of Pilkington and Marcellus (1981) and of Wadhams (1983), who attempted to combine information on pressure-ridge keels, pack ice drift, and observed distributions of gouge depths to estimate required burial depths. The comparison is made for a 76-km-long line from the artificial gravel island "Kopanoar" and the shore. The return periods for impacts are taken to be 10, 100, and 1000 years. The burial depth estimated by our calculations is roughly 3 m less than that by Wadhams. At 25 m water depth our estimates would be only 4.05 and 5.47 m (assuming $\lambda = 3.7$ and 2.7 respectively) if we took $\bar{g}$ to be 20; a value four times that observed. We believe that Wadhams's approach is correct in theory, but lacks appropriate values to use in the theory. Keel depth characteristics in deeper water as recorded by submarines are probably appreciably different from those in water of 50 m or less where gouging is currently taking place. Also, distance drifted per year by the ice cover over a given point on the shelf is poorly known. During gouging the ice is slowed and commonly stopped as the grounded ice tends to stabilize the nearby pack, converting it to fast ice.

The differences between our estimates and those of Pilkington and Marcellus (1981) are less by 1 to 2 m than our differences with Wadhams's estimates. However, we find these discrepancies surprising as their procedures appear essentially identical to ours. The differences apparently result from data differences for estimating the number of new gouges yr^{-1} km^{-1}. For the latter Wadhams used direct observations of keels, while Pilkington and Marcellus indirectly inferred the number of keels from laser measurements of ridge sails. Therefore we expect that Wadhams's numbers are more realistic. Clearly we are far from agreement regarding suitable pipeline burial depths.

IX. CONCLUSION

In this paper we have presented a large amount of data on the statistical characteristics of ice gouges surveyed on the Alaskan shelf of the Beaufort Sea in shallow water (< 38 m). Although at first glance the gouges appear to be rather chaotically distributed, in a statistical sense they are very systematic. Consequently we have used this information to estimate the required burial depths of pipelines that would ensure only one ice impact in a specified number of years.

Further research on the following topics would contribute to the understanding of the geophysics of gouging and to the safe design of seafloor pipelines in regions of active gouging. The weakness of the present study is the paucity of information on the rate of occurrence of new gouges and their characteristics. The U.S. Geological Survey is collecting this kind of

TABLE III. *Burial Depths for a 76-km Pipeline and 1000, a 100, and a 10-yr Return Period.*

Lifetime (yrs)	Water depth (m)	$\bar{g}_1$ (m^{-1})	λ or $(\lambda-1)$ (m)	Burial depth (m)	Source
1000	15	5	5.5	2.54	This paper
		5	4.5	3.06	(Eq. 10)
		10	5.5	2.66	
		10	4.5	3.21	
1000	15	--	--	6.24	Wadhams (1983)
1000	25	5	3.7	3.67	This paper
		5	2.7	4.96	(Eq. 10)
		10	3.7	3.86	
		10	2.7	5.22	
1000	25	--	--	8.10	Wadhams (1983)
100	15	5	5.5	2.12	This paper
		5	4.5	2.54	(Eq. 10)
		10	5.5	2.24	
		10	4.5	2.70	
100	15	--	--	4.4	Pilkington and Marcellus (1981)
100	15	--	--	5.50	Wadhams (1983)
100	25	5	3.7	3.05	This paper
		5	2.7	4.11	(Eq. 10)
		10	3.7	3.24	
		10	2.7	4.36	
100	25	--	--	4.7	Pilkington and Marcellus (1981)
100	25	--	--	7.02	Wadhams (1983)
10	15	5	5.5	1.70	This paper
		5	4.5	2.03	(Eq. 10)
		10	5.5	1.82	
		10	4.5	2.19	
10	15	--	--	4.76	Wadhams (1983)
10	25	5	3.7	2.43	This paper
		5	2.7	3.25	(Eq. 10)
		10	3.7	2.62	
		10	2.7	3.51	
10	25	--	--	5.94	Wadhams (1983)

data in representative inshore regions. Where offshore development is contemplated, such studies should be initiated early. The large variability in yearly gouge rate necessitates several years of observations to obtain meaningful averages; also many offshore regions are not sufficiently ice-free to allow surveys to be made every year.

Systematic regional sampling is also required to reveal changes in the probability density functions of such parameters as gouge depth with changes in location and in environment on the shelf. Current information (Barnes *et al.*, this volume) suggests that there are appreciable changes in such parameters on a regional scale (for instance between the gouge depths in the present study area and those observed off the Mackenzie Delta). Studies should also be carried out to quantify the effects of differences in slope angle and direction and of the nature of bed material on gouging. Such work in conjunction with detailed site-specific studies, as conducted by Reimnitz and Kempema (this volume), would be very useful in evaluating hazards along specific pipeline routes.

Theoretical studies should also be implemented to advance our ability to treat gouging as a stochastic process. For instance a treatment of gouging as a simple covering problem in geometric probability would be useful. If such developments are sufficiently general, they can be applied to different geographic areas by simply changing the values of the input parameters.

An improved understanding of the interactions between pressure ridges and ice-island keels and the seafloor may provide insight into the maximum probable gouge depths for a given sediment type. Until such information is available we can only assume that even apparently "impossibly" deep gouges have a finite probability of occurrence.

Although we have utilized an exponential distribution to describe the relative frequency of occurrence of gouges of different depths because of its simplicity and the fact that pressure-ridge keels can be well described by such a distribution, chi-squared tests of goodness of fit are commonly failed. Therefore attempts should be made to obtain a more satisfactory distribution to describe gouge depths. The same comment can be made about our use of a Poisson distribution to describe N_n in that, as was noted earlier, there are consistently more large N_n values than predicted. We suggest that at least some of these difficulties stem from a lack of adequate treatment of the infilling of the gouges in this and other published reports on gouging. The development of a numerical simulation model that includes a description of both the initial gouging and the subsequent infilling of existing gouges could prove to be illuminating.

ACKNOWLEDGMENTS

We thank the members of the Alaskan OCSEAP for their continued support and patience. We particularly thank Gunter Weller, Dave Norton, and Bill Sackinger for allowing the continuing data-collection effort that made this study possible. The Office of Naval Research supported the first author during a year at the Naval Postgraduate School in the ONR Chair of Arctic Marine Science. He thanks Don Gaver and Pat Jacobs of the NPS

Operations Research Department for their counsel and encouragement. We also profited from the reviews and advice of Ed Phifer, John Kreider, Richard Larrabee, and P. W. Marshall of Shell, Hans Jahns and Albert Wang of Exxon, Paul Teleki of the U.S. Geological Survey, and Austin Kovacs, Malcolm Mellor, and Darryl Calkins of the Army Cold Regions Research and Engineering Laboratory.

REFERENCES

Barnes, P. W. and Reimnitz, E. (1974). *In* "The Coast and Shelf of the Beaufort Sea" (J. C. Reed and J. E. Sater, eds.), p. 439. Arctic Institute of North America, Arlington, VA.

Barnes, P. W., and Reimnitz, E. (1979). *Open-File Rept. 79-848*, U.S. Geol. Surv.

Barnes, P. W., McDowell, D., and Reimnitz, E. (1978) *Open-File Rept. 78-730*, U.S. Geol. Surv.

Fredsoe, J. (1979). *J. Petrol. Technol.*, Oct., 1223.

Hnatiuk, J. and Brown, K. D. (1977). *Proc. 9th Offshore Technol. Conf.*, Vol. 3, p. 519. Houston.

Kindle, E. M. (1924). *Am. J. Sci. 7*, 1223.

Kovacs, A. (1972). *The Oil and Gas J. 70*, 92.

Kovacs, A. and Mellor, M. (1974). *In* "The Coast and Shelf of the Beaufort Sea" (J. C. Reed and J. E. Sater, eds.), p. 113. Arctic Institute of North America, Arlington, VA.

Lewis, C. F. M. (1977a). *Beaufort Sea Tech. Rept. 23 (draft)*. Beaufort Sea Project, Dept. of the Environ. Victoria, B.C.

Lewis, C. F. M. (1977b). *In* "Proceedings POAC 77" (D. B. Muggeridge, ed.), p. 567. Memorial Univ. of Newfoundland, St. Johns.

Pelletier, B. R. and Shearer, J. M. (1972). *Proc. 24th Internat. Geol. Congress*, Sec. 80

Pilkington, G. R. and Marcellus, R. W. (1981). *In* "Proceedings POAC 81," p. 674. Quebec.

Rearic, D. M., Barnes, P. W. and Reimnitz, E. (1981). *Open-File Rept. 81-950*. U.S. Geol. Surv.

Reimnitz, E. and Barnes, P. W. (1974). *In* "The Coast and Shelf of the Beaufort Sea" (J. C. Reed and J. E. Sater, eds.), p. 301. Arctic Institute of North America, Arlington, VA.

Reimnitz, E., Barnes, P. W., Forgatsch, T. C. and Rodeick, C. H. (1972). *Marine Geol. 13*, 323.

Reimnitz, E., Barnes, P. W. and Alpha, T. R. (1973). *Misc. Field Studies Map MF-532*. U.S. Geol. Surv.

Reimnitz, E., Barnes, P. W., Toimil, L. J. and Melchior, J. (1977a). *Geology 5*, 405.

Reimnitz, E., Toimil, L. J. and Barnes, P. W. (1977b). *Proc. 9th Offshore Technol. Conf.*, Vol. 3, p. 513. Houston.

Reimnitz, E., Toimil, L. J. and Barnes, P. W. (1978). *Marine Geol. 28*, 179.

Shearer, J. M. and Blasco, S. M. (1975). *In* "Rept. of Activities, Part A," Geol. Surv. Canada, Paper 75-1A, 483.

Shearer, J. M., MacNab, E. F., Pelletier, B. R. and Smith, T. B. (1971). *Science 174*, 816.

Skinner, B. C. (1971). *U.S. Coast Guard Acad. Rept. RDC GA-23.* U.S. Coast Guard, New London, CT.
Wadhams, P. (1983). *Cold Reg. Sci. and Technol. 6,* 257.
Wadhams, P. and Horne, R. J. (1980). *J. Glaciol. 25,* 401.
Wahlgren, R. V. (1979a).   M.A. Thesis.   Dept. Geog., Carleton Univ., Ottawa.
Wahlgren, R. V. (1979b).   *In* "Current Research, Part B, Geol. Surv. Canada," Paper 79-1b, 51.

DETERMINING DISTRIBUTION PATTERNS OF ICE-BONDED PERMAFROST IN THE U.S. BEAUFORT SEA FROM SEISMIC DATA

K. Gerard Neave

Northern Seismic Analysis
Echo Bay, Ontario

Paul V. Sellmann

U.S. Army Cold Regions Research and Engineering Laboratory
Hanover, New Hampshire

I. INTRODUCTION

The basis for this study is the noticeable increase in seismic velocity that occurs in most materials containing water when they freeze. Applied to the large amounts of seismic data collected for petroleum exploration, this phenomenon makes seismic techniques a reasonable approach for investigating the distribution of ice-bonded subsea permafrost. However, sometimes ice-bonded materials may not be detectable due to their low seismic velocities, which reflect the high temperature and high salt content associated with subsea permafrost. Hunter *et al.* (1976, 1978) showed that when seismic records are available, and their quality and recording parameters are appropriate, permafrost data can be extracted from them. Direct-wave velocities and refraction interpretation of the first returns can give information on velocity structure, which in turn is used to predict the distribution of ice-bonded permafrost and, when resolution is adequate, the depth to the top of the frozen sediments. Reflection interpretations also supplement the depth information when shallow reflectors are strong enough. For the purposes of this study, zones with velocities greater than 2 km s^{-1} were interpreted to be ice-bonded permafrost. This cutoff value was selected based on laboratory studies (King *et al.*, 1982) and comparison of survey data with drill logs near the Sagavanirktok Delta. Earlier investigators have used a higher cutoff value (approximately 2.4 km s^{-1}; Hunter *et al.*, 1976, Morack and Rogers, 1982).

Examples of results from the Alaskan Beaufort Sea were selected to illustrate some common distribution patterns for the high-velocity

THE ALASKAN BEAUFORT SEA:
ECOSYSTEMS AND ENVIRONMENTS

materials in this region. Velocity profiles and maps are included for the coast from Harrison Bay to 40 km east of Prudhoe Bay (Fig. 1). A combination of ice survey and marine records provided coverage that in some cases extended as far as 60 km seaward of the small offshore islands. The distribution of high-velocity material interpreted to be ice-bonded permafrost is extremely complex, but on a regional scale we can organize the results based on the depth to the top of the high-velocity material. Two main categories and possibly a third are suggested.

To help understand the seismic results, we constructed velocity profiles from available laboratory information on velocity versus temperature for various materials and from temperature profiles onshore and offshore. These profiles were used to explain aspects of the velocity gradients and to gauge the depth of ice-bonded material.

II. METHODS

This study interpreted seismic records acquired for petroleum exploration purposes. The first-return data used included nonproprietary monitor records from Geophysical Service Inc. and Western Geophysical Co. as well as data released by British Petroleum. The data for shoreline transitions and other shallow-water coastal areas are from winter surveys conducted over the ice. The remaining shot lines are from marine surveys. A small amount of drilling information was available for control purposes in the eastern part of the study area.

In seismic data processing for petroleum exploration the emphasis is normally on deep targets; the procedures compromise the quality of data from near the surface. To obtain as much information as possible from the records without costly processing, approximately the first 2 seconds of the field tapes was played back with expanded gain and printed in a "wiggletrace" or variable area format, as shown in Fig. 2.

Analysis of the refraction and reflection records used a simple model: a plane homogeneous layer lying on a homogeneous half-space. The equations for calculating depths are available in Grant and West (1965). Ideally the refraction readings yield a depth for the interface and the velocity above and below it. The reflection readings yield the depth of the interface and an average velocity for the upper layer.

A number of error estimates were made to verify that the measurements and analytical methods were reasonable. These estimates were not intended to set strict limits on interpretation error, since difficulties in seismic data acquisition and analysis may occur which cannot be evaluated. For example, signal identification is occasionally a problem. Later arriving reflections can be misinterpreted as refraction events, particularly when the signal-to-noise ratio is low. Errors may also arise from the need to assume plane layers for simplicity in interpretation. Both of these problems are dealt with primarily by looking at the internal consistency of the results where there is duplication in coverage, and in a few cases by verification with independent geophysical techniques and drilling results.

The spatial resolution of the data from this type of study is obviously not as great as can be obtained from a seismic investigation specifically

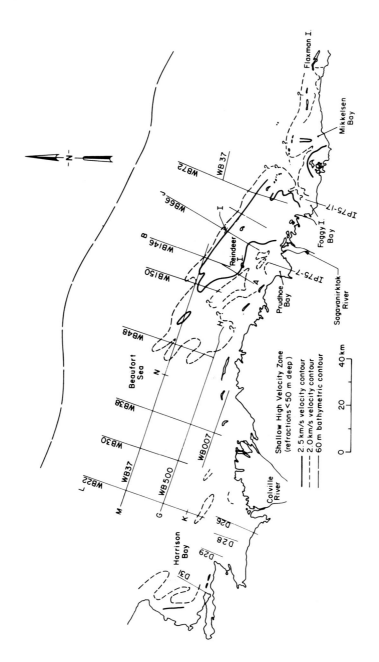

FIGURE 1. Distribution of shallow high-velocity material in part of the Beaufort Sea. The seismic lines shown are used to determine the limits of this material near Prudhoe Bay. Lines used in Harrison Bay are shown in Fig. 6.

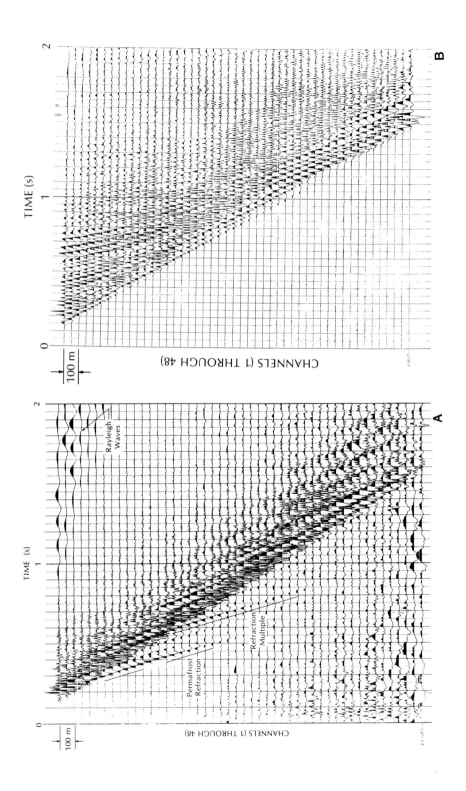

TIME (s)

CHANNELS (1 THROUGH 48)

100 m

Rayleigh Waves

Permafrost Refraction

Refraction Multiple

A

TIME (s)

CHANNELS (1 THROUGH 48)

100 m

2

1

0

B

designed to study offshore permafrost. For refraction and reflection measurements, a number of factors must be considered, including geophone spacing, signal frequency, and complexity of the subsurface. In general, the horizontal extent of a feature that can be detected should be a minimum of three detector spacings, or about 300 m for most of the ice-shooting data, and around 150 m for the marine survey data. The minimum vertical thickness of a detectable high–velocity layer is about half the wavelength of the refracted signal (Sherwood, 1967), or about 50 m for these data. However, Sherwood (1967) showed that shallow layers less than 30 m thick might be detectable. The signals from these would have reduced velocity and amplitude and are known as plate waves.

To simplify the refraction depth determinations, the upper layer velocity was taken as 1.8 km s^{-1} for all profiles out to 15 m water depth. This means that the water layer was combined with the low–velocity bottom sediments to make a single upper layer. Upper-layer velocities were observed to range from 1.6 to 2.0 km s^{-1}; therefore, the error introduced by assuming 1.8 km s^{-1} could be as much as 30% under rare circumstances. For deeper water an upper layer velocity of 1.5 km s^{-1} was used. Therefore, all depths resulting from our data analysis are below sea level.

Refraction velocities and depth determinations from single–ended marine records are subject to errors caused by dipping layers. Our interpretations indicate that dips are normally less than 3%. The corresponding maximum error is approximately 5% in velocity measurements and 2% in depth determinations (Neave and Sellmann, 1982). The assumption of horizontal layers for the reflection interpretation does not result in significant error. The error calculations in Neave and Sellmann (1982) show that a 3% dip usually results in an error of 1% in velocity and 2% in depth.

III. RESULTS

A. Shallow High-Velocity Material

Refraction results from the lines shown in Fig. 1 were used to estimate the horizontal extent and depth to the top of shallow (<50 m) high–velocity material. A shallow high–velocity refractor near Reindeer Island, shown in Fig. 3A (also reported by Rogers and Morack, 1980), has its top between 30 and 50 m below sea level. On the line in Fig. 3A the high–velocity segment (>2 km s^{-1}) is confined to a 17–km-long section that starts adjacent to Reindeer Island. A corresponding velocity profile (Fig. 3B) shows an abrupt increase in velocity at the southern edge of the zone in contrast to a steady decrease in velocity with distance north of the island. Velocities on this line range from approximately 3 to 1.7 km s^{-1} for this seabed refractor.

The shallow high–velocity zone is found in the extensive region shown in Fig. 1 enclosed by the 2–km s^{-1} velocity contour. Its northeastern

FIGURE 2. *Marine seismic record showing clear examples of refractors. A, high velocity; B, low velocity.*

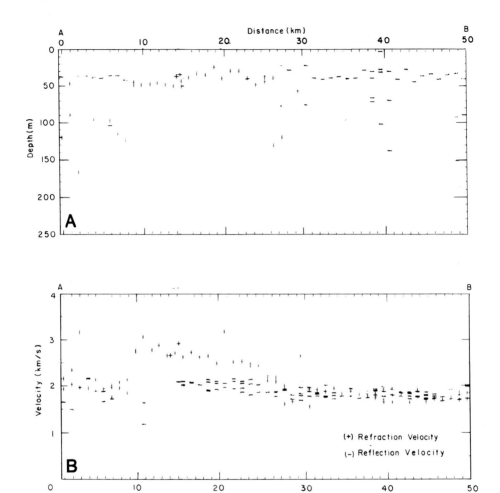

FIGURE 3.   Refraction profiles for line WB 146.   A, depth to the refractors.   Dashes indicate velocity <2 km s^{-1}, crosses indicate velocity >2 km s^{-1}.   B, velocity profile.

boundary roughly parallels the 28-m isobath and the western boundary lies between the Sagavanirktok and Colville Rivers at approximately 149°30' W. longitude. Drilling results (Miller and Bruggers, 1980) extend this zone to the east, on the offshore side of the barrier islands. The high-velocity zone does not exist immediately south of Reindeer Island but instead extends to the southeast to the Sagavanirktok Delta.

A smaller shallow high-velocity layer exists in the western part of Harrison Bay (Fig. 1). An example of the seismic data from this part of the bay is shown in Fig. 4A. A high-velocity unit appears 2 km from the

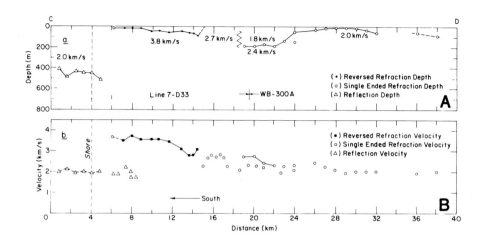

FIGURE 4. *Profiles from the western part of Harrison Bay. Line locations are shown in Fig. 6. A, cross section; B, velocity profile.*

shoreline and continues to the north for 13 km. After a gap of 5 km, a shallow refractor continues to the north end of the line. The general trend is for velocities to decrease away from the shore (Fig. 4B).

The shallow depth of the refractors observed in the western part of Harrison Bay also corresponds with observations made to the west near Lonely, which is in the same geological setting. Harrison and Osterkamp (1981) established a study line off of Lonely, approximately 40 km west of Harrison Bay, and observed shallow ice–bonded permafrost 8 to 15 m below the seabed out to at least 7.8 km from shore.

B. Deep High-Velocity Material

A different seismic-velocity regime was found in the east half of Harrison Bay. A shallow high-velocity unit is not present; however, a deep refractor was observed that dips northward under the bay, starting near the surface at the shoreline (Fig. 5A). This refractor also has a systematic decrease in velocity away from shore (Fig. 5B) from 4 km s^{-1} at the shore to less than 3 km s^{-1} at the northern end of the line.

Velocity profiles and cross sections like those in Figs. 4 and 5 were constructed for all the shot lines in Harrison Bay and are presented elsewhere (Neave and Sellmann, 1982). A contour map of the depth to the refractor (Fig. 6) shows that in the eastern part of the bay the high-velocity layer is at or near the surface at the shore and dips to the northeast under the bay. Beyond this refraction zone a continuation of the high-velocity structure is suggested by deep reflections. The steepest dips, approximately 3%, are near shore off Atiguru Point and just west of the Colville Delta.

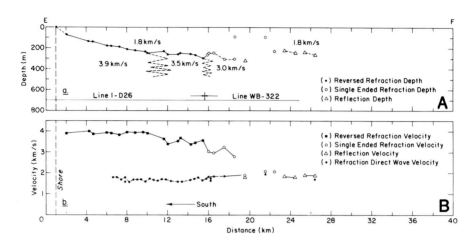

FIGURE 5. Profiles from the eastern part of Harrison Bay. A, cross section; B, velocity profile.

The range of seismic-refraction velocities observed in the Harrison Bay region is shown in Fig. 7. The narrow low-velocity peak at 1.8 km s^{-1} arises from the direct waves and represents the upper layer velocity where there is a simple two-layer subbottom. The broad high-velocity peak at 3.5 km s^{-1} represents the refracting horizons and shows the range of velocities (2.2 to 4.6 km s^{-1}) that is possible for the deeper material.

A contour map of refractor velocities (Fig. 8) also illustrates the difference between the eastern and western parts of Harrison Bay. Velocities west of Atiguru Point are more variable and decrease more rapidly with distance from shore than velocities east of the point.

Outside of Harrison Bay, refraction analysis of ice shooting data located deep refracting layers in all the large bays shown in Fig. 1; examples are shown here from Prudhoe Bay and Foggy Island Bay (Fig. 9). These deep refracting layers only approach the surface at the coastline, where they rise to correspond with the exposed high-velocity permafrost on land.

Reflection analysis for the region beyond the barrier islands indicates that three deep horizontal reflecting horizons are common (Fig. 10). No velocity data are available for these layers as refraction analysis was not possible. The depths of these reflectors are approximately 200, 450, and 750 m, with the upper two occurring at depths that could be related to ice-bonded permafrost. These features are not continuous on all lines but can be found repeatedly across this part of the shelf. An example of good continuity in the reflectors can be seen in Fig. 11A, an east-west section on which the upper (200 m) reflector is a continuous feature for the entire length of the line. The intermediate reflector (450 m) is missing or may rise to 300 m depth on the eastern third of the line. The deep reflector is suggested by a scattering of reflections between 700 and 900 m; however, it is not continuous enough to define a single horizon. In Figs. 11B and 11C,

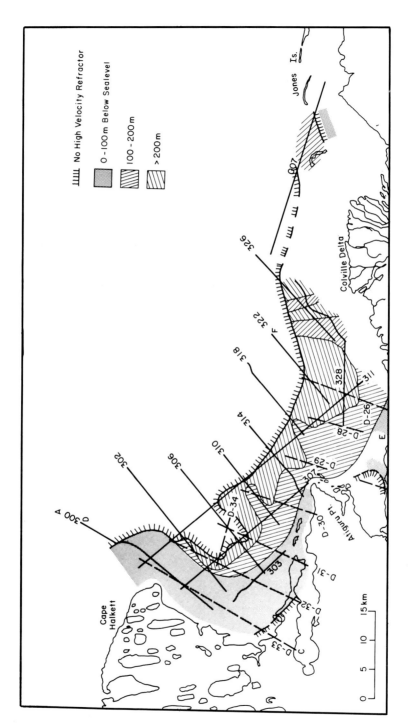

FIGURE 6. *Refraction velocity for the high-velocity material in Harrison Bay.*

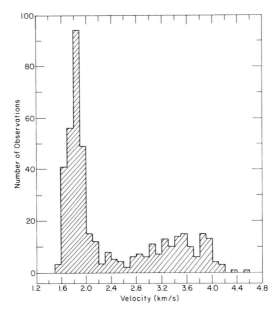

FIGURE 7.   *Velocity histogram from refraction analysis of Harrison Bay records.*

the northern halves of both lines show reflection horizons at 450 and 750 m. In Fig. 11C a segment of the 200-m upper reflector appears at the north end of the line and the intermediate reflector at the south end. The west end of line WB37 is reproduced in Fig. 11D with a good example of a strong continuous reflector at 750 m. The 200-m reflector is well represented on this line, as is a small segment of the 450-m reflector.

There is good correspondence between the deep refractors found near shore and the 200-m reflectors observed on the marine lines beyond the barrier islands. Therefore, there is some evidence that the 200-m reflection is the top of a continuous massive structure, possibly ice-bonded permafrost. The apparent continuity of this structure from the refractions and reflections in the coastal and marine survey data can be shown by comparing the north end of E-F (Fig. 5A) and the south end of K-L (Fig. 11C).

IV. DISCUSSION

A. *Permafrost Interpretations*

The high-velocity materials that were found in the regions shown in Figs. 1 and 6 can be grouped into three general distribution patterns. The first pattern (Fig. 12) represents large areas of shallow high-velocity

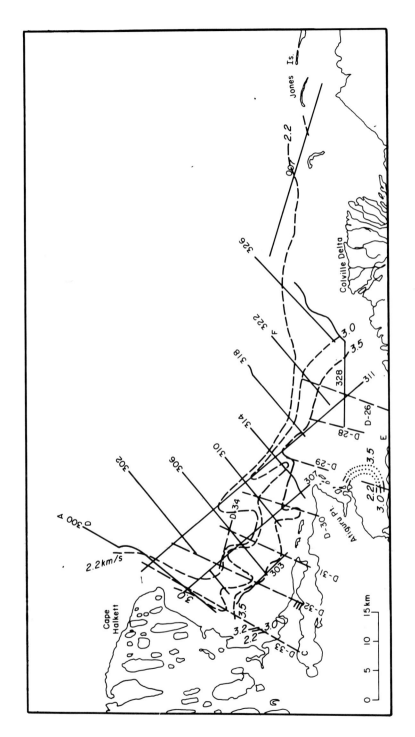

FIGURE 8. Depth below sea level to the top of the high-velocity refractor in Harrison Bay.

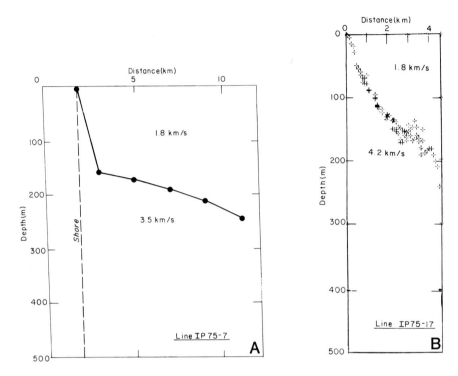

FIGURE 9.   Refraction sections from Prudhoe Bay and Foggy Island Bay.
Depth is measured from sea level. A, Prudhoe Bay. Line IP 75-7 has a flat
refractor just above 200 m. B, Foggy Island Bay. Line IP 75-17 has a layer
dipping at approximately 2.5%. The shore is at distance 0.

material less than 40 m below the seabed. It was observed in the western
part of Harrison Bay and off the Sagavanirktok Delta.   Off the
Sagavanirktok Delta this zone has been confirmed as ice-bonded sediments. The top of ice-bonded permafrost was found at 7 to 24 m below the
seabed in drill and penetrometer holes north of the delta and beyond the
offshore islands to the east of the Sagavanirktok River (Blouin *et al.*, 1979;
U.S. Geological Survey, 1979).
    The second pattern, which covers the majority of the remaining area,
is characterized by a high-velocity refractor that dips steeply offshore and
becomes nearly horizontal at approximately 200 m. Further offshore, a
reflector commonly replaces this refractor that can be detected to the
limit of the study area.   This deep high-velocity zone, extended by a
reflecting horizon, represents the top of relict ice-bonded material, an
interpretation based on the possible link to onshore permafrost, seismic
data and oil well logs from Prudhoe Bay (Osterkamp and Payne, data
1981). Furthermore, the horizontal reflector at approximately 450 m depth
roughly corresponds to the base of the permafrost, found along parts of this
coastline and in the Gull and Niakuk Island wells (Gold and Lachenbruch,

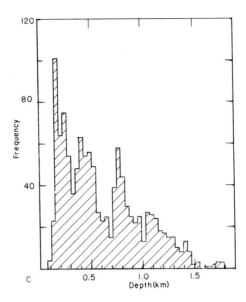

FIGURE 10. *Histogram of depths to subsea reflections for the lines seaward of the barrier islands. Note the three distinct peaks at 200, 450, and 750 m.*

1973, Osterkamp and Payne, 1981). These deep reflectors may also be related to contrast in lithology; however, the limited evidence supports a permafrost interpretation.

The third pattern, consisting of two separate permafrost layers, is suggested from a limited amount of seismic data and information from drilling on and north of the barrier islands (Sellmann and Chamberlain, 1979, Miller and Bruggers, 1980). Layering would be expected in the immediate vicinity of the barrier islands and shoals, or any region where shallow aggrading permafrost may form or be preserved in the marine environment. This shallow ice-bonded sediment would overlie first unbonded then deeper relict ice-bonded materials. The distribution of layered ice-bonded permafrost is largely unknown and in seismic investigations can be confused with the first pattern, because seismic techniques provide limited information beneath a high-velocity surface layer.

B. Simple Velocity Model

In an attempt to better understand velocity distribution offshore due to the melting and temperature changes caused by inundation, we constructed a simple velocity model. Velocity-versus-temperature curves, drawn from laboratory tests on several material types, were used with field temperature data to construct velocity profiles. Simple assumptions were used that did not consider variations in pore water chemistry, material properties, and transgression rate. Our curves for sand, silt, and clay (Fig. 13) were drawn from laboratory studies conducted by King et al. (1982), who tested Mackenzie River samples. Additional data from the Canadian Beaufort Sea are not shown, since they did not include the transition to the thawed state. However, their results from the two sample locations were similar; the slightly lower velocities from the Beaufort samples were attributed to higher salt content. The velocities for the thawed material agree with our field observations as well as those of Rogers and Morack (1978), Morack and Rogers (1982), and Hamilton (1971).

The field temperature data required for the conversion of temperature to velocity were obtained from two sources. The on-land profile was selected from Prudhoe Bay; other coastal profiles in Fig. 14 are shown for

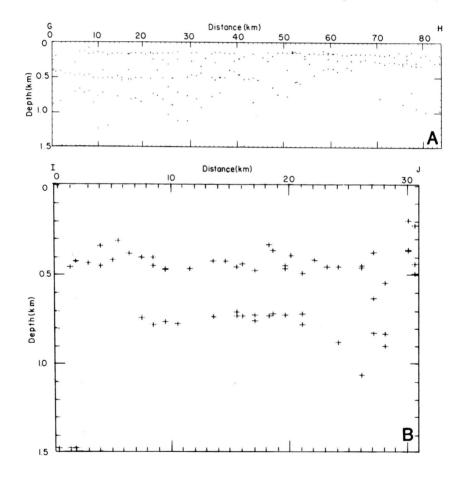

FIGURE 11. Reflection sections showing horizontal structures. A, line G-H with reflectors at approximately 200 and 450 m; B, line I-J with reflectors at approximately 450 and 750 m.

comparison (Gold and Lachenbruch, 1973). For the offshore permafrost sections we used the theoretical profiles developed by Lachenbruch and Marshall (1977), because no field profiles are available, particularly at varying distance from shore. These profiles were useful since their analysis took into consideration the length of time the area was covered by the sea (Fig. 15). These profiles, along with an assumed transgression rate and the laboratory data, permitted construction of the onshore and offshore velocity profiles shows in Figs. 16 and 17. A constant transgression rate of 2 m yr^{-1} was used, suggesting that a location 20 km offshore was inundated 10 000 years ago. Transgression rates can commonly range from 2 to 10 m yr^{-1} in this region (Lewellen, 1977, Hartz, 1978).

The model for the onshore velocities, including various material types, shows decreasing velocity with depth. The velocities for sand remain more uniform to greater depths than for clay and silt. The silt and sand velocities from the model range from 2.8 to 4.3 km s^{-1} in the upper 80% of

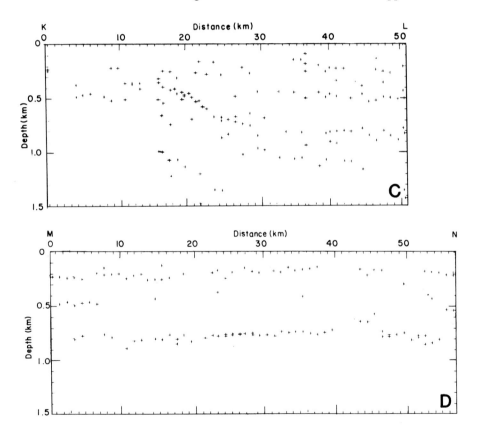

FIGURE 11. (Con'd) Reflection sections showing horizontal structures. C, line K-L with reflectors at about 200 and 450 m; D, west half of line WB 37 (M-N) with reflectors at about 200, 450, and 750 m.

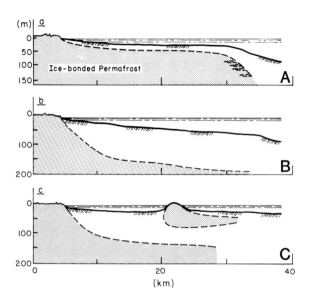

FIGURE 12. Three subsea permafrost distribution patterns interpreted for the region studied in the Beaufort Sea: shallow relict permafrost, deep relict permafrost, and layered ice-bonded permafrost.

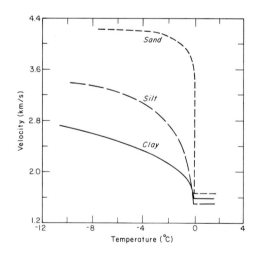

FIGURE 13. Velocity versus temperature curves from laboratory observations, reported by King et al. (1982).

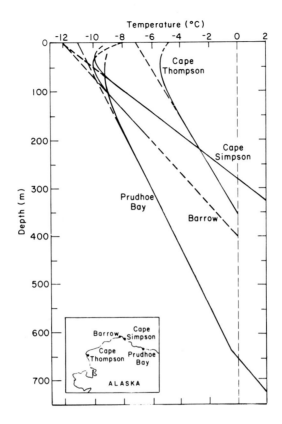

FIGURE 14. *Measured land temperature profiles from the Alaskan arctic coast (solid lines). Dashed lines indicate extrapolations (from Gold and Lachenbruch, 1973).*

the section. This range is close to that observed based on our seismic data analysis (3.3–4.6 km s^{-1}) for the on-land permafrost section in the Prudhoe Bay region. These profiles also help to illustrate the basis for contrasting offshore and onshore velocities.

Anomalous onshore velocities are indicated when the model is compared with data from the southwest corner of Harrison Bay, where reflections show a thick section with an average velocity of 2.0 km s^{-1}. The low-velocity zone may be the result of warming to 400 m depth beneath a lake that once occupied this area.

The simplest offshore velocity models shown in Fig. 17 fit the first two subsea permafrost patterns of Fig. 12; however, each pattern requires a different transgression rage. The 2 m yr^{-1} transgression rate used for the silt and sand models results in velocity profiles that best fit the second pattern (deep permafrost). Shallow offshore permafrost would require a much greater transgression rate. A rate of 10 m yr^{-1}, fast enough to allow shallow ice-bonded permafrost to exist to 10 to 15 km from shore, can be

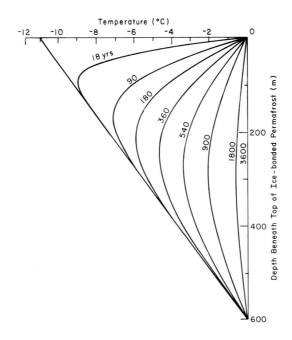

FIGURE 15. *Subsea temperature profiles for the Prudhoe Bay region from a model proposed by Lachenbruch and Marshall (1977). The profiles indicate anticipated temperature modification caused by the sea covering an area for various periods of time.*

modeled from Fig. 17 by multiplying the distance axis by 5. The velocity profiles in Fig. 17 also dramatically help to illustrate the influence of material type on velocity distribution.

In the model, warming is seen to gradually reduce the velocity with distance from shore and explain why high–velocity units in our sections have a corresponding offshore decrease in velocity (Figs. 3B, 4B, and 5B).

V. SUMMARY AND CONCLUSION

According to the traditional definition, based on temperature alone, permafrost is widespread throughout the Beaufort Sea, with material below 0^0 found near the seabed and at depths comparable to those observed on the adajacent land (Lachenbruch and Marshall, 1977, Sellmann and Chamberlain, 1979, Osterkamp and Harrison, 1980, Hartz and Hopkins, 1980, Miller and Bruggers, 1980, Lachenbruch et al., 1982). However, the distribution of ice–bonded material is not as extensive and is variable due to different salt contents, geologic histories, and material types along the coast. In some cases, when temperatures are low enough or when material types are appropriate, ice–bonded permafrost can be detected by the velocity contrast suggested by the laboratory studies shown in Fig. 13. This

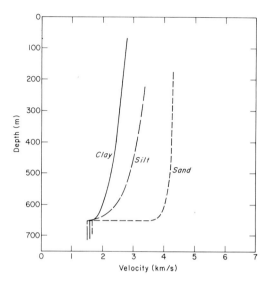

FIGURE 16. Land velocity profiles constructed from data in Fig. 13.

study of seismic records, originally acquired for petroleum exploration, is best suited to defining the minimum regional distribution of ice-bonded permafrost. The deficiencies are that features even tens of meters thick and as great as 100 m in horizontal extent may escape detection. Low velocities associated with warm, frozen, saline sediments and the small amount of control from drilling are additional problems.

Our regional survey indicates that shallow high-velocity zones, in some cases identified as and in others inferred to be ice-bonded permafrost, occur locally in the Beaufort Sea. The largest regions occur off the Sagavanirktok Delta and in the western part of Harrison Bay (Fig. 1). Ice-bonded sediments were found less than 10 m below the seabed near the Sagavanirktok Delta (Miller and Bruggers, 1980).

Shallow ice-bonded permafrost would be expected along large seg-ments of the Beaufort Sea coast because the coast is retreating rapidly by erosion. In places where the shallow high-velocity material is associated with accelerated coastal erosion, this material can contain large ground ice features. Here it represents the upper part of a continuous ice-bonded permafrost section, with the total thickness corresponding closely to that normally found on the adjacent land. These subsurface conditions are expected in the western part of Harrison Bay.

An ice-bonded layer seems to be present at 200 m depth in most areas where the shallow layer was not detected. This identification is largely by inference as drilling control is absent. The major control is the apparent link with the high-velocity permafrost onshore. Associated with this horizon is a reflecting horizon at 450 m depth that may be related to the bottom of ice-bonded permafrost. Our analysis suggests that deep ice-bonded material extends beyond the limits of the lines studied, out to at least 55 km from shore. The deeper permafrost unit apparently can also

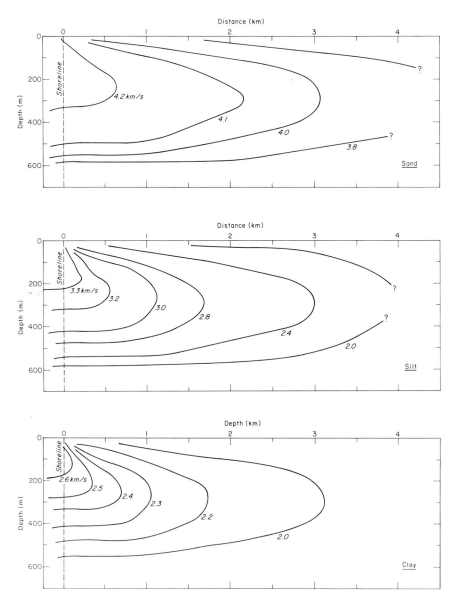

FIGURE 17. Offshore velocity sections constructed from data in Fig. 13 and proposed temperature profiles shown in Fig. 15.

locally underlie a thin, shallow high-velocity layer. The island chains in the eastern part of the study area may contain examples of this situation, where ice-bonded sediments form beneath shoreward-moving islands and shoals, thus remaining as a shallow high-velocity layer to seaward.

The origin of the shallow high-velocity zone near the Sagavanirktok Delta shown in Fig. 1 is not known. It may be related to a more recently degraded land surface, like the western part of Harrison Bay. The fact that the velocities are very high near Prudhoe Bay may be related not only to recent inundation but also to the frozen sand and gravel that occur in this region (Miller and Bruggers, 1980). Sediments can also freeze in the marine environment under ideal conditions, such as when bottom water temperatures are lowered, possibly due to drainage of cold saline water, and when sediments have low salinity and permeability. Therefore, distribution of sediments at the seabed and their properties will have an important influence on the mechanisms that influence the properties of subsea permafrost (Harrison and Osterkamp, 1976; Hartz and Hopkins, 1980).

REFERENCES

Blouin, S., Chamberlain, E., Sellmann, P., and Garfield, D. (1979). *Cold Reg. Sci. Tech.* 1, 3.

Gold, L., and Lachenbruch, A. (1973). *In* "Proc. 2nd Int. Conf. on Permafrost, North American Contribution," p. 3, Nat. Acad. Sci., Washington.

Grant, F. and West, G. (1965). *In* "Interpretation Theory in Applied Geophysics. McGraw-Hill, New York.

Hamilton, E. (1971). *J. geophys. Res.* 76, 579.

Harrison, W., and Osterkamp, T. (1976). *In* "A Coupled Heat and Salt Transport Model for Subsea Permafrost." University of Alaska Geophysical Institute, UAG-R-247.

Harrison, W., and Osterkamp, T. (1981). *In* "Environmental Assessment of the Alaskan Continental Shelf, Annual Reports of Principal Investigators," NOAA-BLM.

Hartz, R. (1978). *In* "USGS Open File Report 78-406."

Hartz, R., and Hopkins, D. (1980). *In* "Environmental Assessment of the Alaskan Continental Shelf, Annual Reports of Principal Investigators" (P. Smith, R. Hartz and D. Hopkins, eds.), Vol. IV, p. 167. NOAA, Boulder.

Hunter, J., Judge, A., MacAulay, H., Good, R., Gagne, R., and Burns, R. (1976). *In* "Beaufort Sea Technical Report," No. 22, 174 p., Geol. Survey of Canada and Earth Physics Branch, Dept. of Energy, Mines and Resources, Victoria, British Columbia.

Hunter, J., Neave, K., MacAulay, H., and Hobson, G. (1978). *In* "Proc. 3rd Int. Conf. on Permafrost," Vol. 1, p. 521. Edmonton.

King, M., Pandit, B., Hunter, J., and Gajtani, M. (1982). *In* "Proc. 4th Canadian Permafrost Conf.," p. 268.

Lachenbruch, A., and Marshall, B. (1977). *U.S. Geol. Surv. Open-File Report 77-395.*

Lachenbruch, A., Sass, J., Marshall, B., and Moses, T. (1982). *J. geophys. Res.* 87, 9301.

Lewellen, R. (1977). *In* "Environmental Assessment of the Alaskan Continental Shelf." Ann. Repts. Vol. 15, p. 491. NOAA, Washington.

Miller, D., and Bruggers, D. (1980). *In* "Proc. 12th Ann. Offshore Tech. Conf.," p. 325.

Morack, J., and Rogers, J. (1982). *In* "Proc. 4th Canadian Permafrost Conf., p. 249.

Neave, K., and Sellmann, P. (1982). *In* "CRREL Report 82-24," U.S. Army Cold Regions Research and Engineering Laboratory, Hanover, NH.

Osterkamp, T., and Harrison, W. (1980). *In* "Environmental Assessment of the Alaskan Continental Shelf, Annual Reports," Vol. IV, p. 497.

Osterkamp, T.E. and Payne M.W. (1981). *Cold Reg. Sci. Tech.* 5., 13.

Rogers, J., and Morack, J., (1978). *In* "Proc. 3rd Int. Conf. on Permafrost," Vol. 1, p. 560.

Rogers, J., and Morack, J., (1980). *J. geophys. Res. 85*, p. 4845.

Sellmann, P., and Chamberlain, E. (1979). *In* "Proc. of the 11th Ann. Offshore Tech. Conf.," p. 1481.

Sherwood, J. (1967). *In* "Seismic Refraction Prospecting" (A. Musgrove, ed.), p. 138, Society of Exploration Geophysicists, Washington.

U.S. Geological Survey (1979). *Data Set AK 17718*. NOAA, Boulder.

ACOUSTIC VELOCITIES OF NEARSHORE MATERIALS IN THE ALASKAN BEAUFORT AND CHUKCHI SEAS

John L. Morack

Physics Department and Geophysical Institute
University of Alaska
Fairbanks, Alaska

James C. Rogers

Electrical Engineering Department
Michigan Technological University
Houghton, Michigan

I. INTRODUCTION

An extensive marine seismic study made to locate ice-bonded subsea permafrost in the Alaskan Beaufort and Chukchi Seas has also yielded information on the acoustic properties of unbonded materials from these areas. Most of the data were taken using air guns as sources and a 350-m, 24-channel hydrophone streamer (Rogers and Morack, 1980a). The data were mostly of a survey nature; however, some were taken in areas adjacent to drill holes where geologic data have been obtained by others. Most of the seismic-refraction lines collected near these drill holes were reversed. A general discussion of marine seismic techniques and analysis of reversed refraction data can be found in Clay and Medwin (1977) and Dobrin (1975).

II. ACOUSTIC VELOCITIES OF MATERIAL TYPES

Materials under the seafloor can be differentiated by various geophysical quantities, but the most commonly used one is the compressional-wave velocity. A significant amount of laboratory and field research has been performed on the many types of unconsolidated materials that underlie the oceans (Hamilton, 1971), and there has been a continuing theoretical interest in their acoustical properties. The general theory of the transmission of sound signals in fluid-filled porous materials was presented by Biot (1941, 1956a, 1956b, 1962a, 1962b) and extended to many of the cases of interest to marine geophysicists by Stoll (1974).

A special feature of the Beaufort Sea is the presence of subsea permafrost. Permafrost is defined as the thermal condition in soil or rock of having temperatures below $0^{\circ}C$ persisting over at least two consecutive winters and the intervening summer (Brown and Kupsch, 1974). Fine-grained water-saturated sediments, especially those containing seawater, may not be actually ice-bonded at several degrees below $0^{\circ}C$, so we will use the term ice-bonded permafrost to indicate that ice bonding is present.

The research of Shackleton and Opdyke (1973) suggests that the world sea level fell to a minimum about 18,000 years ago. During that time, permafrost was formed beneath much of the present continental shelf in the Beaufort Sea, and as the sea level rose due to glacial melting the rising ocean inundated large areas of permafrost. The coastline continues to recede by an average of approximately 1.5 m yr^{-1} along the Alaskan Beaufort Sea (Hopkins and Hartz, 1978); thus, the subsea permafrost that exists today is relict. Models have shown that thousands of years may be required for subsea permafrost to be melted (Lachenbruch, 1957). The presence of permafrost has been confirmed as far as 20 km offshore in the Prudhoe Bay area.

Laboratory work and *in situ* measurements have shown that compressional-wave velocity increases significantly when a material becomes ice-bonded. Roethlisberger (1972) compiled compressional-wave velocities for many northern soil types, some of which are shown in Fig. 1. The figure also includes several measurements made by the authors: 28 measurements made near shore at Point Barrow and 41 made on five offshore islands in the Prudhoe Bay area. In water-saturated sediments, the compressional-wave velocity varies from about 1500 m s^{-1} for silt and clay, which is only slightly higher than that of seawater, to about 1750 m s^{-1} in sand and higher in sandy gravel. Some laboratory measurements (King *et al.*, 1981), given in Fig. 2, show the increase in velocity with cooling below $0^{\circ}C$ for several material samples taken from the Mackenzie River and the Canadian Beaufort Sea. All of these data show the marked increase that occurs when the material becomes ice-bonded. Because compressional-wave velocity alone is inadequate to classify the material type, we have used drill-hole information where possible to identify material types and their state.

III. STUDY AREA WEST OF PRUDHOE BAY

Beginning in 1975 a series of probe and drill holes (Drill Line #1 in Fig. 3) was placed along a line beginning at the foot of the West Dock and heading offshore at a bearing of approximately N. 31° E. (Sellmann and Chamberlain, 1979; Harrison and Osterkamp, 1981). These holes produced detailed information on lithology, temperature, pore water chemistry, blow count data, and other features. A series of seismic-refraction measurements was taken within an area 1 km square just east of these drill holes ("Study Area" in Fig. 3). Non-reversed refraction lines were run in

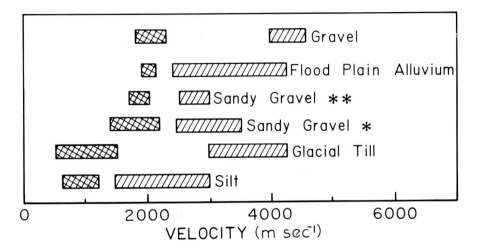

FIGURE 1. *Compressional-wave velocities in unbonded (cross-hatched) and ice-bonded (slashed) materials. The data are from Roethlisberger (1972) except those marked * and **, which were taken by the authors on five offshore islands near Prudhoe Bay and near Point Barrow, respectively.*

the summer of 1979 giving a total of 102 records. Figure 4, a histogram of all velocity measurements made within the area, shows that most seismic lines yielded two velocity ranges: a lower one representing surficial materials and a higher one (>2350 m s^{-1}) representing ice-bonded subsurface materials. Throughout the area the depth of the seawater is less than 2 m, and no refraction at the sea bottom was detected.

The depth to the ice-bonded layer was calculated for each record having a high-velocity refractor, and these data were used to construct the computer-fitted three-dimensional surface shown in Fig. 5. This figure shows an undulating, slightly seaward dipping surface. The least-squares fit to a plane surface (Fig. 6) indicates the average seaward dip of the surface to be 30.9 m km^{-1} or 1.8°. Individual refraction lines, however, give local slopes ranging from -4.5 to + 8.6°. Figure 7 shows the correlation of the best-fit plane surface with the depths as determined from the drilling data. The drill hole nearest shore (OH 300) detected ice-bonded materials at a depth of 2 m, which is 7 m above the plane surface. If indeed the ice-bonded surface suddenly slopes more steeply between 300 m and 350 m from the shoreline, as the drilling indicates, the surface fitted to the seismic data would not show this feature because the water was too shallow (<1 m) to obtain adequate data in this area. The other three drill holes indicated in the figure detected the bonded layer slightly below the best-fit surface; however, the differences are on the order of the standard deviation for the surface. Also, one must keep in mind the roughness of the surface as indicated in Fig. 5 when making this comparison.

The seismic data also displayed a decrease in the higher compressional-wave velocity from a value of approximately 4000 m s^{-1} at the

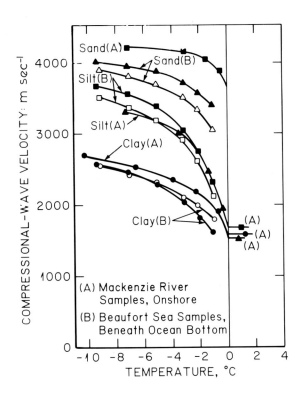

FIGURE 2. Compressional-wave velocities as functions of temperature for three general material types (King et al., 1981).

shoreline to approximately 3000 m s^{-1} at a distance of 1 km from shore. The decrease in velocity, approximately linear with the distance offshore, probably indicates that a degradation of the ice-bonded material surface occurs after the inundation by the ocean along the receding shoreline. No data supporting or contradicting this velocity trend were seen in the drilling data, and the actual degradation process is not yet understood.

The sediment types sampled in drill hole PB 6 are also shown in Fig. 7. Little silt or clay was found in this hole. The upper 2 m (3.5 – 5.5 m depth) of this section is dominated by silty sand. Below 4 m is dense sand and gravel with the exception of a firm, compact silt layer at 11 m. This information is compatible with the seismic velocities found in the unbonded layer. Figure 4 indicates that the velocities in this layer peak at 2000 m s^{-1}, which would be interpreted as sandy gravel. The seismic analysis does not detect the silt layers at 2 m and 11 m because they are too thin, but the overall interpretation of sandy gravel is adequate for most engineering situations.

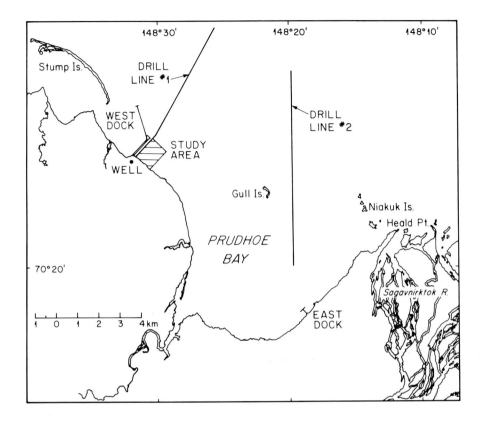

FIGURE 3. *Map of the Prudhoe Bay area showing the locations where detailed studies were made.*

IV. STUDY AREA IN PRUDHOE BAY

The second detailed study area was along a line beginning in Prudhoe Bay and extending offshore into Stefansson Sound (Drill Line #2 in Fig. 3). This area was chosen because of available information from a set of drill and probe holes (Sellmann and Chamberlain, 1979). Figure 8 shows the location of the drill holes and probe stations and the positions of the refraction measurements along the line. Figure 9, a stratigraphic section along the line, shows the locations and depths of the drill and probe holes and the interpretation of material types. In this section silty to sandy materials are underlain by sandy gravel. Ice-bonded materials were not encountered.

Over 20 reversed refraction lines 350 m long were taken along this line during the summer of 1980. Depths to the refractors, determined at each

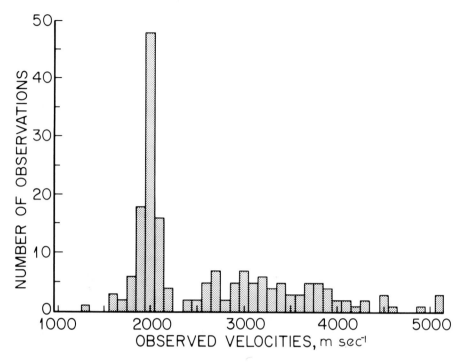

FIGURE 4. *Histogram of seismic velocities measured in the study area near Drill Line #1 (see Fig. 3).*

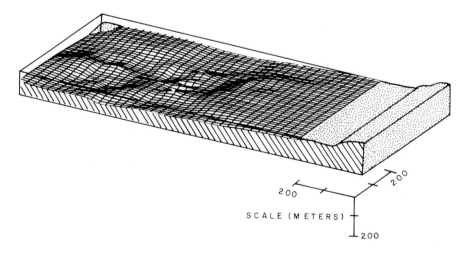

FIGURE 5. *Computer fit to the many high-velocity refractor depth measurements made near Drill Line #1.*

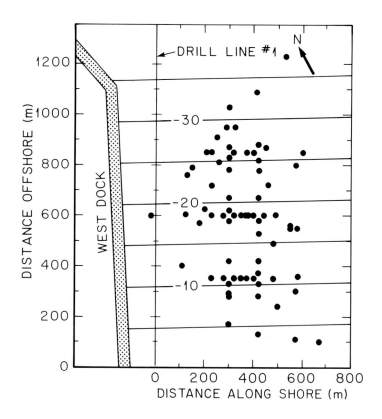

FIGURE 6. *Least-squares fit to a plane surface of the depths of high-velocity refractors (dots) in the study area near Drill Line #1. The depths of the contours are measured in meters from the ocean surface.*

end of the line, are shown on Fig. 10, along with the interpretation from the drilling. The only refractor occurring in the seismic records corresponds to the sandy gravel, and the interface thus plotted agrees with the surface drawn from the drilling data. The refraction data indicate that the surface of the sandy gravel is slightly shallower than that drawn from the drilling data in the region from 3.5 km to 4.5 km. This is plausible since the only probe hole (PH 5) in this region did not intersect the interface.

A plot of the velocities in the upper and lower layers is given in Fig. 11. The upper material has a velocity of approximately 1500 m s^{-1}, typical of water-saturated silty material. The refractor velocities are approximately 2000 m s^{-1} from the beginning of the line to 3.5 km, about 2200 m s^{-1} for the next 2.5 km, and about 2000 m s^{-1} for the rest of the line. These values agree with those expected for sandy gravel. The region from 3.5 to 6.0 km may either contain a higher gravel content or, more likely, be partially ice-bonded. The single measured value of 2500 m s^{-1} is interpreted as ice-bonded material, which is consistent with data shown

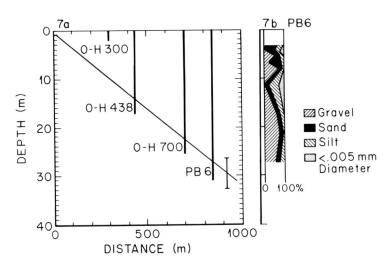

FIGURE 7. Distances along Drill Line #1 showing four drill holes (Sellmann and Chamberlain, 1979; Harrison and Osterkamp, 1981). The depth of the holes indicates the point where ice-bonded materials were encountered. The slope of the plane surface fitted to the refraction data (Fig. 6) is shown as a downward sloping straight line. Graph at right shows sediment textures at PB 6.

later taken along a perpendicular line. Figure 12 shows the calculated angle of the interface along the line. As the sloping interface in Fig. 10 would suggest, these angles are generally slightly negative, but local variations range from +2 to -3° along the line.

These data demonstrate the usefulness of high-resolution reversed refraction data, when interpreted in conjunction with drill holes that provide control information, to give improved mapping of sub-bottom materials in the Beaufort Sea and to allow one to greatly extend the drill-hole information at lower cost.

V. INTERPRETATION OF MATERIAL TYPES ALONG THE NORTH COAST OF ALASKA

Over 500 km of refraction data were collected along the north coast of Alaska in four study areas shown in Fig. 13. Areas A and B, covering the region from Bullen Point 60 km east of Prudhoe Bay to a point west of Cape Halkett 200 km west of Prudhoe Bay, are shown in detail in Figs. 14A and 14B. These data have been analyzed for compressional-wave velocities in the surficial and subsurface materials from which we can interpret what the materials are.

Data from hundreds of measurements taken in the area have been averaged over each small area with similar velocities. Each velocity value

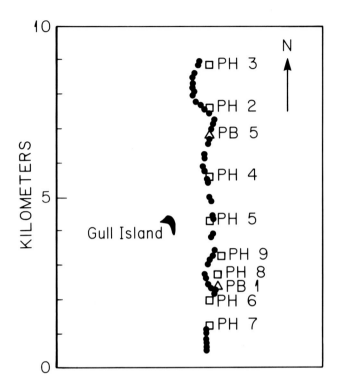

FIGURE 8.   Pattern of probe holes (squares), drill holes (triangles), and reversed refraction lines (circles) along Drill Line #2 (see Fig. 3).

presented on Figs. 14A and B is an average of 10 to 50 values. Average velocities appear on the figures next to the lines along which they were taken: the first value is the velocity in the surficial material, the second that of the subsurface material. In areas with no value given for the second material, either no refractor was observed (broken line) or the materials are ice-bonded with acoustic velocities greater than 2400 m s^{-1} (solid line). The ice-bonded materials are interpreted to be sandy gravel. It is possible that velocities in the range from approximately 1850 to 2400 m s^{-1} are due to partial ice bonding, yet the materials are thermally too fragile to be detected by drilling. From measurements made in the two detailed study areas (Fig. 3), materials with these velocities are interpreted as sandy gravel; however, this may not be correct at large distances from the study areas. The standard deviations of the averages shown are less than 100 m s^{-1} for the velocities of the upper materials and less than 200 ms^{-1} for the lower or refracting materials. Within 300 m of shore, the measured velocities in ice-bonded materials in many places are greater than 4000 m s^{-1}. These values are similar to those measured on the

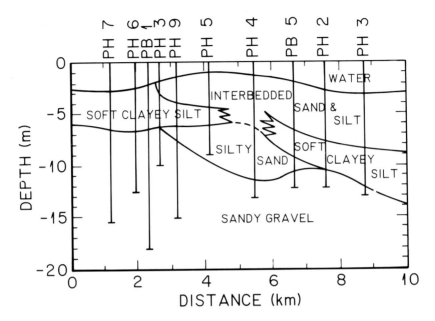

FIGURE 9. *Stratigraphic section along Drill Line #2 showing the locations and depths of the drill and probe holes (Sellmann and Chamberlain, 1979).*

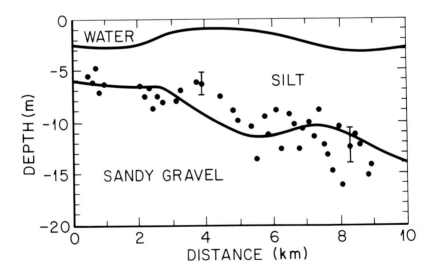

FIGURE 10. *Depths to the first refractor along Drill Line #2 (circles), overlaid on the drilling interpretation of material types. Typical error bars are shown for two data points.*

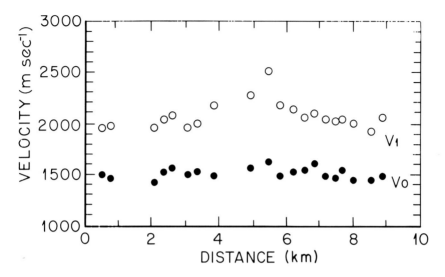

FIGURE 11.   Seismic velocities of the surficial material (V_0) and the subsurface material (V_1) along Drill Line #2.

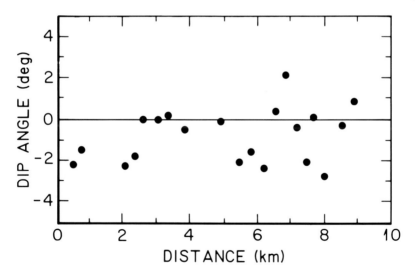

FIGURE 12.   Dip angles of the subsurface refractor along Drill Line #2.

mainland and indicate that the permafrost has not had time to degrade from its onland state. As mentioned earlier, the actual changes that occur in this nearshore region are not well understood. Velocities over 4000 m s^{-1} have also been measured in a few locations (discussed below)

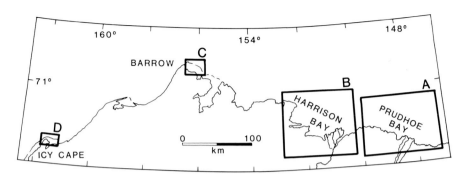

FIGURE 13. *Map of north coast of Alaska showing areas studied.*

quite far from shore. The depths to the refracting layer range from a few meters to over 40 m and vary greatly in each area. Vertical cross sections giving specific depth information for most of the areas are given elsewhere (Rogers and Morack, 1977, 1978, 1980b, 1981, 1982).

Seismic data in the region east and west of Tigvariak Island (Fig. 14A) indicate that the surficial material is silt or clay underlain by sandy gravel that may be partially ice-bonded. Along the mouth of the Sagavanirktok (Sag) River the upper material is silt or clay, underlain by ice-bonded sandy gravel. The center of Prudhoe Bay is silt or clay underlain by sandy gravel; around the seaward perimeter of the bay, however, the surficial materials are sandy gravel underlain by ice-bonded sandy gravel. Seaward of Prudhoe Bay and extending to the barrier islands, the surficial materials are silt or clay underlain by sandy gravel that may be partially ice-bonded. In this area deep ice-bonded materials are known from seismic-reflection data to exist below 50 m depth. Seaward of the barrier islands, the surficial materials exhibit a velocity between those of silty clay and sand. Below this layer lie relatively shallow (less than 20 m deep) ice-bonded materials having velocities over 4000 m s^{-1}, similar to the situation adjacent to the mainland. Offshore and to the west of Stump Island, the surficial materials are silt or clay underlain by sandy gravel that may be partially ice-bonded. However, very near the north shore of Stump Island the surficial materials appear to be sandy gravel. Inside the barrier islands to the west the surficial materials appear to be sandy gravel that may be partially ice-bonded in some areas. Just seaward of the east end of Pingok Island ice-bonded materials with velocities greater than 4000 m s^{-1} were located, similar to the situation occurring seaward of Reindeer Island.

The western half of the study area (Fig. 14B), includes the Colville River mouth and Harrison Bay. The eastern end of this region contains the western end of a long chain of barrier islands. The surficial materials between these islands and the mainland appear to be sandy gravel that may be partially ice-bonded. Seaward of the islands in this area, similar sandy gravel appears to be covered by silt or clay. Between Oliktok Point and Thetis Island, the surficial materials change from sandy gravel to silt or clay underlain by sandy gravel. The surficial materials off the mouth

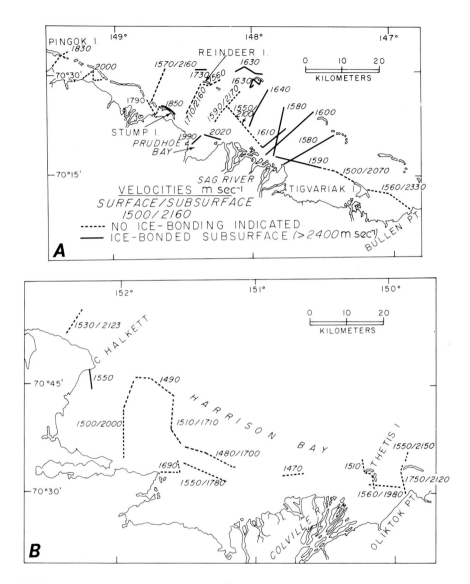

FIGURE 14. Maps showing details of Study Areas A and B. Lines are sites where compressional-wave velocities were taken; solid lines overlie ice-bonded permafrost, dashed lines indicate that no ice-bonded permafrost was found. Velocities are shown in m s^{-1} next to lines: the first number is for the top layer, the second is for the refracting layer (except where there was permafrost or no refracting layer).

of the Colville River are probably silt. The conditions are clearly different in the center and western end of Harrison Bay. The presumed silt or clay overlies materials whose velocities fall between silt or clay and sandy gravel. These velocities probably represent ice-bonded or partially ice-bonded overconsolidated clay. A complicating factor in this region is reports of gas pockets (Sellmann et al., 1981), thus more data are needed before a definite description of material types can be given. The conditions at the extreme western end of Harrison Bay appear to be similar to those near Prudhoe Bay with silt or clay underlain by ice-bonded materials. West of Cape Halkett for approximately 60 km the surficial material is silt or clay underlain by sandy gravel that may be partially ice-bonded. At the western extent of the study area the seismic data indicate that beneath the sandy gravel the materials become ice-bonded at depths less than 50 m.

A large number of monitor seismic records taken in the same general area for petroleum exploration programs have been analyzed for the occurrence of subsea ice-bonded permafrost (Sellman et al., 1980, 1981). Although these data provide lower resolution due to the large hydrophone spacing used, the very long line length employed allows interpretations to a much greater depth. Consequently these data complement our own data. The agreement between the two data sets is excellent in the areas where they overlap. The industry data have made it possible to follow the ice-bonded permafrost interface into areas where it lies beneath the 50-m depth limit of our equipment.

Our data are also in good agreement with the information available from a series of 20 coreholes drilled in 1979 by Harding-Lawson Associates for the United States Geological Survey (U.S. Geological Survey, 1979) in Area A (Fig. 13). Presently no such data are available for Area B.

VI. INTERPRETATION OF MATERIALS IN OTHER AREAS

Seismic data were taken from 1975 to 1977 on the spit just east of Point Barrow, in Elson Lagoon south of Plover Point, and on the Tapkaluk Islands (Area C, Fig. 13). The seismic data indicate that the gravel on the Barrow spit overlies ice-bonded materials with seismic velocities greater than 2400 m s^{-1}. To the east toward the Tapkaluk Islands, velocities characteristic of ice bonding are not observed with the apparatus used. An average seismic velocity of 1690 m s^{-1} in Elson Lagoon indicates material properties that are between those of silt or clay and sandy gravel.

Twenty-one kilometers of refraction lines were run from Icy Cape, approximately 230 km southwest of Point Barrow (Area D, Fig. 13), to a point 13 km to the east along the shoreline. The average seismic velocity of 1570 m s^{-1} for the surficial material indicates silt or clay. An average of 2030 m s^{-1} for the sub-bottom material indicates sandy gravel that may be partially ice-bonded. At locations within 200 m of the shore the values for the surficial material averaged 1780 m s^{-1}, probably indicating a higher gravel content. A few high-velocity (>2400 m s^{-1}) refractors were observed just north of Icy Cape in water depths of 5 m. These may indicate ice-bonded material, although no confirming drill holes exist. An additional 6 km of lines was run inside Kasegaluk Lagoon just to the south of Icy

Cape. These data gave a single velocity for the surficial material averaging 1940 m s^{-1} and indicate sandy gravel that may be partially ice-bonded.

VII. CONCLUSIONS

Much of the sea bottom off the north coast of Alaska consists of silt or clay overlying sandy gravel. In large areas quite near the mainland or islands, the surficial material is sandy gravel. Large areas of the subsurface materials remain ice-bonded from a period when they were subaerial. Two classes of ice-bonded materials have been identified by seismic techniques. The first class has a seismic velocity greater than 3500 m s^{-1} corresponding to onshore velocities. This class is found very near the mainland and in a few locations near islands. The second class, characterized by seismic velocities between 2400 and 3500 m s^{-1}, is found in extensive offshore regions. Seismic velocities in the range from 1850 to 2400 m s^{-1} may indicate partial ice-bonding, although no confirming drill-hole information is available.

ACKNOWLEDGMENTS

This work was partially supported by the National Science Foundation under Grant DPP77-27239. We thank John Engles for his help in using the Surf II computer program to fit some of the seismic data.

REFERENCES

Biot, M.A. (1941). *J. appl. Phys. 12*, 155.
Biot, M.A. (1956a). *J. acoust. Soc. Am. 28*, 168.
Biot, M.A. (1956b). *J. acoust. Soc. Am. 28*, 179.
Biot, M.A. (1962a). *J. appl. Phys. 33*, 1482.
Biot, M.A. (1962b). *J. acoust. Soc. Am. 34*, 1254.
Brown, R.J.E., and Kupsch, W.O. (1974). *Tech. Mem. 111.* Associate Committee on Geotechnical Research, Nat. Res. Council of Canada, Ottawa.
Clay, C.S., and Medwin, H. (1977). "Acoustical Oceanography." John Wiley & Sons, New York.
Dobrin, M.B. (1975). "Introduction to Geophysical Prospecting," 2nd ed. McGraw-Hill, New York.
Hamilton, E.L. (1971). *J. geophys. Res. 76*, 579.
Harrison, W.D., and Osterkamp, T.E. (1981). *In* "Environmental Assessment of Alaskan Continental Shelf," Ann. Rept. NOAA, Rockville, MD. (In Press).
Hopkins, D.M., and Hartz, R.W. (1978). *Open-File Rept. 78-1063.* U.S. Geol. Surv.

King, M.S., Pandit, B.I., Hunter, J.A., and Gajtani, M. (1982). *In* "Proceedings Fourth Canadian Permafrost Conference," p. 268. Nat. Res. Council of Canada, Ottawa.

Lachenbruch, A.H. (1957). *Bull. geol. Soc. Amer. 68*, 1515.

Roethlisberger, H. (1972). *Sci. Eng. Monogr. II-A2a.* U.S. Army Cold Regions Res. and Eng. Lab., Hanover, NH.

Rogers, J.C., and Morack, J.L. (1977). *In* "Environmental Assessment of the Alaskan Continental Shelf," Ann. Rept. Vol.17, p. 467. NOAA, Boulder.

Rogers, J.C., and Morack, J.L. (1978). *In* "Environmental Assessment of the Alaskan Continental Shelf," Ann. Rept. Vol. 11, p. 651. NOAA, Boulder.

Rogers, J.C., and Morack, J.L. (1980a). *J. geophys. Res. 85*, 4845.

Rogers, J.C., and Morack, J.L. (1980b). *In* "Environmental Assessment of the Alaskan Continental Shelf," Ann. Rept. Vol. 5, p. 1. NOAA, Boulder.

Rogers, J.C., and Morack, J.L. (1981). *In* "Environmental Assessment of the Alaskan Continental Shelf," Ann. Rept. NOAA, Boulder. (In Press).

Rogers, J.C., and Morack, J.L. (1982). *In* "Environmental Assessment of the Alaskan Continental Shelf," Ann. Rept. NOAA, Boulder. (In Press).

Sellmann, P.V., and Chamberlain, E.J. (1979). *In* "Proceedings of the 11th Offshore Technology Conference," Vol. 3, p. 1481. Offshore Technol. Conf., Dallas.

Sellmann, P.V., Delaney, A., and Chamberlain, E.J. (1980). *In* "Environmental Assessment of the Alaskan Continental Shelf," Ann. Rept. Vol. 4, p. 125. NOAA, Boulder.

Sellmann, P.V., Neave, K.G., and Chamberlain, E.J. (1981). *In* "Environmental Assessment of the Alaskan Continental Shelf," Ann. Rept. NOAA, Boulder. (In Press).

Shackleton, N.J., and Opdyke, N.E. (1973). *Quat. Res. 3*, 39.

Stoll, R.D. (1974). *In* "Physics of Sound in Marine Sediments" (L. Hampton, ed.), p. 19. Plenum Press, New York.

U. S. Geological Survey (1979). *Data Set AK 17718.* NGSDC, NOAA, Boulder.

SEDIMENT CHARACTERISTICS OF THE LAGOONS OF THE ALASKAN BEAUFORT SEA COAST, AND EVOLUTION OF SIMPSON LAGOON

A. Sathy Naidu
Thomas C. Mowatt

Institute of Marine Science
University of Alaska
Fairbanks, Alaska

Stuart E. Rawlinson

Department of Natural Resources
Division Geological and Geophysical Surveys
Fairbanks, Alaska

Herbert V. Weiss

Department of Chemistry
San Diego State University
San Diego, California

I. INTRODUCTION

During the past two decades there has been a major surge in the study of sediments and sedimentary processes of barrier islands and spits and the adjacent lagoons, bays, or sounds. Much of the impetus for the study has been generated from the recognition by geologists that ancient, lithified representatives of barrier island deposits constitute an important repository for hydrocarbons (Shelton, 1973; Klein, 1975). Characterization of modern lagoonal sediments, and an understanding of their three-dimensional geometry within the framework of various interfingering coastal subfacies, may yield reliable geological criteria to identify ancient analogs of lagoons, and by implication also the associated barriers. For management of fragile coastal regimes, it is critical to understand the overall sedimentological and geomorphological processes of barrier island-lagoon systems. Extensive chains of barrier islands and associated lagoons skirt the arctic coast of Alaska. This paper describes the sediments and depositional settings of the lagoons along the microtidal coast of the Alaskan Beaufort Sea and reports the erosional rates of the adjacent

THE ALASKAN BEAUFORT SEA:
ECOSYSTEMS AND ENVIRONMENTS

275

mainland coast. Finally, stratigraphic evidence is presented showing that some North Slope lagoons are evolved from coalescence of coastal lakes.

II. ENVIRONMENTAL SETTING OF THE LAGOON SYSTEM

Four chains of barrier islands border the Beaufort Sea coastline of Alaska (Fig. 1). A chain extends eastward of the barrier spit at Barrow, two chains stretch westward from west of the Sagavanirktok River and Canning River mouths, and a fourth barrier chain east of the Canning Delta stretches east to Demarcation Point. Associated with these fringing barriers are a number of lagoons; the large ones, from west to east, are Elson Lagoon, Simpson Lagoon-Gwydyr Bay, Stefansson Sound, and Beaufort Lagoon (Fig. 1).

Salient features of the regional climate, seasonal fluvial dynamics, and oceanography of this coastal area have been summarized in a number of publications and also appear elsewhere in this volume (Matthews, 1970; 1981a, b; Wiseman *et al.*, 1973; Walker, 1974; Naidu and Mowatt, 1975a; Dygas and Burrell, 1976; Brower *et al.* 1977; Nummedal, 1979; and Owens *et al.*, 1980; among others). The gross geomorphology of the area was presented by Hartwell (1973) and Owens *et al.* (1980). Dimensions of the lagoons vary widely; widths generally range from a few hundreds of meters to 4 km, whereas lengths may reach 65 km. The lagoons are typically 1 to 3 m deep. Stefansson Sound (Fig. 1) is an extensive lagoon with a maximum depth of 10 m and width of 20 km. The lagoons have tidal communication with the open sea through their lateral open ends or through the interbarrier inlets (or channels), the depths of which range from less than 1 m to about 6 m. Because the lagoon environment along the Beaufort Sea is quite extensive, our sedimentologic and coastal erosion studies were restricted to three type areas. Simpson Lagoon represents a type area off a gently sloping coastal plain adjacent to an extensive deltaic complex, and Beaufort Lagoon represents a type area off a relatively steep coastal plain with less widespread deltation, and Stefansson Sound, the largest lagoon along the Alaskan Beaufort coast, is a distinctive type (Fig. 1).

III. METHODS AND MATERIALS

Surficial sediment samples of the lagoon floor were collected by Ekman or Van Veen grabs. To gain understanding of the evolution of the Simpson Lagoon, four vibracore samples 120 to 170 cm long were taken from the middle of that lagoon, and splits of the cores sectioned at 10-cm continuous intervals were studied (Barnes *et al.*, 1979).

As there were significant small-scale spatial variations in the distribution of sand and gravel around a point on the lagoon shoreline, consideration of a limited number of random samples from such shoreline

FIGURE 1. Area of study and location of sediment grab and vibracore samples from Simpson and Beaufort Lagoons, Prudhoe Bay, and Stefansson Sound.

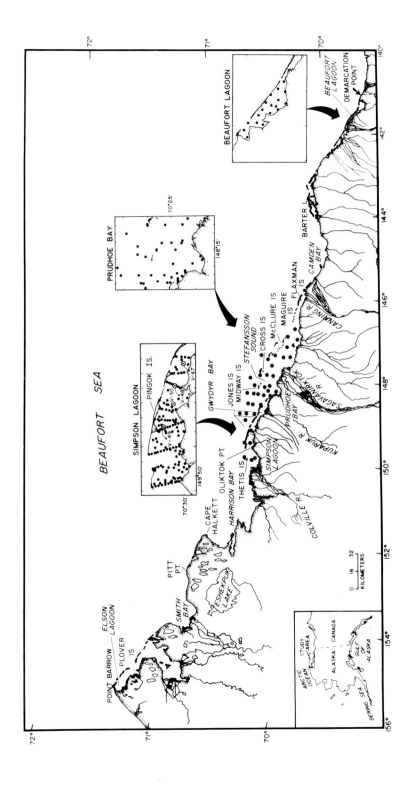

would have been statistically meaningless. Instead, intensive sampling of lagoon shoreline deposits was restricted to a single representative area, the southwest shore of Simpson Lagoon in the vicinity of Oliktok Point.

To estimate sedimentation rates in Simpson Lagoon seven 15- to 33-cm-long core samples were collected from its western and central parts (Fig. 2). All cores were obtained from more than 2 m depth near the center of the lagoon to avoid areas that might have been disturbed by gouging action of sea ice (which is usually 1-2 m thick) or shore-fast ice. The westernmost core sample was retrieved by a Phleger gravity corer and the others by a manually driven coring unit.

To assign time control on a stratigraphic break, one basal peat sample separated from vibracore V-49 (Fig. 3) from under a 150-cm sediment overburden was dated by ^{14}C. To understand sediment distribution, sedimentary facies changes and sediment dynamics, grain-size analysis was performed on sediments by the combined sieve-pipette technique, and conventional statistical parameters of sediment texture were calculated according to Folk (1968). Granulometric data reported by Dygas et al. (1972), Tucker (1973), Barnes (1974), Naidu (1979, 1982), and Naidu and Mowatt (1975a, 1975b, 1976) were collated and integrated for this report.

Depositional rates were derived from ^{210}Pb measurements as a function of core depth, using the approach described by Koide et al. (1973) and method similar to that described by Nittrouer et al. (1979). Coarse fractions (>62 μm) and microfossils of various sections of two of the vibracore samples (V-48 and V-49) were studied under a binocular microscope.

To examine the processes of sediment entrainment in sea ice of lagoons and sounds, samples of ice cores were taken from Stefansson Sound and studied for sediment zonation. Sections of 5 cm were cut from these cores and melted and concentrations of sediment in each of the sections were estimated.

Coastal erosion was estimated from aerial photographs taken 25 years apart in 1955 and 1980. The accuracy of the erosion measurements is dependent upon distortion and resolution of the photographs, measuring techniques, and scale determinations, the third being dependent upon the first two. To minimize the effects of photograph distortion, the measurements were not taken near the edge of the photograph where the distortion is greatest. Taking into account all factors that contribute to error in the erosion measurements, the calculated erosion rates have an estimated precision of ±0.3 m yr^{-1}.

IV. RESULTS

Following the classification scheme of Folk (1954), Simpson Lagoon sediments are very poorly sorted, gravelly, muddy sand to gravelly, sandy mud. In addition, the sediments display very positive- to positive-skewed (fine-skewed) and platykurtic to leptokurtic size distributions. Some general areal variations in the gross texture have been identified previously within Simpson Lagoon (Tucker, 1973; Naidu, 1981). From trend-surface analysis, Tucker (1973) noted a net decrease in the mean size of sediment from the nearshore, more turbulent shallow part to the tranquil deeper central lagoon.   Variations in sand and mud contents (Naidu, 1981) substantiate this trend (Fig. 4).   No distinct regional patterns were observed, however, in the sorting and skewness values.   Naidu (1982) showed that the Beaufort Lagoon sediments have textures quite similar to those of Simpson Lagoon, the only notable difference being the virtual absence of gravel in Beaufort Lagoon.   The substrate of Stefansson Sound varies widely between muddy sand and sandy mud with occasional minor amounts of gravel (Barnes, 1974; Naidu and Mowatt, 1975a, 1975b, 1976). Generally, the sediments are very poorly sorted, and strongly fine-skewed with mean sizes from fine sand to fine silt. Stefansson Sound is, however, widely strewn with boulders up to 1 m across.   In places with high concentrations of boulders a unique ecosystem supporting kelp beds has evolved (Dunton et al., 1982).   The lagoonal deposits of the Alaskan Beaufort Sea have a notable paucity of calcareous bioclastics and microfaunal tests, in contrast to most of the lagoonal sediments of the deltaic regions of temperate and humid climatic belts (Shepard and Moore, 1955; Allen, 1965; Naidu, 1968; Dickinson et al., 1972; Friedman and Sanders, 1978; Reading, 1978; among many others).   Sediments of shoreline deposits of the lagoons, as manifested by the fringing beaches of tundra and barrier islands and the fringing beach of the mainland, are poorly to very poorly sorted gravelly sand, sandy gravel, or gravel with coarse- to fine-skewed and platykurtic size distributions (Naidu and Mowatt, 1975, 1976; Wiseman et al., 1973).

Longitudinal sections of the four vibracores from central Simpson Lagoon are displayed in Fig. 3, and stratigraphic variations of the texture of one vibracore are shown in Table I. The sediments and structures in the four cores vary widely.   In core V-49 the basal section (140-150 cm depth) is mostly finely divided, fluffy peat mixed with fine sand, silt, and occasional fossil remains of the freshwater gastropod Valvata and the pelecypod Psidium.   Overlying this section is a 65-cm sequence of cross-bedded medium sand with a few shells of pelecypods Cyrtodaria kurriana and Serripes groenlandicus, and a low diversity of estuarine or marine foraminiferal and ostracod species[1] intercalated with seeds of the freshwater plants Chara, Hippuris vulgaris L., and Potamogeton (at 120-130 cm below the core top). This sequence is overlain by a 75-cm sequence of sandy silt displaying typical mottled structure in the base and laminated deposits in the rest of the core.   Core V-48, which shows more dramatic stratigraphic changes (Fig. 3) than core V-49, has no apparent freshwater sediment sequence.   The basal 55 cm of the core shows typical mottled structure and contains the ostracod Heterocyprideis sorbyana Jones. There is a 20-cm laminated silty sand sequence intercalated between two layers (60-90 cm and 110-115 cm) of very gravelly sand and pelecypod fragments.   The top 60 cm of this core is homogeneous silty sand (Naidu, 1982).   Stratigraphic analysis on cores V-47 and V-50 has not been

---

[1] Rabilimis septentrionalis Brady; Paracyprideis pseudopuntillata Swain; Eucytheridea bradii Norman; Cytheretta edwardsi Swain; Candona rectangulata; Ilyocypris bradii, Getirectorudeus sorbyaba Jones.

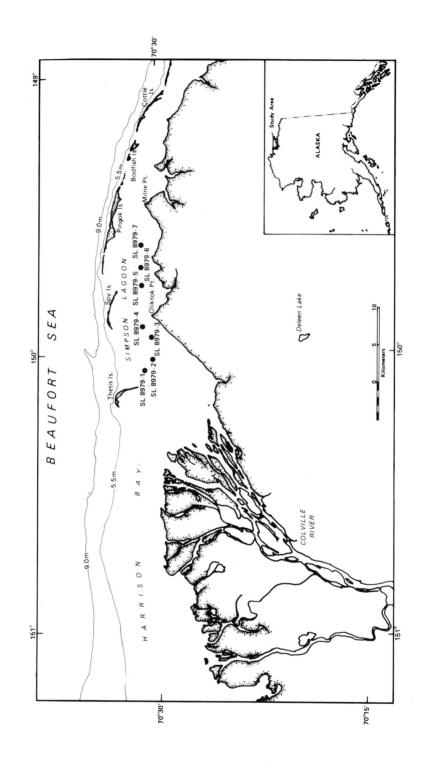

TABLE I. Stratigraphic Variations in Sediment Texture in Vibracore Sample V-49

Core section (cm)	Gravel %	Sand %	Silt %	Clay %	Md	M_z	I	Sk_I	K_G
0-10	-	15.35	78.55	6.10	4.72	5.16	1.46	0.47	1.08
10-20	0.14	9.05	81.31	9.50	5.21	5.54	1.74	0.40	1.38
20-30	-	5.52	86.79	7.69	5.91	5.90	1.38	0.13	1.40
30-40	0.09	25.73	61.03	13.15	5.29	5.38	2.24	-0.02	2.56
40-50	-	48.91	46.38	4.71	4.11	3.98	2.07	0.03	0.92
50-60	0.73	64.26	31.39	3.63	2.78	3.73	2.12	0.58	1.03
60-70	-	30.45	63.11	6.45	4.94	4.77	2.16	-0.02	1.03
70-80	0.54	73.72	23.21	2.54	2.38	3.02	1.82	0.54	1.08
80-90	0.67	99.33	-	-	1.87	1.85	0.49	-0.06	1.10
90-100	0.09	99.91	-	-	2.10	2.15	0.47	-0.15	1.21
100-110	0.12	99.88	-	-	2.39	2.39	0.39	-0.01	0.92
110-120	-	100.00	-	-	2.11	2.05	0.71	-0.17	0.91
120-130	0.46	92.02	5.98	1.53	2.19	2.12	1.00	0.02	1.46
130-150	1.70	51.25	26.13	20.92	3.69	4.18	2.63	0.21	1.22

completed. A laminated layer is present in the top 50 cm of core V-50, whereas the rest of this core and the entire core V-47 are apparently structureless and consist of homogeneous clay-silt-sand.

The sedimentation rates at various sites in Simpson Lagoon, as estimated by the ^{210}Pb method, are presented in Table II. A notable decrease in the sedimentation rate from the western to the central lagoon is suggested.

A wide regional and yearly variation in concentrations of sediments in frazil sea ice of the lagoons and Stefansson Sound is observed (Dr. D. M. Schell, personal commun.). Samples of the sea-ice cores analyzed from Stefansson Sound show a sediment concentration range, computed on the basis of melted ice, of 40 to 730 mg L^{-1} with a mean of 166 mg L^{-1} (Naidu and Larsen, 1980). In addition, Barnes et al. (1982) described sediment zonation that is typically observed in samples of vertical blocks of sea ice. Microscopic examination of entrained sea-ice sediments show various proportions of clay- and silt-size particles, with significant amounts of plant debris, occasional fine sand, and rarely coarse sand and fine gravel.

The ^{14}C date of the basal peat of the vibracore sample V-49 is 4500 years before present (y.B.P.).

Numerous measurements of coastal bluff retreat have been reported for the Beaufort Sea coastline (cited in Hopkins and Hartz, 1978), and new data have been generated for the coastline between Oliktok Point and

FIGURE 2. Locations of core samples for sedimentation rate measurements.

V-47   V-48   V-49   V-50

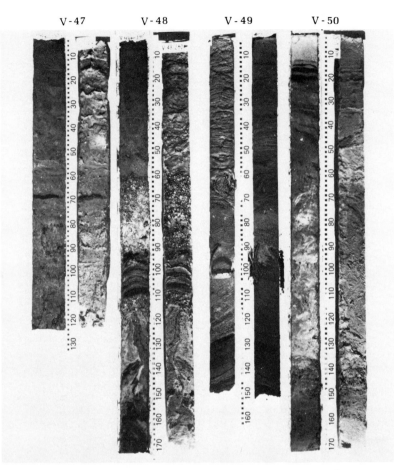

FIGURE 3. Vertical sections of vibracore samples taken from Simpson Lagoon. Photograph by P. W. Barnes, U.S. Geological Survey.

Prudhoe Bay (Fig. 5) using large-scale photographs. The mean value of these new data is 1.1 m yr^{-1}. Figure 5 also shows idealized, hypothetical coastlines from the present to 5000 y.B.P. at 1000-year intervals based on a constant bluff retreat rate of 1.1 m yr^{-1}.

V. DISCUSSION

A. Sedimentary Processes in the Lagoons

Sediment transport in the Beaufort Sea lagoons and barriers is greatly influenced by the seasonal variations in the hydrodynamics and sediment supply. During late winter (November to May) sediment transport and

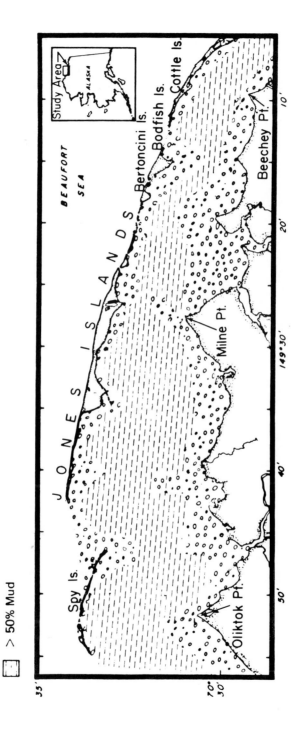

Legend

▨ > 50% Sand

▢ > 50% Mud

FIGURE 4. Textural characteristics in Simpson Lagoon (after Tucker, 1973, and Naidu, 1982).

TABLE II. Sample Sites and Sedimentation Rates Estimated For The Simpson Lagoon Area

Core No.	Lat. (N)	Long. (W)	Water Depth (m)	Sedimentation rate $(cm \cdot yr^{-1})$	Mass Sedimentation rate $(g \cdot cm^{-2} \cdot yr^{-1})$
SL8979-1	70°32'	150°07'	3.9	1.64	2.05
SL8979-2	70°31'	150°01'	3.6	0.82	1.03
SL8979-3	70°31'	149°57'	3.3	0.46	0.58
SL8979-4	70°32'	149°53'	3.0	*	*
SL8979-5	70°32'	149°45'	2.7	0.58	0.73
SL8979-6	70°32'	149°50'	2.6	0.74	0.93
SL8979-7	70°32'	149°35'	2.6	0.52	0.65

*No significant linear exponential decay in ^{210}Pb was noticed, presumably because the core was reworked.

deposition essentially come to a standstill; currents in the lagoons are less than 2 cm s^{-1} (Matthews, 1981). With the possible exception of areas shallower than 2 m and adjacent shore-fast portions, lagoon substrates appear to be generally free from reworking by ice gouging and bioturbation, as shown by preservation of laminated sediment sequences (Fig. 3; Barnes et al., 1979) and by the generally uninterrupted ^{210}Pb profiles exhibited in Simpson Lagoon cores (Naidu, 1982). At spring breakup the ice cover over parts of Simpson Lagoon and Stefansson Sound is subjected to highly turbid overflow from adjacent rivers (Walker, 1974; Reimnitz and Bruder, 1972; Reimnitz and Kempema, 1982). Arnborg et al. (1967) documented that 43% of the annual discharge and 74% of the total inorganic suspended load was discharged from the Colville River during a 3-week period at spring breakup.

Terrigenous sediments in the lagoons appear to have been derived from a number of sources (Owens et al., 1980). Sediments entrained in the fluvial overflow and those derived from longshore transport from river mouths presumably constitute 80% of the contemporary terrigenous debris supplied to the lagoons (Canon, 1978). Occasional wind-driven currents of 40 cm s^{-1} near the lagoon bottom have been reported (Barnes et al., 1977), which conceivably can move sand-size particles. The next important source of sediments is from erosion of tundra cliffs. Washovers resulting from occasional storm surges may also transfer significant volumes of marine and barrier shoreface sediments to the lagoons. Additionally, as discussed later, most of the gravel and the coarse sand in the lagoons is believed to represent either ice-rafted or lag deposits. Eolian transport of fine sand from the adjacent barriers into the lagoons can also be quantitatively important (Reimnitz and Maurer, 1979). However, there is as yet no comprehensive statistical data base available to show what proportions of sediments are transferred into the lagoons from the various sources.

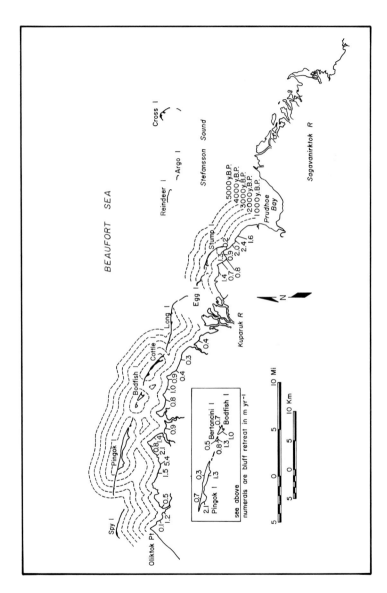

FIGURE 5. Mean annual bluff retreat and hypothetical coastal positions based on a mean coastal retreat rate of 1.1 m yr^{-1} between Oliktok Point and the Sagavanirktok River.

It appears that transport and deposition of sediments into the lagoons by tidal current is unimportant. This is apparent from the absence or, at best, very weak development of flood-tidal deltas in the lagoons and may be attributed to the low tidal range (Matthews, 1970) and the attendant low tidal prism (Nummedal, 1979).

In summer, sedimentary processes in the lagoons and Stefansson Sound are dominated by periodic resuspension and redeposition of sediments. Time–series plots between wind speed and suspensate concentrations exhibit an apparent correlation, and suggest that in 3-m water depths in Simpson Lagoon the threshold wind velocity to induce wave–current resuspension of bottom sediment is about 8 m s^{-1} (Naidu, 1979). Turbid plumes in the lagoons and bays in mid and late summer (late July–September) cannot generally be ascribed to fluvial discharge and erosion of coastal bluffs. No such direct association is apparent from detailed scrutiny of several multi-year satellite images. Except during rare flash floods (Fig. 6), waters low in suspended matter (about 1 mg L^{-1}) are commonly discharged in late summer from the major distributary mouths, while presence of relatively turbid waters can be delineated only at some distance off the mouths (Fig. 6). These observations suggest that in mid and late summer turbidity in coastal waters for the most part is associated with wave-induced resuspension of cohesionless muddy substrates from shallow-water regions. The fine suspended particles are carried westward, which is consistent with results of littoral drift studies, clay mineral dispersals, turbidity plume structure (Fig. 6), and current measurements (Wiseman et al., 1973; Naidu and Mowatt, 1975a, 1983; Dygas and Burrell, 1976; Barnes and Toimil, 1979; Nummedal, 1979; Matthews, 1981a, b; among others).

The presence of bimodal grain size distributions of lagoon sediments can be explained as mixing of two populations of particles from grab sampling; one a better sorted coarse population of gravel and sand and a poorly sorted fine population of silt and clay that was deposited under highly fluctuating but low wave–current energy levels. Because energy levels are low in the central lagoon, the coarse population is probably not a product of contemporary hydraulic sorting. We believe that it was most likely derived from the Gubik Formation from either the present coastal bluffs and tundra islands or their ancient analogs (Naidu, 1982). The coarse fraction, therefore, represents either lag deposits that were reworked from eroded coastal cliffs or material transported by ice rafting. The general paucity of calcareous bioclastics and microfaunal assemblages in the lagoon sediments presumably reflects the low density of the benthic bivalve community (Susan Schonberg, personal commun., 1983) and low microfaunal productivity in the area.

Nummedal (1979) and Owens et al. (1980) discussed transport of coarse-grained particles along the lagoon shorelines of the Alaskan Beaufort Sea. They presented quantitative estimates of potential longshore sediment transport, computed both on theoretical considerations and field observations of wave energy fluxes, and compared them with previous estimates by Short (1973), Short et al. (1974) and Dygas and Burrell (1976). They contended that the net longshore transport of coarse sediments is toward the west at a rate of roughly 10^4 m^3 yr^{-1} along the fringing beach of the barriers, an order of magnitude greater than along the adjacent inner

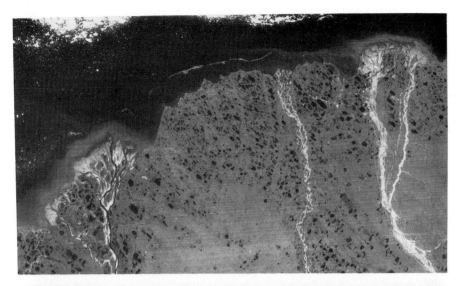

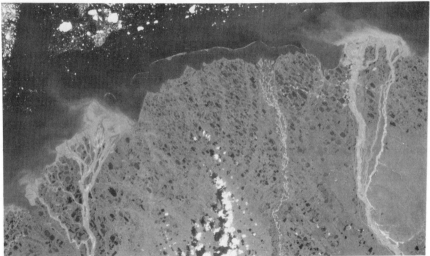

FIGURE 6. LANDSAT images (Band 5) of the Alaskan Beaufort Sea coast. The bottom image (#22374 of July 23, 1981) displays the turbid outflow from the Colville and Sagavanirktok Rivers during a flash flood. The top image (#2915-20483 of July 25, 1977) shows turbid water some distance off the river mouths, suggesting wave resuspension of sediments; river discharge is less turbid.

lagoon beaches. However, almost all of this transport may be accounted for during the three summer months. In addition, sediment transport effects along beaches during storms can be cataclysmic (Hume and Schalk, 1964, 1967); Nummedal (1979) suggested a potential transport of 1500 m³ day^{-1} in Point Barrow adjacent to Elson Lagoon during a 3-day storm.

Sedimentary processes in fall and early winter (September to November) are not well understood for lagoon areas of the Alaskan Beaufort Sea, but they appear to be dominated by sediment entrainment in sea ice. The common presence of sediments in frazil sea ice of the lagoons and sound, in concentrations two to three orders of magnitude higher than in ambient coastal waters (Naidu and Larsen, 1980; Barnes et al., 1982), has been an enigma. Schell (1980) summarized explanations for these high concentrations: one is that storms during the incipient freeze-up period entrain large quantities of sediment into the water column, which subsequently freeze into the sea ice (Barnes et al., 1982). Naidu's (1980) hypothesis outlines a process of sieving the particle-charged seawater through sea ice to explain the sediment concentrations. The presence of occasional coarse sand and fine gravel in the sea ice is most likely related to particle entrainment through anchor ice (E. Reimnitz, oral commun., 1982). During winter, sea ice can hold a substantial amount of fine-grained sediments, which can be released into the lagoon during spring and summer. There is no way yet to estimate the flux of this sediment into the lagoons of the Alaska Beaufort Sea and its effect on the sediment budget.

The ^{210}Pb stratigraphy has shed some light on the contemporary depositional processes in Simpson Lagoon (Naidu, 1982). The generally continuous ^{210}Pb linear profiles down the cores suggest that, unlike on the open shelf (Barnes, 1974; Reimnitz and Barnes, 1974), sediments in the central lagoon are not intensely reworked by ice gouging or bioturbation. Although there may be occasional wave-induced resuspension of sediment the Simpson Lagoon is a depositional sink. There is a regional variation in the sedimentation rates (Table II); the higher rate in the western end of the lagoon (Core SL8979-1) must be ascribed to the larger flux of sediment locally from the Colville River (Naidu, 1982), whereas the lower rate in the central lagoon is presumably related to sediment from the adjacent Kuparak River (Reimnitz and Kempema, 1982; Naidu, 1982). The sedimentation rate of 0.46-1.64 cm yr^{-1} determined by the ^{210}Pb method on contemporary lagoon sediments (Table II) does not match the long-term deposition rate of 0.03 cm yr^{-1} based on the 4500 y.B.P. ^{14}C date of the peat at 150 cm in core V-49. Apparently sedimentation has varied greatly in Simpson Lagoon over the past 4500 years.

B. Evolution of Simpson Lagoon

The recent paleogeographic history of the Simpson Lagoon area has been quite complex. Progressive changes in sediment type, structure, and coarse fractions in two core samples retrieved from Simpson Lagoon (cores V-48 and V-49 in Fig. 1) and coastal retreat measurements, allow us to construct a model of the lagoon's evolution.

Basal sections of core V-49 suggest that Simpson Lagoon, west and south of the tundra islands, was recently a low-lying tundra coastal plain with a number of freshwater lakes and ponds like what now lies to the south. This is substantiated by the presence in the basal layer of core V-49 of the freshwater gastropod *Valvata* and the pelecypod *Psidium*. The accumulation of plant debris in these lakes is represented as a basal peat in the core. We believe that the peat is an autochthonous lake deposit and not part of a paleotundra mat dislodged from coastal bluffs and rafted to the present site because of the fluffy nature of the peat and excellent preservation of the gastropod shells. In contrast, representatives of ancient tundra mats would exhibit, as they do now, a tightly intertwined fibrous mesh. As suggested by the ^{14}C date of the basal peat, the coastal plain with lakes dominated the area around 4500 y.B.P. Subsequently, the lakes were occasionally breached by the sea and remained connected to it through tidal channels, as implied by the presence of estuarine faunal residues over the basal peat. It is most likely that the breached lakes progressively coalesced and eventually were inundated by the postglacial rise in sea level over the past 5000 years (Nelson and Bray, 1970; Kraft, 1971; among many others). Pingok, Bertoncini, and Bodfish tundra islands, which presumably represent Pleistocene relict coastal highlands, separated the inundated lakes from the open sea to form the present Simpson Lagoon. We maintain that the upper 70 cm of muddy sediments in core 49 are contemporary lagoonal deposits, and the cross-bedded sequences of medium sand between 70 and 145 cm in core 49, characterized by well-sorted, fine-skewed, and mesokurtic size distributions (Table I), most probably represent the beach facies (Friedman, 1961) of the ancient shallow lagoon. Thus, the results of our stratigraphic study complement the geomorphic evidence earlier presented by Wiseman *et al.* (1973) on the evolution of the Simpson Lagoon from coastal lakes.

We do not know fully how the rest of the lagoons along the Alaskan Beaufort Sea coast have evolved. The Simpson Lagoon and a few other lagoonal areas that are delineated by the tundra remnant islands and the mainland coast have quite a different geomorphic setting than the lagoons associated with barrier islands, whose evolutions are presumably closely tied. Changes in the configuration of some of the barrier islands and spits between 1906 and 1972 (Wiseman *et al.*, 1973; Barnes *et al.*, 1977) and stratigraphic variations displayed in a few core samples from barriers (Lewellen, 1977; Lewis and Forbes, 1974) indicate a net landward migration of the barrier islands along the Beaufort coast. This implies the transgressive nature of the barriers (Short, 1979; Nummedal, 1979), and that the lagoons adjacent to them have formed in a different manner than that of the Simpson Lagoon. It appears that Holocene barrier islands and, thus, the lagoons associated with them can form by different ways (refer to Schwartz, 1973 and Nummedal, 1982 for a review on the subject). We believe that for a firm understanding of the evolution of these lagoons adjacent to barrier islands, it is crucial first to reconstruct the Holocene stratigraphy of lagoons on a wide regional base.

ACKNOWLEDGMENTS

This research was funded partly by the U.S. Geological Survey, Menlo Park (contract No. 14-08-0001-14827), by the State of Alaska appropriation to the Institute of Marine Science, University of Alaska, Fairbanks, Office of Alaska Sea Grant Program, and the U.S. Environmental Protection Agency (Grant R801124-03).

J. A. Dygas, R. W. Tucker, M. D. Sweeney and J. Helmericks helped in sample collection, and P. W. Barnes, E. Reimnitz and T. Osterkamp also provided samples. We have profited from a number of discussions with D.M. Hopkins on various aspects of the study. R. Nelson, K. McDougall, E. Brouwers and L. Marincovich identified microfossils. Textural analyses were by Weislawa Wajda, and some illustrations by S. B. Hardy. C. Rawlinson helped at various stages of manuscript preparation. Special thanks to Peter Barnes, Scott Briggs and two anonymous reviewers for review and suggestions on the manuscript.

REFERENCES

Allen, J.R.L. (1965). *Amer. Assoc. Petrol. Geol. Bull. 49,* 547.

Arnborg, L., Walker, H.J., and Deippo, J. (1967). *Geografiska Annaler 49,* 131.

Barnes, P.W. (1974). *In* "An Ecological Survey in the Beaufort Sea" (Huffort *et al.,* eds.), p. 183-227. U.S. Coast Guard Oceanographic Report CG363-64.

Barnes, P.W., Reimnitz, E., Drake, D. and Toimil, L. (1977). U.S. Geol. Surv. *Open-File Rept. 77-477.*

Barnes, P.W., Reimnitz, E. and Fox, D. (1982). *J. Sed. Pet. 52,* 493.

Barnes, P.W., Reimnitz, E. and McDowell, D. (1977). U.S. Geol. Surv. *Open-File Rept. 77-477.*

Barnes, P.W., Reimnitz, E. and Toimil, L. (1979). U.S. Geol. Surv. *Open-File Rept. 79-351.*

Barnes, P.W. and Toimil, L. (1979). U.S. Geol. Surv. *Misc. Field Inv. Map MF-1125.*

Brower, W.A. Jr., Searby, H.W., Wise, J.L., Diaze, H.F. and Prechtel, A.S. (1977). "Climatic Atlas of the Outer Continental Shelf Waters and Coastal Regions of Alaska," Vol. III, Chukchi-Beaufort Sea, Arctic Environmental Information and Data Center, Anchorage.

Canon, J. (1978). *In* "Coastal Geology and Geomorphology," Arctic Project Office Bull. 20, Geophysical Inst., Univ. Alaska, Fairbanks.

Dickinson, K.A., Berryhill, H.L. and Holmes, D.W. (1972). *In* "Recognition of Ancient Sedimentary Environments" (J.K. Rigby and Wm. K. Hamblin, eds.), p. 192-214. Soc. Econ. Paleont. Mineralogists Spec. Pub. 16, Tulsa, Oklahoma.

Dunton, K., Reimnitz, E. and Schonberg, V. (1982). *Arctic 35,* 465.

Dygas, J.A. and Burrell, D.C. (1976). *In* "Assessment of the Arctic Marine Environment" (D.W. Hood and D.C. Burrell, eds.), p. 263-285. Inst. Marine Sci., Univ. Alaska Occas. Pub. 4, Fairbanks.

Dygas, J.A., Tucker, R. and Burrell, D.C. (1972). *In* "Baseline Data Study of the Alaskan Arctic Aquatic Environment" (P.J. Kinney *et al.,* eds.), p. 62-121. Inst. Marine Sci., Univ. Alaska Rep. R-72-3, Fairbanks.

Folk, R.L. (1954). *J. Geol. 62,* 344.

Folk, R.L. (1968). "Petrology of Sedimentary Rocks." Hemphills, Austin.

Friedman, G.M. (1961). *J. Sed. Pet. 31*, 514.

Friedman, G.M. and Sanders, J.E. (1978). "Principles of Sedimentology." Wiley, New York.

Hartwell, A.D. (1973). *Arctic 26*, 244.

Hopkins, D.M. and Hartz, R.W. (1978). U.S. Geol. Surv. *Open-File Rept. 78-1063.*

Hume, J.D. and Schalk, M. (1964). *Am. J. Sci. 262*, 267-273.

Hume, J.D. and Schalk, M. (1967). *Arctic 20*, 86.

Klein, G.D. (1975). "Sandstone Depositional Models for Exploration for Fossil Fuels." Continuing Education Publishing Company. Champaign, IL.

Koide, M., Bruland, K.W. and Goldberg, E.D. (1973). *Geochim. Cosmochim. Acta 37*, 1171.

Kraft, J.C. (1971). *Geol. Soc. Amer. Bull. 82*, 2131.

Lewellen, R.I. (1977). *In* "Research Techniques in Coastal Environments" (H.J. Walker, ed.), Louisiana State Univ., Baton Rouge.

Lewis, C.P. and Forbes, D.L. (1974). Geol. Survey Canada, *Rept. 74-29.*

Matthews, J.B. (1970). *The Northern Engr. 2(2)*, 12.

Matthews, J.B. (1981a). *J. Geophys. Res. 86*, 6643.

Matthews, J.B. (1981b). *Ocean Management 6*, 223.

Naidu, A.S. (1968). Ph.D. Thesis, Andhra University, Waltair.

Naidu, A.S. (1979). *In* "Environmental Assessment of the Alaskan Continental Shelf", p. 3. Ann. Rept. of the Principal Investigators, March 1980, Vol. 7. National Oceanic and Atmospheric Administration, Boulder.

Naidu, A.S. (1980). Ann. Rept. National Oceanic and Atmospheric Administration, Boulder.

Naidu, A.S. (1981). *In* "Environmental Assessment of the Alaskan Continental Shelf," Ann. Rept. of the Principal Investigators, March 1981. National Oceanic and Atmospheric Administration. Boulder.

Naidu, A.S. (1982). *In* "Final Report," p. 114. National Oceanic and Atmospheric Administration, Boulder.

Naidu, A.S. and Larsen, L.H. (1980). *In* "Beaufort Sea Winter Watch" (D.M. Schell, ed.), p. 21-23. Arctic Project Office, Geophysical Institute, University of Alaska, Fairbanks.

Naidu, A.S. and Mowatt, T.C. (1975b). *In* "Deltas: Models for Sub-surface Exploration" (M.L.S. Broussard, ed.), p. 284. Houston Geological Society, Houston.

Naidu, A.S. and Mowatt, T.C. (1976). *In* "Recent and Ancient Sedimentary Environments in Alaska," (T.C. Miller, ed.), p. D1. Alaska Geological Society, Anchorage.

Naidu, A.S. and Mowatt, T.C. (1983). *Geol. Soc. Amer. Bull. 94*, 841.

Nelson, H.F. and Bray, E.E. (1970). *In* "Deltaic Sedimentation: Modern and Ancient" (J.P. Morgan, ed.), p. 48. Society of Economic Paleontologists and Mineralogists, Tulsa.

Nittrouer, C.A., Sternberg, R.W., Carpenter, R. and Bennett, J.T. (1979). *Mar. Geol. 31*, 297.

Nummedal, D. (1979). *In* "Proceedings 5th Conference on Port and Ocean Engineering Under Ice Arctic Conditions," Trondheim, Vol. II, p. 845-858.

Nummedal, D. (1982). "Barrier Islands." Tech. Rept. 82-3, Coastal Research Group, Louisiana State Univ., Baton Rouge.

Owens, E.H., Harper, J.R., and Nummedal, D. (1980). Proc. 17th Intn'tl. Coastal Engr. Conf. ASCE/Sydney, 1344.

Reading, H.G. (1978). "Sedimentary Environments and Facies." Elsevier, New York.

Reimnitz, E. and Barnes, P.W. (1974). *In* "The Coast and Shelf of the Beaufort Sea" (J.D. Reed and J.E. Sather, eds.), p. 301. Arctic Institute of North America, Arlington.

Reimnitz, E. and Bruder, K.F. (1972). *Geol. Soc. Amer. Bull. 83*, 861.

Reimnitz, E. and Kempema, E.W. (1982). U.S. Geol. Surv. *Open File Rept. 82-588.*

Reimnitz, E. and Maurer, D.K. (1979). *Geology 7*, 507-510.

Schell, D.M. (1980). "Beaufort Sea Winter Watch. Ecological Processes in the Nearshore Environment and Sediment-Laden Sea Ice: Concepts, Problems, and Approaches." Arctic Project Office, Geophysical Institute, University of Alaska, Fairbanks.

Schwartz, M.L. (1973). "Barrier Islands: Benchmark Papers in Geology." Dowden, Hutchinson and Ross, Stroudsburg, PA.

Shelton, J.W. (1973). *Univ. Oklahoma Bull. 118.*

Shepard, F.P. and Moore, D.G. (1955). *Amer. Assoc. Petrol. Geol. Bull. 39*, p. 1463.

Short, A.D. (1973). Ph.D. Thesis, Louisiana State Univ., Baton Rouge.

Short, A.D. (1979). *Geol. Soc. Amer. Bull 90(11)*, 77.

Short, A.D., Coleman, J.M. and Wright, L.D. (1974). *In* "The Coast and Shelf of the Beaufort Sea" (J.D. Reed and J.E. Sater, eds.), p. 477. Arctic Institute of North America, Arlington.

Tucker, R.W. (1973). M.S. Thesis, University of Alaska, Fairbanks.

Walker, H.J. (1974). *In* "The Coast and Shelf of the Beaufort Sea" (J.D. Reed and J.E. Sater, eds.), p. 513. Arctic Institute of North America, Arlington.

Wiseman, W.J. Jr., Coleman, J.M., Shu, S.A., Short, A.S., Suhayda, J.N., Walters, C.D. Jr. and Wright, L.D. (1973). "Alaskan Arctic Processes and Morphology." Coastal Studies Inst. Tech. Rep. 149. Louisiana State University, Baton Rouge.

Biological Interactions

PHYTOPLANKTON ABUNDANCE, CHLOROPHYLL a, AND PRIMARY PRODUCTIVITY IN THE WESTERN BEAUFORT SEA

Rita Horner

Seattle, Washington

I. INTRODUCTION

Little was known about phytoplankton populations in the western Beaufort Sea before the early 1970's, although some early collections of diatoms were made during the Canadian Arctic Expedition of 1913-1918 (Mann, 1925). With the discovery of oil on the North Slope in 1968, several studies were designed to provide baseline information on the ecology of the marine and freshwater ecosystems with work being done primarily in shallow, nearshore environments in summer (Alexander *et al.*, 1974; Coyle, 1974; Horner *et al.*, 1974). Here I will discuss phytoplankton associations, chlorophyll *a*, and relative phytoplankton production in the area between the 20-m depth contour and the edge of the ice in summer.

II. MATERIALS AND METHODS

Sampling was done during five interdisciplinary cruises on U.S. Coast Guard icebreakers in 1973-74 and 1976-78; all sampling was done from late July to mid-September. The study area was the Beaufort Sea seaward from the 20-m depth contour and extending from Point Barrow ($156°30'W$) to near Demarcation Point ($141°W$) (Fig. 1). Station locations were determined to some extent by ice conditions and the sampling requirements of other projects on the cruises. Water samples were collected with 5-L Niskin bottles evenly spaced to provide coverage from the surface to the bottom in shallow water or to 100 m in deeper areas.

Phytoplankton standing stock samples were preserved in glass jars using 5-10 mL formaldehyde (4%) buffered with sodium acetate. Samples from 1973 and 1976-78 were analyzed using an inverted microscope technique (Utermöhl, 1931).

Water for plant pigment analysis was filtered through 47 mm, 0.45 μm Millipore filters (Strickland and Parsons, 1968). In 1973-74, chlorophyll *a* concentrations were determined using trichromatic methods (Strickland and Parsons, 1968) and calculated using SCOR/UNESCO equations (UNESCO, 1966). Phaeopigment concentrations were not determined. In 1976-78, chlorophyll *a* and phaeopigment concentrations were determined using fluorometric methods (Strickland and Parsons, 1968).

THE ALASKAN BEAUFORT SEA:
ECOSYSTEMS AND ENVIRONMENTS

295

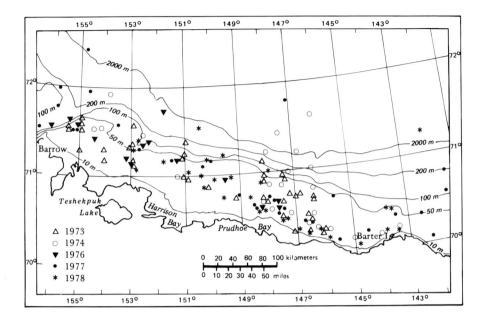

FIGURE 1. Phytoplankton sampling sites in the Beaufort Sea.

Primary productivity measurements in 1976-78 were made in 60-mL reagent bottles with two light and one dark bottle per depth. Five μ Ci ^{14}C, as sodium bicarbonate, was added to each bottle, aluminum foil was wrapped around the dark bottles, and the samples were incubated in a laboratory sink under a bank of cool white fluorescent lights. Light levels averaged about 40-50 $\mu E\ m^{-2}\ s^{-1}$. Temperature in the sink was maintained near 3°C with continuously running seawater and was monitored throughout the 3-4-hour incubation period. Following incubation, the samples were filtered onto 25 mm, 0.45 μm Millipore filters, rinsed with 5 mL 0.01N HCl and 5 mL filtered seawater, and placed in scintillation vials. Radioactive uptake was measured using liquid scintillation techniques in a Packard Tri-Carb Scintillation Spectrometer with Aquasol as the scintillation cocktail. Carbon uptake was calculated using the equation in Strickland and Parsons (1968).

Salinity samples were stored in tightly closed 250-mL polyethylene bottles and analyzed once a week using a Beckman RS-7A induction salinometer. Temperatures, taken with reversing thermometers, were provided by the U.S. Coast Guard in 1973-74. In 1976-78, thermometer readings were corrected following U.S. Naval Oceanographic Office Publ. 607 (1968).

A more detailed description of methods and extensive data tables will be found in Horner (1981a). All 1976-78 data, including cell numbers by species, were submitted to the National Oceanographic Data Center, Washington, D.C.

III. RESULTS AND DISCUSSION

Ice conditions were variable between the summers of sampling. In 1973, severe ice conditions in late July near Point Barrow prevented the icebreaker from entering the Beaufort Sea for several days. Later, ice or fog prevented some sampling. Ice conditions in 1974 were particularly heavy from 146° to 148°W. In 1976 and 1977, heavy ice near Point Barrow again prevented the icebreaker from entering the Beaufort Sea for several days. Heavy ice in 1976 prevented sampling east of Prudhoe Bay, while ice conditions in 1977 were relatively light except at a few stations well offshore and in the area from 147° to 148°W. In 1978, heavy ice was present off Harrison and Prudhoe Bays and east to near 146°W.

Salinity in the upper 50 m was usually about 28-32°/oo, but in all years, at stations where heavy ice was present, there was a layer of low-salinity water (<20°/oo) about 5 m thick at the surface. This low-salinity layer was present at stations east of Harrison Bay in 1973 and 1974, at nearly all stations in 1976, and at a few stations in 1977 and 1978.

Temperature in the upper 50 m ranged from -1.70° to +8.59°C. A layer of warmer water (0.1 to 8.59°C) extended from 0 to 60 m depending on water depth and spread eastward from Point Barrow to near 152°W in all years. This water was probably advected from the Bering Sea (Johnson, 1956; Hufford, 1973; Paquette and Bourke, 1974). Salinity in this layer was generally 28-32°/oo, but surface salinity was often considerably lower, down to 8.35°/oo on one occasion. Warm surface water was present east of 152°W in 1977 and 1978, particularly in the area between Harrison and Prudhoe Bays (151°-146°W) and north of Barter Island (144°W).

Ninety-four species and 18 additional taxonomic categories of phytoplankton, such as unidentified species and groups of species, were identified from the Beaufort Sea (Table I). Not all species were found each year. Because of the large number of species and the large number of samples, phytoplankton species were grouped into four categories based on taxonomic affinities and numerical abundance: *Chaetoceros* spp., other diatoms, dinoflagellates, and flagellates. The genus *Chaetoceros*, consisting of at least 20 species identified from the Beaufort Sea, was the most important group of diatoms based on number of species and number of cells. The other diatoms category consisted of all diatom species except *Chaetoceros*, with the number of species in this category varying depending on the year. Dinoflagellates were never very numerous, but were given category status because they are a major taxonomic group in the phytoplankton. The flagellates category consisted of identified and unidentified cells. Most of the flagellates could not be identified and were grouped into size classes with cells <10 μm in diameter usually being the most abundant.

Results of the quantitative analyses of phytoplankton samples collected in 1973 and 1976-78 are presented in Table II. The 1974 data have not been included because only chlorophyll *a* values are available; chlorophyll *a* and carbon uptake data for 1977 stations where standing stock was not determined have also not been included.

The most abundant species in these samples were widely distributed across the western Beaufort Sea. Small species of the genus *Chaetoceros*

TABLE I. Phytoplankton species found in the Beaufort Sea[a]

Bacillariophyceae
  Amphiprora hyperborea Grunow
  Asterionella kariana Grunow
  Bacterosira fragilis Gran
  Biddulphia aurita (Lyngbye) Brébisson & Godey
  Chaetoceros atlanticus Cleve
  Chaetoceros borealis Bailey
  Chaetoceros ceratosporum Ostenfeld
  Chaetoceros compressus Lauder
  Chaetoceros concavicornis Mangin
  Chaetoceros danicus Cleve
  Chaetoceros debilis Cleve
  Chaetoceros fragilis Meunier
  Chaetoceros furcellatus Bailey
  Chaetoceros gracilis Schütt
  Chaetoceros karianus Grunow
  Chaetoceros mitra (Bailey) Cleve
  Chaetoceros radicans Schütt
  Chaetoceros septentrionalis Østrup
  Chaetoceros socialis Lauder
  Chaetoceros subsecundus (Grunow) Hustedt
  Chaetoceros subtilis Cleve
  Chaetoceros teres Cleve
  Chaetoceros wighami Brightwell
  Chaetoceros spp.
  Coscinodiscus centralis Ehrenberg
  Coscinodiscus curvatulus Grunow
  Coscinodiscus oculus-iridis Ehrenberg
  Coscinodiscus spp.
  Cylindrotheca closterium (Ehrenberg) Reimann & Lewin
  Detonula confervacea (Cleve) Gran
  Eucampia zoodiacus Ehrenberg
  Gomphonema spp.
  Leptocylindrus danicus Cleve
  Leptocylindrus minimus Gran
  Licmophora sp.
  Melosira arctica (Ehrenberg) Dickie
  Melosira jurgensii Agardh
  Melosira moniliformis (O.F. Müller) Agardh
  Navicula pelagica Cleve
  Navicula transitans Cleve
  Navicula spp.
  Nitzschia cylindrus (Grunow) Hasle
  Nitzschia delicatissima Cleve
  Nitzschia frigida Grunow
  Nitzschia grunowii Hasle
  Nitzschia seriata Cleve
  Nitzschia spp.

(continued)

Porosira glacialis (Gran) Jørgensen
Rhizosolenia alata Brightwell
Rhizosolenia hebatata Bailey
Skeletonema costatum (Greville) Cleve
Stauroneis granii Jørgensen
Thalassionema nitzschioides Hustedt
Thalassiosira anguste lineata (A. Schmidt) Fryxell & Hasle
Thalassiosira antarctica Comber
Thalassiosira decipiens (Grunow) Jørgensen
Thalassiosira gravida Cleve
Thalassiosira hyalina (Grunow) Gran
Thalassiosira lacustris (Grunow) Hasle
Thalassiosira nordenskioeldii Cleve
Thalassiosira spp.
Thalassiothrix frauenfeldii Grunow
Unidentified diatoms, mostly pennate diatom species

Dinophyceae
Amphidinium longum Lohmann
Ceratium arcticum (Ehrenberg) Cleve
Ceratium longipes (Bailey) Gran
Dinophysis acuta Ehrenberg
Dinophysis norvegica Claparède & Lachmann
Dinophysis rotundata Claparède & Lachmann
Dinophysis sphaerica Stein
Gonyaulax spinifera (Claparède & Lachmann) Diesing
Gonyaulax spp.
Gymnodinium lohmanni Paulsen
Gymnodinium spp.
Micracanthodinium setiferum (Lohmann) Deflandre
Oxytoxum spp.
Peridiniella catenata (Levander) Balech
Protoceratium reticulatum (Claparède & Lachmann) Bütschlii
Protoperidinium belgicum (Wulff) Balech
Protoperidinium bipes (Paulsen) Balech
Protoperidinium brevipes (Paulsen) Balech
Protoperidinium conicum (Gran) Balech
Protoperidinium depressum (Bailey) Balech
Protoperidinium grenlandicum (Woloszynska) Balech
Protoperidinium pallidum (Ostenfeld) Balech
Protoperidinium pellucidum Bergh
Protoperidinium spp.
Scrippsiella faeroense (Paulsen) Balech & Soares
Scrippsiella trochoidea (Stein) Loeblich III

Haptophyta
Phaeocystis pouchetii (Hariot) Lagerheim

Cryptophyta
Chroomonas spp.
Cryptomonas spp.

(continued)

(Table I, continued)

Chrysophyta
   Calycomonas gracilis Lohmann
   Calycomonas ovalis Wulff
   Dinobryon balticum (Schütt) Lemmermann
   Dinobryon petiolatum Willén
   Distephanus speculum (Ehrenberg) Haeckel
   Ebria tripartita (Schumann) Lemmermann
   Pelagococcus subviridis Norris

Chlorophyta
   Platymonas spp.
   Pterosperma spp.

Euglenophyta
   Dinematomonas litorale (Perty) Silva
   Eutreptiella braarudii Throndsen

Choanoflagellates
   Crinolina aperta (Leadbeater) Thomsen
   Diaphanoeca grandis Ellis
   Parvicorbicula socialis (Meunier) Deflandre
   Unidentified species

Undentified flagellates from several taxonomic categories

Organisms of unknown affinities
   Hexasterias problematica Cleve
   Piropsis polita Meunier
   Radiosperma corbiferum Meunier

[a] *Names and authors are those given in Hustedt (1930, 1959-1962); Hendey (1974); Parke and Dixon (1976); Balech (1974, 1977); and Schiller (1933, 1937).*

were the most abundant diatoms present during the sampling periods. These small species, often measuring about 6 μm along the apical axis, are difficult to identify to species on the basis of vegetative cells alone. Resting spores, known for some of these species, make identification easier, but they were not always present in these samples.

Other abundant diatom species included *Nitzschia grunowii* (probably also including *N. cylindrus*), *N. delicatissima, Thalassiosira gravida, T. nordenskioeldii, Cylindrotheca closterium, Eucampia zoodiacus,* and *Bacterosira fragilis,* but not all species were found all years. Of these species, *N. grunowii* is known primarily from northern coastal regions (Hasle, 1965). It is a component of the sea ice community in early spring and may also occur commonly in the water column in spring (Horner, 1981b).

Thalassiosira nordenskioeldii is a common and abundant spring bloom species in cold water (Durbin, 1974), probably because of its short

TABLE II. *Maximum phytoplankton abundances by major category, primary productivity, and chlorophyll a at Beaufort Sea stations, 1973, 1976-78[a]*

Station Lat (N)	Station Long (W)	Max Sample Depth	Standing Stock Chaetoceros Cells	d_{max}	Diatoms Cells	d_{max}	Flagellates Cells	d_{max}	Dinoflagellates Cells	d_{max}	Prim Prod mg m^{-3} d_{max}	Chlorophyll a mg m^{-3}	d_{max}
1973													
71°28'	155°26'	20	10.0	10	6.6	0	12.6	10	0.6	10		1.67	10
71°34'	155°30'	60	1.3	60	13.8	60	17.1	0	1.8	40		2.35	60
71°26'	153°08'	60	19.1	30	77.8	40	21.0	0	2.7	10		5.12	40
71°11'	152°59'	20	16.2	15	28.5	15	15.8	0	1.9	5		5.52	15
71°08'	152°59'	20	10.2	20	21.2	20	26.6	0	1.6	15		4.13	15
71°35'	152°59'	120	7.2	30	65.4	50	9.8	10	4.8	10		4.30	30
71°20'	153°05'	75	30.8	30	45.6	30	18.4	0	5.0	10		5.38	30
70°56'	150°55'	20	110.8	15	41.3	20	7.6	0	0.6	0		8.30	20
71°02'	151°00'	19	88.2	19	28.6	10	6.6	0	4.2	19		8.77	19
71°11'	151°16'	20	5.9	15	6.7	20	28.1	0	4.6	5		2.46	15
71°16'	151°04'	90	44.6	30	28.4	75	10.4	0	4.0	90		3.20	30
71°23'	151°00'	100	13.4	40	45.4	75	56.3	0	2.0	40		4.58	50
70°51'	150°10'	20	382.7	10	67.0	0	7.8	0	1.9	0		7.84	20
70°43'	149°11'	15	24.2	10	33.8	15	20.6	5	1.1	5		2.99	10
70°42'	148°00'	20	240.1	10	14.3	20	81.6	5	4.1	0		4.28	20
70°41'	148°07'	30	299.5	30	13.8	15	30.7	0	2.8	0		4.12	30
70°42'	148°10'	30	215.5	30	14.6	30	8.5	5	1.4	10		2.99	30
70°56'	148°10'	39	138.7	20	13.0	10	11.0	0	2.8	20		3.38	39
71°02'	148°12'	49	149.4	49	12.2	30	10.7	5	3.2	15		4.53	49
71°09'	148°09'	100	87.3	20	34.2	75	12.0	0	2.2	20		3.59	75
70°57'	147°30'	50	57.3	30	8.4	50	24.0	0	2.2	5		1.22	50
70°52'	147°30'	40	109.4	15	12.2	15	17.8	0	1.7	15		2.17	40
70°32'	147°30'	20	67.6	10	19.0	20	10.6	0	1.1	0		2.77	20
70°23'	147°07'	15	63.4	15	64.9	15	8.2	0	1.0	0		3.55	15
70°19'	146°33'	20	171.7	10	49.4	20	10.0	0	1.0	5		4.09	10

[a] Cell numbers = cells L^{-1} x 10^4; d_{max} = depth in meters at which maximum value was found. Primary productivity was not measured in 1973.

Table II continued

Station Lat (N)	Long (W)	Max Sample Depth	Standing Stock — *Chaetoceros* Cells	d_{max}	Diatoms Cells	d_{max}	Flagellates Cells	d_{max}	Dinoflagellates Cells	d_{max}	Prim Prod mg m^{-3}	d_{max}	Chlorophyll mg m^{-3}	d_{max}
70°06'	149°54'	20	4.2	20	3.8	10	24.2	5	2.8	15	0.21	20	0.57	15
71°04'	150°53'	20	5.4	20	3.0	20	26.4	0	4.2	10	0.13	20	0.55	20
71°11'	151°51'	20	3.2	15	1.2	0	25.2	0	4.2	0	0.17	15	0.42	15
71°05'	152°51'	20	39.0	15	7.0	20	23.6	0	3.8	10	0.60	20	0.72	20
71°20'	152°48'	50	8.8	30	10.6	20	21.4	5	4.4	15	0.18	10	0.21	40
71°22'	152°41'	90	6.6	45	8.2	20	20.2	5	7.0	15	0.49	45	1.03	15
71°34'	150°27'	75	5.4	45	13.4	5	23.0	0	2.2	5	0.23	0	0.27	15
70°36'	147°39'	18	70.4	18	29.6	9	24.2	15	0.4	9	0.47	9	2.86	6
70°29'	147°23'	21	164.4	9	69.2	9	21.0	12	1.4	15	0.90	12	1.11	18
70°22'	146°52'	18	155.6	12	77.8	12	24.8	12	0.6	12	1.05	15	3.45	18
70°34'	145°52'	18	154.8	15	95.4	12	40.2	0	0.4	9	1.93	12	4.42	18
70°13'	143°23'	18	167.8	15	56.2	15	24.8	18	0.2	3	1.52	15	2.65	15
69°59'	142°15'	15	189.6	15	25.0	15	20.8	15	0.2	6	0.90	15	0.84	15
70°58'	142°21'	45									0.11	30	0.26	30
69°45'	141°18'	15	81.6	12	3.2	6	14.4	6	0.4	6	0.28	15	0.55	12
70°28'	143°33'	35	32.0	5	6.4	5	11.4	5	1.0	35	0.14	35	0.27	35
70°29'	143°42'	55	19.8	20	1.4	0	16.8	0	0.6	10	0.16	30	0.23	30
70°15'	143°40'	25	233.8	20	269.2	25	17.8	20	0.6	12	2.18	20	5.46	20
70°08'	144°48'	15	6.4	9	11.2	15	11.8	3	0.2	6	0.28	15	0.92	15
70°18'	146°31'	18	370.2	18	35.2	18	40.4	18	0.6	18	2.56	18	6.11	18
70°28'	147°26'	21	148.2	18	20.8	21	13.4	21	0.8	18	1.50	21	3.35	21
71°01'	147°57'	45	4.8	30	17.8	15	75.8	15	2.6	20	2.93	45	0.94	15
70°45'	148°34'	21	113.2	12	23.6	18	100.6	0	4.4	15	1.08	21	2.34	21
70°36'	148°00'	18	70.8	18	16.4	18	52.6	0	1.8	9	1.09	15	2.28	18
70°47'	149°30'	18	71.0	3	15.8	18	93.8	0	1.6	9	0.98	18	1.76	15
71°13'	149°38'	60	8.6	5	46.4	5	17.6	0	2.2	9	0.24	20	0.51	20
70°52'	150°16'	22	44.8	9	14.6	12	23.2	0	2.8	9	0.68	22	1.17	17
71°01'	150°25'	21	59.0	0	26.1	12	19.0	9	2.2	15	0.49	18	0.98	18

Table II continued

Lat (N)	Long (W)	Max Sample Depth	Chaetoceros Cells	d_{max}	Diatoms Cells	d_{max}	Flagellates Cells	d_{max}	Dinoflagellates Cells	d_{max}	Prim Prod mg m^{-3}	d_{max}	Chlorophyll a mg m^{-3}	d_{max}
70°13'	146°01'	20	44.6	10	63.9	5	15.8	0	1.2	0			3.45	10
70°21'	146°00'	20	46.6	15	21.9	20	7.9	0	1.1	0			1.80	15
70°24'	146°31'	20	41.5	20	20.2	20	10.8	0	0.9	0			2.04	20
70°39'	146°31'	30	72.9	15	10.9	15	11.9	0	1.4	10			1.97	30
70°40'	146°29'	40	108.6	30	8.7	20	9.8	0	1.6	15			2.99	30
70°44'	146°45'	30	75.4	30	7.8	15	4.6	5	1.2	10			1.97	30
70°42'	147°04'	39	43.8	14	9.6	4	14.5	0	1.6	0			1.72	39
70°42'	147°10'	30	41.0	20	6.7	10	8.4	0	2.4	15			1.81	30
70°39'	147°33'	20	40.6	20	7.7	5	4.1	10	1.4	20			1.21	20
71°09'	154°03'	15	0.7	15	9.6	10	18.3	5	4.5	0			1.86	15
71°15'	154°05'	20	0.2	0	6.6	20	9.4	10	7.6	5			0.81	5
71°19'	155°00'	13	0.4	10	5.2	13	16.1	10	4.7	0			1.33	10
71°28'	154°55'	20	0.6	20	5.4	15	11.0	5	5.3	5			1.55	5
71°33'	155°00'	30	0.1	10	9.0	30	10.0	0	3.7	0			1.42	5
71°36'	154°57'	50	0.2	50	17.8	10	7.8	5	3.3	10			1.90	50
1976														
71°31'	156°09'	100	46.8	0	2.3	0	69.8	0	2.6	0	2.81	50	3.10	50
71°31'	155°05'	25	222.9	0	3.1	20	31.7	0	3.1	0	0.54	15	0.59	10
71°11'	153°09'	20	473.2	20	26.8	10	61.8	0	1.6	0	0.55	5	0.62	0
70°36'	148°12'	15	305.3	10	38.3	10	16.9	5	0.9	10	4.36	10	5.38	15
70°32'	147°33'	20	316.2	10	61.7	15	63.6	0	4.3	0	5.98	10	3.80	10
70°39'	147°37'	15	15.9	5	2.9	5	29.5	5	1.8	5	3.95	10	3.36	10
70°57'	149°33'	25	8.7	10	2.0	10	7.1	10	1.6	15	0.45	10	0.71	5
71°08'	151°19'	15									2.45	10	3.90	10
71°43'	151°47'	100									0.72	20	0.74	20
71°22'	152°20'	75	8.6	5	61.1	15	52.8	0	6.1	30	0.97	40	1.98	15
71°19'	152°32'	45	3.7	15	71.3	15	21.7	0	2.3	0	0.77	15	1.03	15

Standing Stock

303

Table II continued

Station Lat (N)	Long (W)	Max Sample Depth	Chaetoceros Cells	d_{max}	Diatoms Cells	d_{max}	Flagellates Cells	d_{max}	Dinoflagellates Cells	d_{max}	Prim. Prod mg m^{-3}	d_{max}	Chlorophyll mg m^{-3}	d_{max}
71°08'	152°57'	15	117.5	10	8.4	15	59.4	0	1.4	0	2.25	10	2.20	10
71°23'	154°21'	25									0.48	15	0.69	10
71°36'	155°32'	30	113.2	20	17.8	10	143.7	0	6.8	0	1.24	20	1.28	20
1977														
71°46'	155°51'	100	293.1	75	28.5	45	12.5	60	1.6	10	1.02	45	1.85	60
71°57'	154°33'	175	156.2	45	16.0	125	68.3	45	1.9	75	0.66	45	0.98	60
72°24'	154°37'	100	33.4	100	6.9	20	68.8	20	0.3	30	0.23	20	0.24	100
71°35'	153°29'	45	69.3	20	6.7	20	15.4	25	4.8	10	0.56	20	0.75	30
71°18'	152°43'	50	160.3	5	42.4	20	32.3	50	0.8	50	2.26	20	2.12	0
71°10'	151°30'	20	224.5	5	62.4	10	11.7	5	1.4	20	2.65	10	4.40	10
71°05'	150°23'	25	111.8	5	23.7	10	13.6	5	0.8	10	1.37	10	1.85	5
71°10'	150°04'	45	165.0	10	82.9	10	6.6	5	1.0	20	3.21	10	3.12	5
70°38'	148°28'	18	207.2	6	99.2	3	12.3	3	0.4	6	4.12	12	5.29	9
70°42'	147°59'	25	315.0	5	141.2	0	5.6	5	0.8	5			9.99	5
70°40'	147°48'	30	708.2	12	113.0	5	5.9	5	0.5	5			7.77	5
70°33'	147°24'	25	598.2	12	208.9	15	8.2	0	1.2	12	7.35	15	8.74	3
70°25'	146°41'	25	658.4	25	104.2	25	7.4	9	1.4	3	3.65	20	4.69	20
69°49'	141°31'	25	781.8	12	100.6	9	49.0	15	4.4	12	10.35	6	18.84	12
70°04'	142°14'	30	278.2	30	91.8	12	41.6	6	1.8	6	2.93	6	8.90	30
70°21'	143°24'	35	1223.2	35	31.4	20	16.0	15	1.4	25	5.12	25	11.03	25
1978														
71°11'	150°14'	25	33.4	35	3.6	15	24.6	5	1.8	35	0.22	25	0.56	25
70°59'	149°17'	35	23.6	15	5.0	35	28.6	3	1.4	20	0.18	35	0.40	15
70°20'	146°05'	25	6.0	25	1.2	20	28.2	15	0.2	10	0.11	10	0.33	25
70°36'	148°20'	20	68.6	12	19.2	6	38.2	0	3.2	9	0.27	6	1.30	6
70°55'	148°11'	35	17.2	20	7.0	5	82.8	0	1.0	0	0.12	20	0.38	5

generation time at low temperatures (Baars, 1982). It is able to grow at low light intensities under the ice and is also favored by continuous light conditions at low temperatures in spring and summer (Baars, 1982). *Thalassiosira gravida*, known from about 35° to 70°N (Hasle, 1976), is a common neritic, boreal species. *Bacterosira fragilis* is a common spring species in arctic neritic regions, and resting spores are sometimes abundant.

Leptocylindrus minimus was the only species in these samples that had not been reported previously from the Beaufort Sea. It was found only in water identified as being from the Bering Sea. *Cylindrotheca closterium*, a common neritic species present throughout the area, was most often found in the warm Bering Sea water.

Microflagellates were abundant all years and were most common in surface samples, perhaps because they are better able to adapt to higher light intensity and lower salinity. It was not possible to determine in preserved samples if these organisms were photosynthetic, although chlorophyll *a* and carbon uptake were low at stations and depths where microflagellates were abundant. However, large numbers of very small (<0.2 μm) photosynthetic flagellates have recently been reported over large areas of the Atlantic Ocean (Johnson and Sieburth, 1982), suggesting that additional studies are needed to determine the photosynthetic capabilities of the Beaufort Sea organisms.

Dinoflagellates were nearly always present, but were never very abundant, usually being less than 5% of the total population. On occasion, however, as at five stations off Smith Bay (154°W) in mid-August 1973, they made up 10-25% of the population at some depths. In temperate waters, dinoflagellates usually become more numerous in fall when nutrient concentrations are low, and this is apparently the situation in the Beaufort Sea as well. Species present included *Ceratium arcticum*, *C. longipes*, *Dinophysis acuta*, *D. sphaerica*, *Gonyaulax spinifera*, *Gymnodinium lohmanni*, *Peridiniella catenata*, *Protoperidinium brevipes*, *P. bipes*, *P. pallidum*, and *Scrippsiella trochoidea*. All of these species are known from arctic-subarctic regions (Paulsen, 1908; Schiller, 1933, 1937; Horner, 1969).

Samples from these cruises can be grouped into five regions to show trends in horizontal and vertical distribution of the four major taxonomic categories (Fig. 2). The regions from west to east are Point Barrow (155°-156°W), Pitt Point (152°-153°W), Harrison Bay (149°-150°30'W), Prudhoe Bay (147°30'-148°30'W), and Barter Island (143°-144°W). In general in the Point Barrow area, flagellates were the most abundant organisms, especially at the surface. *Chaetoceros* spp. and other diatoms were sometimes numerous below 5 m (Fig. 2a).

Off Pitt Point, other diatoms and flagellates were the most abundant organisms, with flagellates being dominant at the surface and other diatoms dominating from 15 to 50 m (Fig. 2b). *Cylindrotheca closterium* and *Leptocylindrus minimus* were abundant in warm water at two stations in this area in 1976. Dinoflagellates were relatively common in this area in 1978. Flagellates were sometimes numerous below the surface in 1978.

Near Harrison Bay, *Chaetoceros* spp. and flagellates dominated the phytoplankton. Flagellates were most abundant at the surface, and *Chaetoceros* spp. were dominant below 5 m (Fig. 2c). Other diatoms were common below the surface at some stations while dinoflagellates were more abundant in this region in 1978.

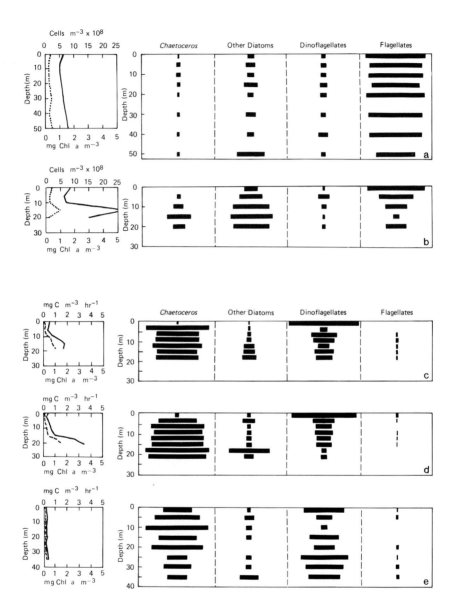

FIGURE 2. Regional vertical distributions of chlorophyll a, cells density or primary production, and phytoplankton species by major category. a) 71°34'N, 155°30'W near Pt. Barrow. b) 71°11'N, 152°59'W off Pitt Point. c) 70°47'N, 149°30'W off Harrison Bay. d) 70°28'N, 147°26'W off Prudhoe Bay. e) 70°28'N, 143°33'W near Barter Island. Solid line is chlorophyll a; dotted line is cells m^{-3}x10^8 in 2a and b, primary productivity in c, d, and e.

Near Prudhoe Bay, *Chaetoceros* spp. were most abundant below the surface with highest numbers usually at 5-20 m (Fig. 2d). Microflagellates were dominant at the surface. Other diatoms were usually present and sometimes numerous below the surface. Dinoflagellates were sometimes common in 1978.

Off Barter Island, *Chaetoceros* spp. and flagellates were the most abundant organisms, with both categories being common from the surface to the bottom. Dinoflagellates were usually not very abundant in this area (Fig. 2e).

In general, highest chlorophyll *a* and carbon uptake occurred at depths where *Chaetoceros* spp. or other diatoms were dominant (Fig. 2). At stations and depths where chlorophyll *a* and carbon uptake were low, flagellates were a major portion of the populations. Flagellates were dominant at some depths where productivity was highest for a station, but then maximum productivity was usually <1 mg C m^{-3} h^{-1}. High carbon uptake usually occurred at or near the same depths as high chlorophyll *a* concentrations, but high carbon uptake and high standing stock did not always occur at the same depth. This was especially true when microflagellates were the most abundant organisms. The same species were usually abundant at the depth of greatest productivity except where *Leptocylindrus minimus* and *Cylindrotheca closterium* were dominant.

Primary productivity was not measured in 1973-74, but rates during 1976-78 were variable with two and three-fold differences between years. Chlorophyll *a* concentrations were also variable, although high values were generally 6-8 mg m^{-3}. Highest production and standing stock occurred in 1977, the year with the least amount of ice cover, while lowest production and standing stock occurred in 1978, when ice cover was greatest.

Harrison and Prudhoe Bays are the only areas of the western Beaufort Sea where earlier studies have been done (Alexander, 1974; Horner *et al.*, 1974), and these were in shallow water, generally shoreward of the barrier islands and the 20-m depth contour. Results, however, were similar to those found during this study with flagellates being most numerous at the surface and diatoms at depth. Primary productivity and chlorophyll *a* were also highest at depths where diatoms were the most abundant organisms.

In the southern Beaufort Sea of western Canada, Hsiao *et al.* (1977) found diatoms dominating in areas close to shore and near river mouths with flagellates more important offshore beyond the Mackenzie River plume. They attributed this distribution to higher nutrient concentrations, lower light levels, and higher temperatures in the coastal waters. Primary productivity and standing stock also decreased with increasing distance from shore and river mouths for the same reasons. This is in contrast to results from the western Beaufort Sea where diatoms, including *Chaetoceros* spp., and flagellates predominated over the whole area with distribution generally determined on a vertical rather than horizontal scale. This suggests that light levels were of more importance than salinity or temperature in determining distribution.

Some estimates of annual production in the Beaufort Sea have been made primarily from data collected in the shallow, nearshore areas. Alexander (1974) estimated annual production of the water column in Harrison Bay and Simpson Lagoon to be about 10-15 g C m^{-2}. Horner *et al.* (1974) calculated annual production for the water column inside Prudhoe

Bay to be <10 g C m^{-2} and 13-23 g C m^{-2} in the lagoon system. Cold-water oceanic phytoplankton are apparently rarely if ever nutrient-limited (Round, 1981), and recent evidence from the Canadian Arctic (Harrison *et al.*, 1982) indicates that light is probably more important than nutrient concentrations in regulating growth of phytoplankton in northern latitudes.

To my knowledge, there are no data based on sufficiently intensive sampling to indicate the occurrence of a spring bloom in the offshore Beaufort Sea. Based on data from the nearshore Chukchi Sea at Barrow (Horner, 1969), it has been assumed that a spring bloom occurs during and just after ice breakup when light levels increase and nutrient concentrations are high. Species common in the spring bloom at Barrow include *Nitzschia grunowii*, *Thalassiosira* spp., *Porosira glacialis*, and *Bacterosira fragilis*, while most *Chaetoceros* spp. are more indicative of summer populations (Horner, 1969). All of these species were present in Beaufort Sea samples collected during these cruises, with *Chaetoceros* spp. usually being the most abundant diatoms. Because of the timing of these cruises in August and September, however, any spring bloom may have been missed, although in some areas ice was either still present or still in the process of breaking up and spring species were still present.

Perhaps there is no large, single pulse of phytoplankton early in the growing season as is known in more temperate regions, but instead a more gradual and moderate increase extending from the time of ice breakup when adequate light becomes available to early September when light again becomes a limiting factor with a gradual species change from dominance by typical spring species to summer species as light levels and nutrient concentrations change. It is also possible that zooplankton grazing may control phytoplankton production to the point where no bloom occurs, as in the northeast Pacific (Frost *et al.*, 1983). Alternatively, heavy ice conditions might limit light reaching the water column until August-September with the spring bloom occurring then rather than earlier. This was suggested for Prudhoe Bay in 1975 when typical spring species were common in samples collected in September and nutrient concentrations were high (English and Horner, 1977), but factors controlling phytoplankton production might be quite different in nearshore areas such as Prudhoe Bay and the offshore areas considered here.

IV. SUMMARY

The present study provides a framework on which future studies of phytoplankton in the Beaufort Sea can be based. Dominant species of diatoms and dinoflagellates are now known along with relative cell numbers. The importance of microflagellates has been shown, but these have not been identified, nor is it known if they are photosynthetic or what their role in the Beaufort Sea food web might be. Chlorophyll *a* concentrations and carbon uptake have been shown to be dependent on species present, with higher values occurring where diatoms are abundant, and on the greater availability of light when the ice disappears. There is currently no real evidence for a major spring bloom, but reasons for this are unclear and may be related to light levels, nutrient concentrations, or zooplankton grazing.

ACKNOWLEDGMENTS

D. Wolfe, R. Bowman, J. Krezoski, and T. Kaperak provided field assistance; W. S. Grant provided field assistance and counted the 1973 phytoplankton standing stock samples. The captains and crews of the icebreakers did a superb job of providing ship support, especially J. McClelland, R. Tuxhorn, and D. Sobeck. The 1973 cruise was part of the U.S. Coast Guard's Western Beaufort Sea Ecological Cruises program (WEBSEC). This study was funded by the Office of Naval Research grant N00014-67-A-0317-003 in 1973-74 and by the Bureau of Land Management through interagency agreement with the National Oceanic and Atmospheric Administration in 1976-78.

REFERENCES

Alexander, V. (1974). *In* "The Coast and Shelf of the Beaufort Sea" (J.C. Reed and J.E. Sater, eds.), p. 609. The Arctic Institute of North America, Arlington.

Alexander, V., Burrell, D.C., Chang, J., Cooney, R.T., Coulon, C., Crane, J.J., Dygas, J.A., Hall, G.E., Kinney, P.J., Kogl, D., Mowatt, T.C., Naidu, A.S., Osterkamp, T.E., Schell, D.M., Seifert, R.D., and Tucker, R.W. (1974). Univ. Alaska, Inst. Mar. Sci. Rep. R74-1. Fairbanks.

Baars, J.W.M. (1982). *Mar. Biol. 68*, 343

Balech, E. (1974). *Revta Mus. argent. Cienc. nat. Bernardino Rivadavia Inst. nac. Invest. Cienc. nat. Hidrobiol. 4*, 1.

Balech, E. (1977). *Revta Mus. argent. Cienc. nat. Bernardino Rivadavia Inst. nac Invest. Cienc. nat. Hidrobiol 5*, 115.

Coyle, K.O. (1974). M.S. Thesis, Univ. Alaska, Fairbanks.

Durbin, E.G. (1974). *J. Phycol. 10*, 220.

English, T.S., and Horner, R. (1977). *In* "Environmental Assessment of the Alaskan Continental Shelf," Ann. Rept., Vol. 9, p. 275. NOAA, Boulder.

Frost, B.W., Landry, M.R., and Hassett, R.P. (1983). *Deep-Sea Res. 30 (IA)*, 1.

Harrison, W.G., Platt, T., and Irwin, B. (1982). *Can. J. Fish. Aquat. Sci. 39*, 335.

Hasle, G.R. (1965). *Norske Videnskaps-Akad. I. Mat-Nat. K1. n.s. 21*, 1.

Hasle, G.R. (1976). *Deep-Sea Res. 23*, 319.

Hendey, N.I. (1974). *J. mar. biol. Ass. U.K. 54*, 277.

Horner, R. (1969). Ph.D. Dissertation, Univ. Washington, Seattle.

Horner, R. (1981a). *In* "Environmental Assessment of the Alaskan Continental Shelf," Final Rept., Vol. 13, p. 65. NOAA, Boulder.

Horner, R. (1981b). *In* "Proceedings of the Sixth Symposium on Recent and Fossil Diatoms" (R. Ross, ed.), p. 359. Otto Koeltz, Koenigstein.

Horner, R., Coyle, K.O., and Redburn, D.R. (1974). Univ. Alaska, Inst. Mar. Sci. Rep. R74-2.

Hsiao, S.I.C., Foy, M.G., and Kittle, D.W. (1977). *Can. J. Bot. 55*, 685.

Hufford, G.L. (1973). *J. geophys. Res. 78*, 2702.

Hustedt, F. (1930). *In* "Kryptogamen-Flora von Deutschland, Österreich und der Schweiz" (L. Rabenhorst, ed.), Vol. 7, No. 1.

Hustedt, F. (1959-62). *In* "Kryptogamen-Flora von Deutschland, Österreich und der Schweiz" (L. Rabenhorst, ed), Vol. 7, No. 2.

Johnson, M.W. (1956). The Arctic Institute of North America Tech. Paper No. 1. Washington, D.C.

Johnson, P.W., and Sieburth, J.McN. (1982). *J. Phycol. 18*, 318.

Mann, A. (1925). *Rep. Can. arct. Exped. IV(F)*, 1.

Paquette, R.G., and Bourke, R.H. (1974). *J. mar. Res. 32*, 195.

Parke, M., and Dixon, P.S. (1976). *J. mar. biol. Ass. U.K. 56*, 527.

Paulsen, O. (1908). Nordisches Plankton, Botanischer Teil, Part 18, 1. Lipsius & Tischer, Kiel.

Round, F.E. (1981). The Ecology of Algae. Cambridge Univ. Press, Cambridge.

Schiller, J. (1933). *In* "Kryptogamen-Flora von Deutschland, Österreich und der Schweiz" (L. Rabenhorst, ed.), Vol. 10, No. 1.

Schiller, J. (1937). *In* "Kryptogamen-Flora von Deutschland, Österreich und der Schweiz" (L. Rabenhorst, ed.), Vol. 10, No. 2.

Strickland, J.D.H., and Parsons, T.R. (1968). *Bull. Fish. Res. Board Can.* 167.

UNESCO (1966). "Monographs on Oceanographic Methodology, No. 1. Determinations of Photosynthetic Pigments in Sea-water." UNESCO, Paris.

U.S. Naval Oceanographic Office (1968). "Instructional Manual for Obtaining Oceanographic Data." Publ. 607. Washington, D.C.

Utermöhl, H. (1931). *Verh. int. Verein. theor. angew. Limnol. 5*, 567.

AN ANNUAL CARBON BUDGET FOR
AN ARCTIC KELP COMMUNITY

Kenneth H. Dunton

Institute of Water Resources
University of Alaska
Fairbanks, Alaska

I. INTRODUCTION

Most of what we know concerning the growth of seaweeds and their productivity in coastal systems is the result of studies in temperate climates. These studies have documented that marine algae make substantial contributions of carbon to coastal systems (Clendenning, 1971; Mann, 1972; Johnston et al., 1977; Dieckmann, 1980). In some cases, the productivity of the seaweeds matches or exceeds the highest estimated productivity of the phytoplankton. For example, in St. Margaret's Bay, Nova Scotia, the annual carbon input of the seaweeds averaged over the entire bay was 603 g m^{-2}, three times the phytoplankton production (Mann, 1972).

In arctic regions, the productivity of benthic macroalgae has been largely ignored and considered insignificant. This is due both to the limited exploration of the arctic shelf for benthic macroalgae and the nature of the physical environment, which in general is not favorable for the attachment and growth of macroalgae. But the recent discovery of a large area of kelp associated with a diverse and abundant invertebrate population off the north coast of Alaska (Dunton et al., 1982), and the work of Chapman and Lindley (1981) on the productivity of *Laminaria solidungula* in a kelp bed in the Canadian High Arctic, indicate that this carbon source may be more important than previously thought.

A complete annual carbon budget for an arctic kelp community has not yet been achieved. Although measurements of growth and productivity have been reported for the predominant alga *L. solidungula*, the effect of herbivores on standing crop have not been considered, and with the exception of phytoplankton production, the presence of other carbon inputs has not been ascertained. In temperate regions, these factors have been addressed by several investigators who have assessed the role of kelp carbon in the environment (Miller et al., 1971; Vadas, 1977; Carter, 1982; Newell et al., 1982).

THE ALASKAN BEAUFORT SEA:
ECOSYSTEMS AND ENVIRONMENTS

311

In the Canadian High Arctic, Chapman and Lindley (1981) found that the annual productivity of an arctic kelp bed dominated by *L. solidungula* was about 20 g C m^{-2}, roughly equivalent to phytoplankton production on an area basis in open water. Growth was greatest in the late winter and early spring, and was correlated with the concentration of inorganic nitrogen, as in most perennial kelps (Chapman and Lindley, 1980). Dunton *et al.* (1982) also studied the growth of *L. solidungula* during the same period and latitude as Chapman and Lindley (1980, 1981) but in the Alaskan Beaufort Sea. In an area known as the Boulder Patch in Stefansson Sound, Dunton *et al.* estimated annual productivity at 7 g C m^{-2}, but under substantially different light regimes. In the Boulder Patch, *L. solidungula* is subject to nine-month periods of complete darkness caused by the presence of a turbid ice canopy. The kelp relies on stored food materials to complete over 90 % of its annual linear growth (Dunton *et al.*, 1982).

The patchy occurrence of turbid ice in the Stefansson Sound Boulder Patch, however, exposes areas of seafloor to light throughout the fall and spring, resulting in changes in the growth and productivity of *L. solidungula*. The annual consumption of kelp by herbivores in the Boulder Patch had also not been estimated quantitatively. In temperate kelp communities, about 10% of the annual productivity is utilized directly by herbivores (Miller *et al.*, 1971; Newell *et al.*, 1982). The remaining 90% is released in dissolved or particulate form, and is a potential food source for filter-feeding animals by conversion into bacterial biomass (Lucas *et al.*, 1981; Stuart *et al.*, 1981).

In this report I summarize the sources of carbon available to consumers in the Boulder Patch and estimate the fraction of kelp carbon derived from *Laminaria solidungula* that is consumed by a common herbivore. Quantitative variations in the annual carbon input of *L. solidungula*—caused by differences in winter light availability—are presented and compared to the carbon contributions made by benthic microalgae, ice algae, and phytoplankton.

II. STUDY AREA

The Boulder Patch lies in Stefansson Sound, 20 km northeast of Prudhoe Bay in the Beaufort Sea off Alaska (Fig. 1), and consists of boulders and cobbles that occur in patches of various sizes and densities. Field studies were conducted at an acoustically marked site, dive site 11 (DS-11; 70°19.25'N, 147°35.1'W), on the eastern side of the Boulder Patch (Fig. 1). The rocks provide a substrate for a diverse assortment of invertebrates and several species of algae (Dunton *et al.*, 1982). The predominant brown alga is *Laminaria solidungula*, which constitutes over 90% of the brown algal biomass. The area covered by the Boulder Patch is approximately 2.03 x 10^7 m^2. Rock cover at DS-11 is estimated at 42%.

Water depths in the Boulder Patch range from 3 to 9 m. Freeze-up is usually complete by mid-October, and breakup begins in late June or early July. Ice thickness reaches a maximum of 2 m in early May before

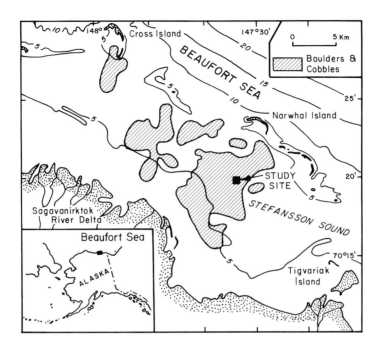

FIGURE 1. *The location of the Boulder Patch and dive site 11 (DS-11) in Stefansson Sound. Depth contours are shown in meters.*

deterioration of the ice canopy begins. Bottom water temperatures range from a nearly constant $-1.9^{o}C$ under the sea ice to $7^{o}C$ during the summer open-water period. Salinity ranges from 14 to 35 $^{o}/oo$ (Barnes et al., 1977).

III. MATERIALS AND METHODS

During ice-covered periods, divers entered the water through holes cut in the ice; a heated insulated hut placed over the dive hole functioned as a dive shelter and laboratory. Biological sampling and *in situ* experiments were conducted at approximately three-month intervals by a team of divers between July 1978 and November 1981. All plants and animals used in this study were haphazardly collected or tagged *in situ.*

A. Linear Growth and Productivity of L. solidungula

Linear growth in *Laminaria solidungula* was followed by punching holes in the base of the blade, above the meristematic region (Chapman and

Craigie, 1977). Since *L. solidungula* constitutes over 90 percent of the brown algal biomass, I focused on the productivity of this plant. The frond of *L. solidungula* is divided into distinct ovate blades of different sizes by constrictions which form annually. The growth of a new basal blade starts in November and continues until the following November, allowing an accurate measurement of the plant's annual production. Here I define a growth year (gwyr) as the period beginning 15 November one year and ending on 15 November the following year.

The formation of a constriction in November followed by the growth of a new blade segment made it possible to monitor linear growth in untagged plants subjected to different light conditions at another site. The distance from the junction of stipe and frond to the first constriction gave an accurate measurement of total linear growth since the preceding November. Linear growth in previous years was also assessed by measuring the distance between successive constrictions up the frond. In May 1979 two sites (DS-11, located under turbid ice near the tagged plants) and DS-11A (200 m distant under clear ice) were chosen for the comparative study. Over 40 plants were collected at each site for linear measurement in May and again following ice breakup in late July 1979.

To establish the relationship between basal (new) blade segment length and tissue biomass, about 20 plants were randomly collected in late April and again in early August 1980. In addition, length-to-biomass relationships were established for entire blades from a total of 87 plants collected in November 1979, 1980, and 1981. Production-to-biomass (P:B) ratios were obtained from plants collected in August at both sites and November 1979 at DS-11. The weight of the first blade segment (annual production) in each plant was divided by the weight of the remaining part of the blade (initial biomass). This fractional value was used to compute annual productivity for the growth year 1979 (gwyr 1979).

Net kelp carbon production (g m^{-2} yr^{-1}) was calculated by multiplying the P:B ratio by a mean standing crop of 137 g m^{-2}, the percent dry weight of wet weight (17%), and percent carbon of dry weight (31%). The standing crop of the kelp and the conversion factors are from Dunton *et al.* (1982). Because growth of the stipe rarely exceeds 0.5 cm year^{-1}, it was considered negligible in comparison to annual blade production.

B. Biomass and Productivity of Microalgae

Quantitative sampling of the biota by scraping and airlifting all organisms in 0.05-m^2 areas on rock surfaces resulted in the collection of chained diatoms. *In situ* observations confirmed that these diatoms were attached to epilithic organisms and thus easy to collect without significant loss using an airlift. Water flow in the airlift is created in a 1-m long tube by the expansion of compressed air introduced at the bottom of the tube.

In the laboratory, filamentous diatoms were sorted and removed from other material, blotted to remove excess water, and weighed to the nearest 0.001 g. Biomass of the benthic microalgae is expressed per m^2 of rock substrate and is based on 50 replicate samples collected randomly over a three-year period. Carbon content of the microalgae was determined using a Leco TC-12 automatic carbon determinator.

Biomass and productivity data for ice algae and phytoplankton were obtained from Schell et al. (1982), from his site BP in Stefansson Sound, which is close to DS-11.

C. Macroalgal Consumption by Herbivores

The chiton Amicula vestita is an active herbivore in the Boulder Patch, and grazes almost exclusively on L. solidungula. The animal leaves distinct grazing scars on blade and stipe tissues of the plant. Examination of over 100 plants collected in 1980 showed 42% with recent grazing scars on stipes or blades. Biomass and density of A. vestita in the Boulder Patch have been reported by Dunton et al. (1982). Since animals smaller than about 1 g in wet weight have been observed to feed primarily on detritus, not kelp, the raw data were reanalyzed to determine the biomass of chitons greater than 1 g in wet weight per m^2. The mean biomass of all animals heavier than 1 g wet weight was also calculated.

Feeding experiments were conducted in the laboratory to determine the amount of tissue consumed by A. vestita. Three chitons were placed in separate small mesh cages, 17 x 12 x 12 cm, and given preweighed fresh blades from L. solidungula. The cages were placed in a 200-l aquarium kept in darkness at a constant temperature of $1°C$ ± 1.5. Three feeding experiments were conducted lasting 67, 51, and 58 days. At the end of each experiment, the amount of kelp consumed by each chiton was determined by weighing the remaining blade. Decreases in the wet weight of the blade tissue (due to respiratory loss of carbon) was small, accounting for less than 10% of the mean daily uptake of kelp tissue by the chitons.

D. Light Energy

Photon flux density was measured with a LI-185 quantum radiometer/photometer with a LI-192S underwater quantum sensor (LI-COR, Inc., Lincoln, NE).

IV. RESULTS

A. Linear Growth in L. solidungula Beneath Turbid and Clean Ice

The light available for photosynthesis between October and July is dependent on the conditions of ice formation, so the effective day length varies from year to year. In Stefansson Sound, considerable amounts of sediment are entrapped in the ice canopy during its formation in October, resulting in what is referred to as "turbid ice" (Reimnitz and Dunton, 1979; Dunton et al., 1982). Turbid ice blocks light transmission completely, even during periods of 24-hour daylight. For example, photon flux density on the bottom at DS-11 in May 1979 under turbid ice was below the detection limits of the LI-COR instrument (less than about 0.05 $\mu E s^{-1} m^{-2}$).

Despite the widespread distribution of turbid ice in Stefansson Sound, "windows" of clean ice allowed some light to penetrate to the bottom. Photon flux density in May 1979 at one such location about 200 m from DS-11 (DS-11A) ranged from 1.8 to 3.5 $\mu E s^{-1} m^{-2}$. Photon flux density under a similar ice canopy in April was 2.4 $\mu E s^{-1} m^{-2}$, which represented 0.57% of the total incident illumination at the upper surface of the ice (D. M. Schell, personal commun.).

The growth of *L. solidungula* over the past three years at two locations, DS-11 and DS-11A, with differing ice cover during the 1978-79 winter, is shown in Fig. 2. With the exception of the latest growth cycle at DS-11A, where plants were exposed to light under the ice canopy in the spring of 1979, linear growth between years and between sites is not significantly different (analysis of variance (ANOVA) completely randomized design, Student-Newman-Keuls-test (SNK), $p > 0.05$). This similarity in growth suggests that turbid ice was present at both locations in the winter seasons of 1976-1977 and 1977-1978. The greater growth of plants under clean ice in 1979 is significant (SNK, $p < 0.05$) and suggests

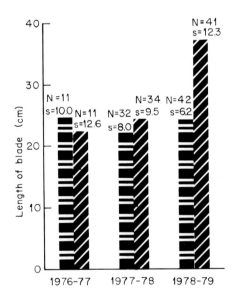

FIGURE 2. *Annual linear growth in L. solidungula at two adjacent sites since 1976-1977. The number of plants sampled (N) and the standard deviation (s) are listed above the bars. The difference in linear growth between the two sites in 1978-1979 is significant (ANOVA, p <.05). DS-11 is indicated by the left bar, DS-11A by the right bar.*

that these plants are actively photosynthesizing under the ice canopy. The mean length of the basal blades in plants that had grown under clean ice at DS-11A was 37.7 cm compared to 24.1 cm under turbid ice at DS-11.

B.   Productivity in *L. solidungula* Exposed to Light Under the Ice Canopy

The productivities of *L. solidungula* exposed to winter light at DS-11A and winter darkness at DS-11 in 1979 were determined by correlating blade lengths with biomass. Figure 3 shows the relationship between basal blade segment length and biomass in plants collected in late April and early August 1980. Measured first blade segments at DS-11 and DS-11A in 1979 were transformed to biomass using this correlation. Figure 4 shows the relationship between total blade length and biomass in plants collected in November 1979, 1980, and 1981. A similar relationship between blade length X and biomass Y, expressed by the equation $Y = 0.03X^{1.67}$ was obtained by Chapman (1981; personal commun.) for *L. solidungula* in the Canadian arctic. Using a regression equation derived from Fig. 4, the initial blade biomasses of plants collected at DS-11A and DS-11 in 1979

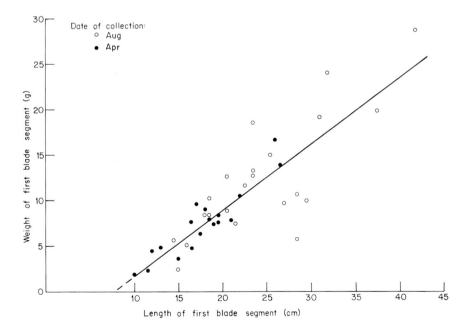

FIGURE 3.   The relationship of basal blade segment length and biomass in *L. solidungula* plants haphazardly collected in late April (open circles) and early August (solid circles) at DS-11 in 1980; correlation coefficient, 0.85.

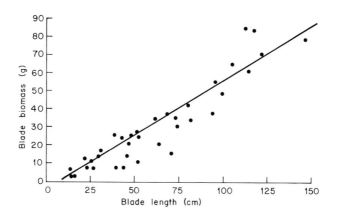

FIGURE 4.   Total plant blade biomass as a function of total blade length
(all blade segments and eroded tip) in plants collected in November 1979,
1980 and 1981 (not all data points shown); correlation coefficient, 0.88.

were calculated from measured lengths.   From these data, I obtained an
average production (basal blade segment biomass)-to-biomass (initial blade
weight of entire plant) ratio (P:B) for each plant.

    The productivity of the plants collected at DS-11 and DS-11A based on
length-to-biomass correlations is shown in Table I. Mean production was

TABLE I. Mean Production (g ± 95% confidence limits) and Production-to-
Biomass Ratios (P:B) of Plants Collected at Sites DS-11 (turbid
ice canopy) and DS-11A (clean ice canopy) in the 1979 Growth
Year.

Site	Mean production		P:B ratio	
	15 Nov 1978 to 22 May 1979	15 Nov 1978 to 22 July 1979	15 Nov 1978 to 19 July 1979	15 Nov 1978 to 15 Nov 1979
DS-11	10.12 ± 1.16	11.98 ± 1.28	0.80 ± 0.27	0.91 ± 0.22
DS-11A	21.77 ± 1.86	21.42 ± 2.64	1.26 ± 0.44	--

about 60% greater in *L. solidungula* at DS-11A than at DS-11. For the periods from November to May and from November to July, mean biomass production at DS-11 and DS-11A was significantly different (p <0.05). Differences in P:B ratios at the two sites for the period from November to July were not significant (0.05 <p <0.10), mainly due to anomalously high P:B ratios for some plants at DS-11. Plants at DS-11 produced additional tissue between May and July, but no tissue production was noted for plants at DS-11A during the same period. The lack of growth in plants at DS-11A may be explained by a shift to food storage starting in May, when nutrient concentrations begin to drop, but I have no tissue-density data to support this hypothesis. The mean production-to-biomass (P:B) ratio by late July was 1.26 at DS-11A compared to 0.80 at DS-11. The P:B ratio for the entire growth year (gwyr 1979) at DS-11 was 0.91, an increase of 14% from July. Since I did not return to DS-11A in November, it was not possible to calculate the P:B for gwyr 1979 at this site.

C. Consumption of Kelp by the Chiton Amicula vestita

Amicula vestita appears to be the only herbivore in the Boulder Patch that noticeably consumes kelp blades (Fig. 5). The animal has a mean density of 16 individuals m^{-2} and a biomass of 11 g m^{-2} on rock substrata at DS-11 (Dunton et al., 1982). Of this biomass, 86% is represented by animals whose individual wet weights are greater than 1 g. The mean biomass of this group is 2.11 g.

The daily rate of consumption of kelp blades by the same three chitons during three different periods (Fig. 6) varied from 13 to 46 mg of fresh kelp per day and was related to live body biomass. The mean annual consumption-to-biomass ratios (C:B) for the three animals were similar (3.82, 3.34, and 2.73), giving an average of 3.3. No changes in the wet weights of the animals were found.

These data allow a calculation of the annual consumption of kelp by *A. vestita*. The biomass of *A. vestita* per m^2 of seafloor is 4 g m^{-2} (11 g m^{-2} multiplied by 0.86, the percentage of animals greater than 1 g wet weight, and 0.42, the percentage rock cover at DS-11). From Fig. 6, the C:B ratio of an average chiton (wet weight, 2.11 g) is 3.71. Assuming that carbon makes up 5.3% of a kelp's wet weight (Dunton et al., 1982), the annual ingestion of kelp by *A. vestita* is roughly 0.8 g C m^{-2}.

D. Benthic Microalgal Production

Benthic diatoms appeared in 62% of the benthic samples collected. Mean biomass was 3.4 g m^{-2}. The percentage dry weight to wet weight of the diatoms was 21%, and the percentage carbon of dry weight was 8%. Based on these data, the mean standing crop of benthic diatoms is approximately 0.6 g C m^{-2}. I found no correlation between biomass and month of collection over the three-year period. Filamentous diatoms consisted primarily of *Amphipleura*. These were most common on bryozoans, hydroids, and branched and foliose red algae, such as *Rhodomela confervoides*, *Phyllophora truncata*, and *Phycodrys rubens* (Fig. 7).

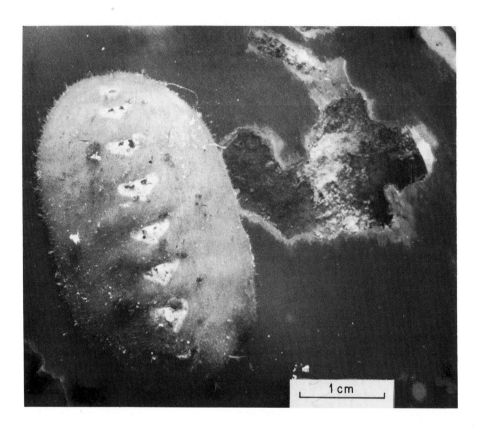

FIGURE 5. The chiton <u>Amicula vestita</u> is a common and obvious grazer of
Laminaria solidungula in the Boulder Patch.   Areas which have been
previously grazed are visible to right of the chiton.

V. DISCUSSION

A. Annual Kelp Carbon Production

Beneath clean ice, and in response to increasing day length in the late
winter and early spring, linear growth in   L. solidungula is about 55%
greater than under turbid ice, resulting in a substantial increase in the
standing crop of these plants.   The mean P:B ratio of plants exposed to
light (under clean ice) measured in July was 1.26, compared to a ratio of
0.80 for plants exposed to dark (under turbid ice). The measured increase
of the P:B ratio in dark-exposed plants between July and November was

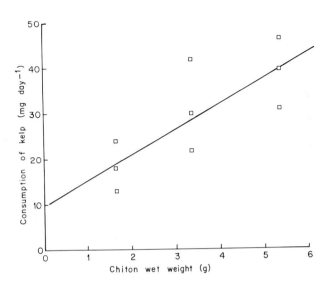

FIGURE 6. Relationship between herbivore biomass and consumption of kelp (mg day^{-1}) for _Amicula vestita_. Each point represents mean consumption over time periods ranging from 51 to 67 days; correlation coefficient, 0.79.

about 14%. Applying this increase to plants at DS–11A (subject to the same late summer and early fall light conditions), an annual P:B ratio of 1.44 was obtained. Using this P:B ratio, the productivity of the kelp was about 10.4 g C m^{-2}yr^{-1}. This compares to about 6.6 g C m^{-2}yr^{-1} under normal winter conditions of darkness under the ice canopy. Since turbid ice is patchily distributed in Stefansson Sound, it is likely that the productivity of the kelp falls between the minimum and maximum values listed above.

The P:B ratios of _L. solidungula_ in the Boulder Patch varied between 0.9 and 1.4; this is considerably lower than other species of _Laminaria_ in more temperate regions. In St. Margaret's Bay, Nova Scotia, for example, the P:B ratio of _L. longicruris_ varies between 7.2 and 10.7 (Mann, 1972), almost an order of magnitude greater. In a kelp bed off the coast of South Africa near Cape Town, the P:B ratio of _L. pallida_ is about four times that of _L. solidungula_ (Newell et al., 1982). But despite the much lower production in the Boulder Patch compared to kelp beds in temperate regions, the proportional split between phytoplankton and kelp production remains nearly the same. In the Beaufort Sea, the presence of an isolated source of kelp carbon provides a unique opportunity to assess its importance to both invertebrate and vertebrate animals.

A portion of this annual production is consumed directly by _Amicula vestita_. On a m^2 basis, _Amicula_ consumes nearly four times its weight in fresh kelp annually, which likely accounts for most (but not all) of its

FIGURE 7.   A cluster of filamentous diatoms attached to the hydroid
Sertularia.

dietary intake.   The kelp carbon utilized by *Amicula* (0.8 g m^{-2}yr^{-1})
represents between 8 and 12% of annual production by *L. solidungula,*
depending on the character of the ice canopy.   Off the coast of Nova
Scotia, Miller *et al.* (1971) found that the annual consumption of kelp by

herbivores (sea urchins and snails) was about 9% of total kelp production. Newell *et al.* (1982) calculated that urchins consumed about 12% of total kelp production off the west coast of South Africa. The fraction of kelp not consumed by herbivores enters the foodweb in dissolved or particulate form, where it becomes a direct or a microbially disguised indirect source of carbon to benthic consumers.

B. *An Annual Carbon Budget for the Boulder Patch*

In years characterized by the presence of a turbid ice canopy, only three sources of marine carbon are available to consumers in the Boulder Patch: kelp, phytoplankton, and benthic microalgae. The spring bloom of ice microalgae, which is common in most arctic coastal areas (Alexander, 1975; Hsiao, 1980) does not occur because of the lack of light caused by the presence of turbid ice. But in the absence of turbid ice in Stefansson Sound, productivity of the ice algae is about 5.2 g m^{-2}yr^{-1} (Schell *et al.*, 1982).

The productivity of benthic microalgae in the absence of turbid ice is about 0.4 g C m^{-2}yr^{-1}, based on a P:B ratio of 7 (Chapman, 1981). Although it seems unlikely that this P:B ratio would hold for arctic regions, Matheke and Horner (1974) reported high productivities for benthic microalgae near Barrow, in which chlorophyll *a* concentrations increased nearly an order of magnitude during a three-month summer period. Since measured productivity rates in the benthic microalgae are undetectable under the ice canopy (Horner and Schrader, 1981), carbon production occurs only during the summer open-water months, between July and October. In the presence of turbid ice, which is accompanied by extremely high water turbidity during breakup and freeze-up, I assume that annual benthic microalgal productivity is cut by about 25 percent. The carbon contribution made by the phytoplankton is 5.0 g m^{-2}yr^{-1} in the Boulder Patch (Schell *et al.*, 1982) and is largely unaffected by turbid ice. Based on these data, it is possible to compare the sources of carbon and their magnitudes, either in the presence or absence of turbid ice (Fig. 8).

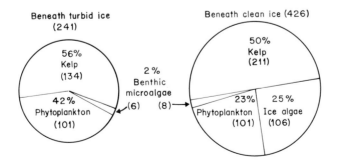

FIGURE 8. *The sources of carbon and their inputs (g x 10^6) to the Stefansson Sound Boulder Patch in years characterized by either turbid ice or clean ice (total area 2.03 x 10^7 m^{-2}).*

In both cases, kelp is clearly a major source of carbon to consumers in the Boulder Patch. Carbon derived from phytoplankton, ice algae, and benthic microalgae balance the input from kelp, but the benthic microalgal component is unimportant. Annual growth increments in *L. solidungula* support observations that turbid ice is a frequent occurrence in the Boulder Patch (Dunton *et al.*, 1982). Thus, it appears that 98% of the carbon available to consumers is derived from either kelp or phytoplankton in most years. Particulate organic matter (the fragmented portion) derived from kelp is about 60% (Newell *et al.*, 1982) of the total released to the environment. The importance of this fraction to consumers in an isolated community dominated by filter feeders, which make up 65% of the total standing stock (Dunton *et al.*, 1982), is now being investigated.

Recent work has shown that the carbon in the particulate fraction is converted into bacterial biomass with an efficiency of between 8% and 11% (Stuart *et al.*, 1981; Robinson *et al.*, 1982). But it is not known to what extent consumers depend on the protein enrichment of fragmented kelp by bacterial mineralization, although Newell *et al.* (1982) reported that bacterial degradation of detritus is important to a kelp community on the west coast of South Africa. Findlay and Tenore (1982) recently demonstrated that a polychaete detritivore derived a major portion of its nitrogen directly from seaweed detritus, not the microbial population that colonized the detritus. It is possible that in the Boulder Patch, the low carbon flux (in comparison to temperate latitudes) may force many consumers to develop a high assimilatory efficiency for kelp detritus. Consumers so adapted would also be able to utilize a year-round source of kelp carbon, especially during the period of total ice cover, when phytoplankton production virtually ceases.

ACKNOWLEDGMENTS

D. M. Schell gave valuable support and criticism in the synthesis of this paper and supplied the phytoplankton and ice algae productivity data. This study is part of the Boulder Patch research program carried out jointly by the University of Alaska, Western Washington University, and the United States Geological Survey. I thank S. V. Schonberg, D. Parrish, and P. Ziemann for many valuable discussions, A. C. Broad and G. Walker for supplying field equipment, and P. Plesha, J. Hanes, G. Cinkovich, J. Olson, and D. Pope for diving assistance. My thanks also to A. C. Paulson, W. J. North, and three anonymous referees for outstanding comments on the manuscript, and to L. Flood for typing the manuscript.

REFERENCES

Alexander, V. (1975). *In* "The Coast and Shelf of the Beaufort Sea" (J. C. Reed and J. E. Slater, eds.), p. 609. Arctic Institute of North America, Arlington, VA.

Barnes, P. W., Reimnitz, E., and McDowell, D. (1977). *Open-File Rept. 77-477*, U.S. Geol. Surv.
Carter, R. A. (1982). *Mar. Ecol. Prog. Ser. 8*, 9.
Chapman, A. R. O. (1981). *Mar. Biol. 62*, 307.
Chapman, A. R. O., and Craigie, J. S. (1977). *Mar. Biol. 40*, 197.
Chapman, A. R. O., and Lindley, J. E. (1980). *Mar. Biol. 57*, 1.
_____ (1981). *In* "10th International Seaweed Symposium, Proceedings" (T. Levring, ed.), p. 247. Walter de Gruyter, Berlin.
Clendenning, K. A. (1971). *In* "The Biology of Giant Kelp Beds (Macrocystis) in California" (W. J. North, ed.), p. 259. Nova Hedwigia, Vol. 32.
Dieckmann, G. S. (1980). *Botanica Mar. 23*, 579.
Dunton, K. H., Reimnitz, E., and Schonberg, S. (1982). *Arctic 35*, 465.
Findlay, S., and Tenore, K. (1982). *Science 218(4570)*, 371.
Horner, R. A., and Schrader, G. C. (1981). *In* "Environmental Assessment of the Alaskan Continental Shelf, Final Reports" Vol. 13, p. 65. NOAA, Boulder.
Hsiao, S. I. C. (1980). *Arctic 33*, 768.
Johnston, C. S., Jones, R. G., and Hunt, R. D. (1977). *Helgoländer Wiss. Meeresunters. 30*, 527.
Lucas, M. I., Newell, R. D., and Velimirov, B. (1981). *Mar. Ecol. Prog. Ser. 4*, 43.
Mann, K. H. (1972). *Mar. Biol. 14*, 199.
Matheke, G. E. M., and Horner, R. (1974). *J. Fish. Res. Board Can. 31*, 1779.
Miller, R. J., Mann, K. H., and Scarratt, D. J. (1971). *J. Fish. Res. Board Can. 28*, 1733.
Newell, R. C., Field, J. G., and Griffiths, C. L. (1982). *Mar. Ecol. Prog. Ser. 8*, 103.
Reimnitz, E., and Dunton, K. H. (1979). Diving observations of the soft ice layer under the fast ice at DS-11 in the Stefansson Sound Boulder Patch. *In* "Environmental Assessment of the Alaskan Continental Shelf: Principal Investigators' Reports for the Year Ending March 31, 1979." OCSEAP, Boulder. 20 p.
Robinson, J. D., Mann, K. H., and Novitsky, J. A. (1982). *Limnol. Oceanogr. 27*, 1072.
Schell, D. M., Ziemann, P. J., Parrish, D. M., Dunton, K. H., and Brown, E. J. (1982). *In* "Environmental Assessment of the Alaskan Continental Shelf Final Reports." NOAA, Boulder. (In press.)
Stuart, V., Lucas, M. I., and Newell, R. C. (1981). *Mar. Ecol. Prog. Ser. 4*, 337.
Vadas, R. L. (1977). *Ecol. Monogr. 47*, 337.

BACTERIAL POPULATIONS OF THE BEAUFORT SEA[1]

Ronald M. Atlas

Department of Biology
University of Louisville
Louisville, Kentucky

Robert P. Griffiths

Department of Microbiology
Oregon State University
Corvallis, Oregon

I. INTRODUCTION

Prior to 1975 there was little information on the bacteria of the Beaufort Sea. Since that time several studies have been conducted on many aspects of the ecology of the Beaufort Sea, including bacteriology. Like other polar seas, the Beaufort Sea is ice-covered for much or all of the year and has continuous sunlight during the summer and a continuous darkness during the winter. Bacteria living in arctic ecosystems must be adapted to these conditions. The studies discussed here consider the bacteria that occur in the Beaufort Sea with respect to their numbers, diversity, physiological adaptations, metabolic activities, and the potential impact of oil and gas development on them.

II. METHODS

A. Sample Collection

Samples were collected at stations in the western Beaufort Sea during late summer (August-September) 1975 and 1976 and during late winter

---

[1] *Published as technical paper No. 7071 Oregon Agricultural Experiment Station.*

(April) 1976 (see Kaneko et al., 1977, for sampling locations). The April sampling was considered to be a late winter sampling as the sea was still covered with thick ice and the spring phytoplankton bloom had not yet occurred. Samples also were collected during summer (September) 1977 and 1978. Surface water samples (1 m depth) were collected with a Niskin butterfly sterile water sampler. Surface sediment samples were collected with a bottom grab during 1975-1976 and with a Smith MacIntyre grab sampler during 1977 and 1978. Ice samples were collected by chipping out pieces of surface ice and placing them in a sterile container. The ice was allowed to melt and maintained at temperatures below 5°C. Winter water and sediment samples were collected through holes drilled in the ice. All samples were placed on ice in sterile containers using aseptic techniques and returned to the laboratory for processing within a few hours (1-10 hr) of collection. The shortest processing times were during summer 1976 and summer 1978 when processing was done aboard the sampling vessel; longer processing times occurred when samples were collected by helicopter or small craft and flown to the Naval Arctic Research Laboratory at Barrow for processing. The longest processing times occurred during the 1977 cruise when sediment samples were shipped on ice to Oregon State University, where they were analyzed after storage times of up to 4 weeks.

B. Enumeration of Microorganisms

Enumerations of bacterial populations were done using both direct count (total count), plate count (viable count), and most probable number (hydrocarbon utilizer) procedures. For direct counts, samples were preserved in 50% formaldehyde. Samples were filtered through 0.2-μm cellulose nitrate black filters and stained with acridine orange according to the procedure of Daley and Hobbie (1975). Samples were viewed under an epifluorescence microscope with a BG-12 exciter filter and 0-530 barrier filter. Ten fields per filter and two filters per sample were viewed and the counts averaged.

For viable plate counts, surface spread inoculations from serial dilutions were used. For some sea ice and water samples, bacteria were concentrated by filtration through 0.45-μm filters, which were immediately placed onto agar plates. Marine agar 2216E (Difco) was used to enumerate viable heterotrophic microorganisms. Media and dilution blanks (Rila sea salt solution) were cooled to 4°C before platings were performed. Replicate plates were incubated aerobically at 4°C for 3 weeks and colonies were counted with a binocular microscope so that even slowly growing bacteria were enumerated. All platings were done in triplicate and the average reported. The choice of media and incubation conditions for the enumeration of viable bacteria was made after testing six different media, several different incubation conditions, and various incubation times with 50 water and sediment samples collected in the Beaufort Sea; marine agar 2216 always gave the highest counts of colony-forming units.

A Most Probable Number (MPN) procedure was used to estimate numbers of hydrocarbon–utilizing microorganisms (Atlas, 1979). Dilutions of samples were added to 30-ml stoppered serum vials containing 1 ml autoclaved Bushnell Haas broth (Difco) at 5X concentration, and 30 μl filtered (0.2 μm filter), sterilized Cook Inlet crude oil spiked with $1\text{-}^{14}\text{C}$ n-hexadecane (s.p. act. = 0.9 $\mu\text{Ci ml}^{-1}$ oil). A 3-tube MPN procedure was used. Following incubation at 5°C for 3 weeks, the solutions were treated with concentrated KOH to stop microbial activity, then acidified with concentrated HCl and the $^{14}\text{CO}_2$ recovered by purging the vials with air and trapping the $^{14}\text{CO}_2$ in 10 ml Oxifluor CO_2 (New England Nuclear). Counting was with a liquid scintillation counter. Background-corrected counts greater than or equal to 2 times control were considered positive; counts less than 2 times control were considered negative. The most probable number of hydrocarbon-degrading microorganisms was determined from the appropriate MPN tables.

C. Characterization of Bacterial Populations

Approximately 300 phenotypic characteristics were determined for each isolate, including: morphologies of both cells and colonies, physiological characteristics such as tolerance to temperature, salt, and pH, and biochemical tests such as determination of a variety of enzymatic activities, and nutritional characteristics such as the abilities to utilize a large number of biochemically diverse substrates. Details of the test procedures have been described previously (Kaneko et al., 1979). Data were coded and processed for computer analysis (Krichevsky, 1979) and subjected to cluster analysis to determine taxonomic groupings (phenotypic clusters) using the Jaccard coefficient (S_j) and average linkage clustering (TAXON computer program) (Sneath and Sokal, 1973; Walczak and Krichevsky, 1980). Feature frequencies for each phenotypic characteristic were calculated using the computer program FREAK (Walczak, 1979). The resultant data from the feature frequency and cluster analyses were used to calculate taxonomic diversity, physiological tolerance, and nutritional utilization indices of the populations in the bacterial community as described below.

D. Physiological Tolerance Indices

Physiological tolerance indices describe the capacity of the bacterial community to tolerate deviations from ambient conditions of temperature, salinity, and pH (Hauxhurst et al., 1981). Ambient conditions were considered as 5°C, 3% NaCl, and pH 8, which approximate both environmental and isolation conditions. Physiological tolerance indices were calculated for temperature (P_T), salinity (P_S), and pH (P_H). The basic formula used to calculate the physiological tolerance indices is:

$$P_x = \Sigma Gx_i \, n^{-1}$$

where P_x is the tolerance index for x, Gx_i represents the proportion of the population indicated by the number of isolates capable of growth at the specified condition of parameter x, and n represents the number of specified conditions of x examined. The tolerance indices range from 0 to 1. An index near 0 represents stenotolerance, with growth restricted to ambient conditions. An index near 1 represents eurytolerance, indicating that the indigenous populations can grow under a wide range of conditions.

E. Nutritional Utilization Indices

Indices were calculated to assess the nutritional versatility of bacterial communities. Separate indices were calculated for carbohydrates, alcohols, carboxylic acids, amino acids, and hydrocarbons. A total nutritional utilization index also was calculated to describe the overall nutritional diversity. Each index was calculated by summing the number of substrates that could be utilized by any member population and dividing by the total number of substrates within that class. Details of the substrates employed in these procedures have been reported previously (Hauxhurst *et al.*, 1981). An index N_x of 1, for any substrate class x, indicates that all substrates included in that class could be utilized by some member population of the bacterial community. An N_x near 0 indicates a lack of versatility.

F. Taxonomic Diversity Index

The number of taxonomic groups and the number of individuals within each group, determined by the cluster analyses, were used to calculate the Shannon diversity index H' (Kaneko *et al.*, 1977; Shannon, 1948). The formula

$$H' = CN^{-1}(N\log_{10} N - \Sigma n_i \log_{10} n_i)$$

was used, where $C = 3.3219$, N = total number of individuals, and n_i = total number of individuals in the ith taxonomic grouping (Kaneko *et al.*, 1977; Lloyd *et al.*, 1968). A Shannon diversity index above 3.0 is considered high, whereas an index under 3.0 for bacterial communities generally is indicative of environmental stress. Analyses of variance and the Duncan multiple mean comparison procedure were used to assess statistical significance. $P < 0.05$ was considered as indicating a significant difference.

G. Nitrogen Fixation

The method used to measure nitrogen fixation rates in the sediments was a modification of the acetylene blocking technique described by Stewart (1967). Details of the procedures used have been described elsewhere (Haines *et al.*, 1981).

H. Carbon Assimilation and Adenylate Concentrations

Relative microbial activities were measured by the incorporation and respiration of ^{14}C-labeled glucose and glutamic acid by microorganisms

(Griffiths *et al.*, 1977; 1978). Substrate utilization was measured at one substrate concentration. The resulting relative microbial activity data are highly correlated with potential heterotrophic activity (V_{max}) estimates made by the multiconcentration method (Griffiths *et al.*, 1972). The sediments were diluted 1,000 fold with sterile seawater prior to analyses. The methods used to measure ATP, ADP, and AMP were essentially those of Bulleid (1977). These analyses have been described previously (Griffiths *et al.*, 1981). Organisms larger than 2 mm in diameter were excluded.

III. RESULTS

A. Enumeration

Enumeration of bacterial populations in the Beaufort Sea indicates that bacterial numbers decline somewhat during winter, especially in surface waters (Table I). As in other marine ecosystems, numbers of bacteria are highest in sediment, lower in water, and lowest in ice; also, the numbers of viable bacteria are several orders of magnitude lower than the total numbers of bacteria directly counted. Numbers of hydrocarbon-utilizing bacteria represent only a small proportion of the total population,

TABLE I. *Counts of Bacteria ± Standard Deviations (g^{-1} dry wt.).*

	Date	Number of samples	Direct count	Viable count	Hydrocarbon utilizers
Ice	Winter 1976		$9.9 \pm 8.2 \times 10^4$	$6.6 \pm 1.8 \times 10^1$	-
Water	Summer 1975	40	$8.2 \pm 7.2 \times 10^5$	$9.6 \pm 4.8 \times 10^3$	-
	Winter 1976	20	$1.8 \pm 1.3 \times 10^5$	$6.1 \pm 7.0 \times 10^2$	-
	Summer 1976	20	$5.2 \pm 3.9 \times 10^5$	$5.0 \pm 3.2 \times 10^4$	-
	Summer 1978	50	$6.7 \pm 4.9 \times 10^5$	$3.5 \pm 2.9 \times 10^4$	$2.6 \pm 1.3 \times 10^1$
Sediment	Summer 1975	30	$6.2 \pm 1.1 \times 10^8$	$2.0 \pm 1.1 \times 10^6$	-
	Winter 1976	20	$3.7 \pm 1.0 \times 10^8$	$2.5 \pm 1.9 \times 10^5$	-
	Summer 1976	20	$2.1 \pm 0.9 \times 10^9$	$8.3 \pm 6.7 \times 10^6$	-
	Summer 1978	40	$1.6 \pm 0.8 \times 10^9$	$5.3 \pm 3.2 \times 10^6$	$2.5 \pm 2.2 \times 10^4$

a fraction of 1 percent. Compared to other Alaska continental shelf regions, the numbers of viable bacteria in surface waters are significantly higher in the Beaufort Sea. For example, the average number of viable bacteria enumerated in the northern Bering Sea surface waters during several spring-summer sampling cruises was 6.9×10^2, which is lower by over an order of magnitude than for comparable Beaufort Sea surface water samples. No such differences, however, appear in direct counts of water or sediment, nor in viable counts of sediment bacterial populations. The evidence suggests that the numbers of viable bacteria in the Beaufort Sea, at least those enumerated on marine agar, are an order of magnitude higher than the numbers of viable bacterial in nearby subpolar seas. Results using the INT method (Zimmermann et al., 1978) indicate that 1 percent or less of the bacteria are enumerated by acridine orange direct microscopy, suggesting that most viable bacteria are enumerated on marine agar. Additionally, counts of oligoheterotrophic bacteria are consistently several orders of magnitude lower than those for the copiotrophic bacteria enumerated on marine agar.

B. Taxonomic Diversity

The dominant bacterial populations isolated from the Beaufort Sea can be placed into seven categories (Kaneko et al., 1979). They appear to be taxonomically different species from those found in temperate marine ecosystems. *Flavobacterium* and *Microcyclus* species are dominant in surface waters during summer. *Vibrio* species represent a major taxonomic group in both sediment and water. *Acinetobacter* and *Alcaligenes* species occur in smaller proportions. Many of the Gram-negative pleomorphic bacteria could not be assigned to previously defined genera.

A relatively high degree of taxonomic diversity was characteristic of the marine bacterial communities of the Beaufort Sea (Table II). Taxonomic diversity was significantly greater in sediment than in water communities and significantly greater in summer than winter. Similar diversities have been found in subarctic marine ecosystems in sediment (Hauxhurst et al., 1981), but there somewhat higher diversities occur in

TABLE II. Taxonomic Diversities of Bacterial Populations.

Location	Date	Shannon diversity index H'
Ice	Winter 1976	3.0
Water	Summer 1975	2.6
	Winter 1976	2.1
	Summer 1976	2.6
	Summer 1978	2.6
Sediment	Summer 1975	3.5
	Winter 1976	3.6
	Summer 1976	4.0
	Summer 1978	4.1

TABLE III. Physiological Tolerance Indices of Surface Water and Sediment Bacterial Communities in Beaufort Sea Ecosystems Sampled in 1976.

Index	Summer		Winter	
	Water	Sediment	Water	Sediment
P_T (temperature)	0.60	0.55	0.56	0.50
P_H (pH)	0.62	0.72	0.65	0.70
P_S (salinity)	0.26	0.35	0.30	0.35

surface waters than were found in the Beaufort Sea (Kaneko et al., 1978). Also, while there are seasonal shifts in bacterial diversity in Beaufort Sea surface water communities, in the subarctic no comparable variations have been observed. The lower taxonomic diversity in arctic waters during winter may reflect the stress from arctic conditions, including the limited substrates available from phytoplankton.

Definite geographic trends in the diversity of bacterial communities were also observed. During summer, diversity is greatest in the western Beaufort Sea, whereas during winter it is lowest there. Community diversity in a given habitat was found to be steady from year to year at a given season, although the individual populations varied from one year to the next. It appears that there is a maximum taxonomic diversity for a bacterial community occupying a given habitat, but that different populations can occupy the niches of that ecosystem.

C. Physiological Tolerance Indices

The arctic populations are somewhat less tolerant of temperature fluctuations than subarctic populations, particularly with respect to tolerance of high temperatures. Most Beaufort Sea bacteria, however, are not true psychrophiles; psychrotrophs, capable of growth at temperatures of $25^{\circ}C$, make up over 85% of the bacterial populations in these ecosystems. Relatively high physiological tolerance indices, nevertheless, are characteristic of Beaufort Sea bacterial communities (Table III) despite the relatively low annual variations in temperature, salinity, and pH in these ecosystems. The indigenous bacterial populations are quite tolerant of fluctuations in temperature, salinity, and pH beyond the limits to which they are exposed in nature.

D. Nutritional Utilization Indices

The nutritional utilization indices suggest that the bacterial populations of the Beaufort Sea are relatively versatile (Table IV).

However, hydrocarbons are not metabolized by the dominant populations of bacteria in water or in sediment. Sediment populations are capable of growing on more organic substrates (50 %) than water populations. The major seasonal difference in the nutritional utilization indices occurred in the abilities of surface-water populations to utilize carbohydrates. During summer, carbohydrates were the most readily utilized substrates, followed by carboxylic and amino acids. During winter, carboxylic and amino acids were utilized by a greater proportion of the community than were carbohydrates. This shift in nutritional capabilities presumably reflects a shift in the available food resources.

TABLE IV. Nutritional Utilization Indices of Surface Water and Sediment Bacterial Communities in Beaufort Sea Ecosystems Sampled in 1976.

Index	Summer		Winter	
	Water	Sediment	Water	Sediment
N_c (carbohydrates)	0.55	0.67	0.40	0.56
N_a (alcohols)	0.29	0.56	0.31	0.50
N_{ca} (carboxylic acids)	0.37	0.69	0.45	0.65
N_{aa} (amino acids)	0.32	0.40	0.38	0.42
N_h (hydrocarbons)	0.00	0.00	0.00	0.00
N_T (all substrates)	0.31	0.46	0.31	0.45

E. Carbon Cycling and Adenylate Concentrations

Distinct patterns of relative microbial activity were observed in samples collected during two summer cruises. The sediment samples collected in the western Beaufort Sea in 1977 showed much greater activity than those collected to the east (Fig. 1). In 1978, water and sediment samples were collected from locations close to shore from the Colville River to the U.S.-Canada border. The relative microbial activity was greatest in the waters associated with the major river plumes (Fig. 2) and in the sediments associated with river plumes, most notably off the Colville River (Fig. 3).

In addition to these geographical patterns we observed seasonal changes in relative microbial activity. In earlier studies in the Beaufort Sea, there was a statistically significant rise in relative microbial activities observed in the water samples from late winter to summer (Griffiths et al., 1978). This pattern, confirmed in Elson Lagoon waters (Table V), appeared whether either glucose or glutamate was used as the test substrate. A summer rise was also noted in total adenylate concentrations (Table VI). These differences were much greater than those observed in subarctic sediments collected in and near Kasitsna Bay (Cook Inlet) (Fig. 4).

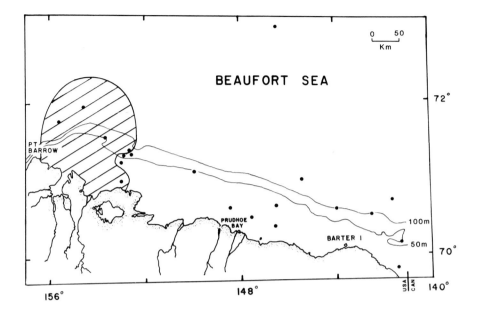

FIGURE 1. *Locations of sample collection during September 1977 showing area of elevated glucose uptake rates observed in sediments. The shaded area represents those stations where uptake rates were equal to or greater than the mean.*

In the earlier Beaufort Sea studies, we observed maximum heterotrophic potential values (microbial activities) in marine waters that were comparable to those in more temperate climates (Griffiths *et al.*, 1978). That comparison was made on the basis of glutamate uptake kinetic experiments. In the present study, we compared uptake rates using single concentrations of both glucose and glutamate in both arctic and subarctic offshore marine waters (Table VII). There was no significant difference in

TABLE V. *Seasonal Variation in Glucose and Glutamate Uptake in Elson Lagoon Waters (ng l^{-1} hr^{-1}).*

Date	Glucose	Glutamate
January 1978	0.4 ± 0.1	3.8 ± 1.0
April 1978	0.2 ± 0.1	0.5 ± 0.1
August 1978	21 ± 21	20 ± 15
January 1979	0.2 ± 0	1.7 ± 1.5

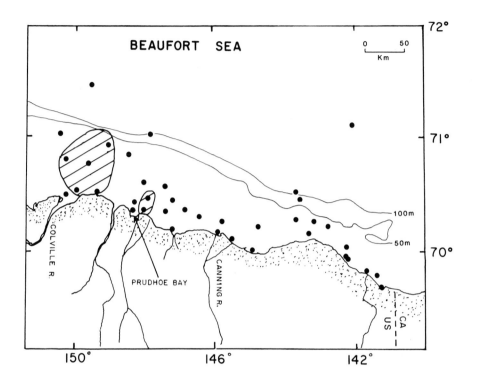

FIGURE 2. *Locations of sample collection during August 1978 showing area of elevated glutamate uptake rates observed in water samples. The shaded area represents those stations where uptake rates were equal to or greater than the mean.*

TABLE   VI. *Total   Adenylate   Concentrations   (nmol   g^{-1})   Observed   in Sediments   Collected   in   Elson   Lagoon   (Beaufort   Sea)   and Kasitsna Bay (Cook Inlet).*

| | | Total Adenylates | | |
Location	Date	Mean	Range	n
Elson Lagoon	8/78	158	16-352	6
Elson Lagoon	1/79	16	13-23	4
Kasitsna Bay	2/79	4.4	0.1-7.3	9
Kasitsna Bay	4/79	6.9	4.2-8.5	4
Kasitsna Bay	7/79	7.7	3.3-12.9	14

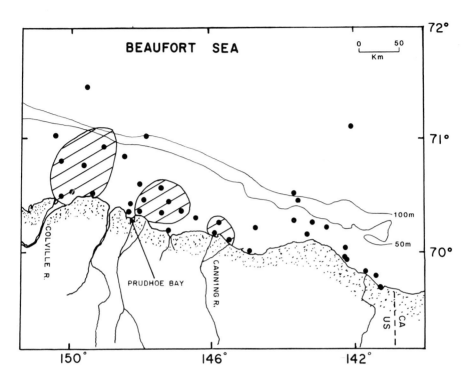

FIGURE 3. Locations of sample collection during August 1978 showing area of elevated glutamate uptake rates observed in sediments. The shaded area represents those stations where uptake rates were equal to or greater than the mean.

TABLE VII. Microbial Uptake Rates Observed in Arctic and Subarctic Marine Waters (ng l^{-1} hr^{-1}).

Location	Date	Glucose			Glutamate		
		Mean	Range	n	Mean	Range	n
Beaufort Sea	8/76	5	1-13	18	8	0.5-24	18
Beaufort Sea	8/78	7	1-44	40	14	1-24	40
Norton Sound	7/79	12	1-110	47	19	2-24	47
Bristol Bay	8/80	--	--	--	6	1-20	28
Bristol Bay	5/81	--	--	--	15	2-80	51
Cook Inlet	10/76	3	0.1-33	17	5	0.2-53	19
Cook Inlet	4/77	2	0.2-12	31	8	0.5-67	32
Cook Inlet	11/77	1	0.2-7	59	4	0.4-61	56
Cook Inlet	4/78	10	0.1-153	67	13	0.2-72	77

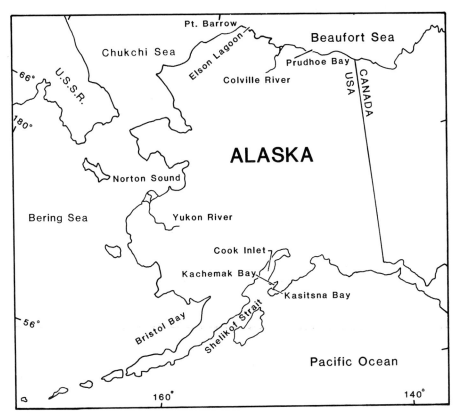

FIGURE 4. Map of Alaska showing study areas and locations discussed in the text.

glutamate uptake rates between arctic and subarctic waters, but there was one in glucose uptake rates between Beaufort Sea and Norton Sound water samples. The same pattern was observed between arctic and subarctic marine sediments (Table VIII). The elevated glucose uptake in the waters and sediments of Norton Sound probably reflects the input of terrestrial carbon from the Yukon River rather than a basic difference in carbon cycling.

F. Nitrogen Fixation

During the 1978 Beaufort Sea summer cruise, we measured nitrogen fixation (acetylene reduction) rates in all sediment samples collected. There was no geographical trend in these data, nor were there significant differences between arctic and subarctic sediments except between the Beaufort Sea and Shelikof Strait sediments (Table IX). This difference most probably reflected properties of the detrital carbon settling into the sediments of the Shelikof Strait rather than a difference in these two

TABLE VIII. *Microbial Uptake Rates Observed in Arctic and Subarctic Marine Sediments (ng g^{-1} hr^{-2}).*

Location	Date	Glucose			Glutamate		
		Mean	Range	n	Mean	Range	n
Beaufort Sea	8/76	4	1-15	11	80	20-180	11
Beaufort Sea	8/78	9	1-24	30	96	7-162	30
Norton Sound	7/79	28	0.1-154	32	127	3-1063	32
Bristol Bay	5/81	-	-	-	215	149-345	23
Cook Inlet	10/76	12	1-56	8	220	70-1190	13
Cook Inlet	4/77	8	1-18	9	217	80-370	9
Cook Inlet	11/77	4	0.4-22	27	63	20-252	27
Cook Inlet	4/78	4	0.1-38	25	89	5-595	26

regions, as no comparable difference occurred between Cook Inlet and Beaufort Sea nitrogen fixation rates.

IV. DISCUSSION

The enumeration data present a paradox. Why are numbers of viable bacteria higher in the Beaufort Sea than in subarctic seas overlying the Alaska continental shelf? The current studies do not provide any definitive answers, but perhaps at the low temperatures at the surface of the Beaufort Sea bacteria survive for prolonged periods.

The taxonomic diversity studies support general ecological theory as evidenced, for example, by the lower diversities in surface waters during winter when the communities are particularly stressed. Despite the high physiological versatilities of the populations, the taxonomic diversity reflects population shifts due to environmental variations. Benthic communities, which are exposed to less environmental variability, have higher diversities than surface-water communities and do not exhibit seasonal variability. Our studies indicate that bacterial communities maintain a high state of diversity unless severely stressed. This appears to be true of most ecosystems including arctic marine ecosystems, where one might have predicted a greater degree of specialization, that is, less versatility. Beaufort Sea ecosystems appear to have a specified number of niches that vary seasonally in surface waters. The formation of coastal ice removes bacterial populations from surface waters, and after the spring melt there is an annual succession to reestablish the surface-water bacterial community. Different populations are included in the community in different years, but the same level of community diversity is achieved each summer. It is likely that random recruitment determines which populations successfully occupy the niches of this ecosystem and that factors such as phytoplankton productivity (substrate availability) and

TABLE IX. Nitrogen Fixation Rates Observed in Arctic and Subarctic Marine Sediments (ng $g^{-1} h^{-1}$).

Location	Date	Mean	Range	n
Beaufort Sea	8/78	0.20	0-1.40	25
Norton Sound	7/79	0.33	0-0.88	33
Bristol Bay	1/81	0.10	0-0.30	13
Bristol Bay	5/81	0.20	0.05-0.34	19
Cook Inlet	4/77	0.23	0-0.90	11
Cook Inlet	11/77	0.46	0.05-1.70	15
Cook Inlet	4/78	0.31	0.10-1.10	18
Shelikof Strait	11/77	1.60	0.3-4.4	11
Shelikof Strait	4/78	0.51	0.30-1.10	10
Shelikof Strait	4/79	0.68	0.45-0.95	7

temperature (abiotic factors) determine the structure of the stable climax community that develops.

The bacterial populations in the surface waters exhibit definite adaptations that enhance their ability to survive (Kaneko et al., 1979), including predominance of pigmented bacteria during summer when populations are exposed to continuous sunlight; pigmentation protects bacteria against ionizing radiation. As with taxonomic diversity, the physiological tolerance indices attest to the diversity and versatility of the community. It is interesting that the bacterial populations of the Beaufort Sea maintain the ability to tolerate conditions to which they are not exposed. Psychrophiles do occur in the Beaufort Sea in higher proportions than in subarctic waters, but psychrotrophs dominate even in this polar sea. The salinity tolerance indices suggest that surface-water populations are better adapted to low salinities and that benthic bacterial populations are adapted to higher salinities. Benthic communities are exposed to hypersaline waters, produced when ice forms, that sink to the bottom. In contrast, surface communities are exposed to low-salinity waters in the spring from ice melt and river runoff.

The nutritional utilization indices suggest that these ecosystems are phytoplankton-supported, particularly during summer. This is evidenced by the large number of carbohydrates that can be utilized by the bacterial populations. During winter there is a shift toward usage of non-carbohydrate substrates, including carboxylic and amino acids, suggestive of a shift to a detrital food web. The potential for utilization of particular classes of substrates, as expressed by the nutritional utilization indices, presumably reflects the natural patterns of substrate availability and usage in Beaufort Sea ecosystems. This hypothesis is substantiated by a recent report by Griffiths et al. (1982c) which shows that the ratio of glucose to glutamate uptake rates changes during a marine phytoplankton bloom. During periods of little primary productivity, this ratio is close to 0.1;

however, during the height of the bloom, it increases to 1.0. This shift appears to reflect a change in the type of organic nutrients available to the microbial community. During a bloom, the phytoplankton release carbohydrates such as glucose, which are used by the bacteria present.

Our seasonal study of glucose and glutamate uptake in Elson Lagoon waters can be interpreted in much the same way (Table V). In August, the uptake of both glucose and glutamate was high and the ratio near one. In both January samplings, the uptake of both substrates was reduced relative to the August values, and the ratio of glucose to glutamate uptake had also been reduced. These data suggest that at this time of the year, there is proportionately more glutamate than glucose being released from the degradation of detritus. During the April sampling period, the glutamate uptake rates were further reduced. These samples presumably were collected prior to the spring under-ice phytoplankton bloom. Therefore, the availability of both carbohydrates and amino acids should have been minimal for the year with readily degradable detritus nearly exhausted.

The inability of the dominant populations to metabolize hydrocarbons indicates that these are not significant natural substrates and that a relatively long period of adaptation may be needed for the indigenous bacterial communities of the Beaufort Sea to respond to inputs of hydrocarbons resulting from offshore oil and gas development in this region. Likewise, there are few hydrocarbon utilizers. Indeed, studies have shown that petroleum biodegradation in Beaufort Sea ecosystems will be slow and that petroleum pollutants will persist (Atlas, 1978; Atlas et al., 1978; Horowitz and Atlas, 1978; Haines and Atlas, 1982).

The elevated microbial activity levels associated with the river plumes are not unique to this region. We have observed similar patterns associated with the major rivers of Cook Inlet (Griffiths et al., 1981), Norton Sound, and Bristol Bay (Griffiths and Morita, 1981a, b). Valdes and Albright (1981) concluded that the elevated activity in the Fraser River plume of British Columbia was the result of nutrient mixing from the parent waters.

Another feature of the North Slope river plumes is of potential ecological significance. The respiration uptake ratios in the plume waters were lower than in the surrounding waters (Griffiths et al., in press), indicating that in the plume waters proportionately more of the organic material taken up by the microorganisms is converted into microbial biomass. Since there are both elevated microbial activities and reduced respiration percentages in these waters, the microbial biomass production should be very high and may represent a significant source of food for higher trophic levels. This could be an important mechanism by which the soluble organic material leaching from the vegetation of the North Slope is made available to nearshore marine organisms.

The seasonal microbial activity variations are apparently not unique to this region. In a similar study of subarctic waters (Cook Inlet), the same magnitude of seasonal change was observed (Griffiths et al., 1982c). Other investigators have reported variations of the same magnitude in more temperate climates (Carney and Colwell, 1976; Hobbie and Rublee, 1977; Delattre et al., 1979).

Unlike the water column, the arctic sediments showed much greater seasonal changes than those observed in the subarctic. The microbial activities observed in August were higher than in January by a factor of 6 and 60, respectively, during 1978 and 1979. During a similar seasonal study of Cook Inlet sediments, the microbial activities observed in August were only 3 times that observed in January (Griffiths and Morita, 1981a).

This larger variation is probably related to seasonal organic carbon input. In Kachemak Bay (Cook Inlet), terrestrial carbon is transported into the bay throughout the year.   There is also a large population of macrophytes such as *Laminaria* sp. which store polysaccharides during the summer and use this material for plant growth during the winter months (Lees, 1978). This is not the case in the southern Beaufort Sea as there is no terrestrial runoff during much of the year and there are very few macrophytes that could provide new detrital carbon during the winter months.   Carney and Colwell (1976) concluded that the observed seasonal changes were caused by temperature; however, our temperature-effects studies on water samples collected near Point Barrow showed that while temperature changes could cause a doubling of activity in the summer months compared to winter, they could not account for the magnitude of change observed in Beaufort Sea sediments (Griffiths *et al.*, 1978).

During winter, when there is low light availability and no input of terrestrial carbon, all organisms present should be entirely dependent on detrital carbon as defined by Fenchel and Jorgensen (1977). It is currently thought that the bacterial biomass is the most important link between detrital carbon and the rest of the detrital food chain (Pomeroy, 1974). Assuming that uptake measurements reflect the potential production of bacterial biomass, the reduced uptake rates observed during winter in the Beaufort Sea should reflect a much reduced food source for higher trophic levels.   The seasonal differences observed in arctic and subarctic marine sediments then should be reflected in the qualitative and quantitative differences in the benthic infauna of these two regions.   There is some indirect evidence that this may be the case.   The seasonal differences observed in total adenylates were much greater in arctic than in subarctic sediments, suggesting that the total biomass in organisms smaller than 2 mm was much reduced in these sediments during the winter.   This observation is what we would predict from the relative microbial activities observed in these same samples since both bacteria and the organisms that feed on them would be included in the adenylate measurements.

During our studies of Alaska marine sediments, we also examined the effects of crude oil perturbation on a number of microbial functions (Haines *et al.*, 1981; Griffiths *et al.*, 1981, 1982a, 1982b). We found that relative microbial activity decreased and respiration percentages increased in both arctic (Elson Lagoon) and subarctic (Kasitsna Bay) marine sediments that had been perturbed by crude oil.   In the subarctic sediments these changes took place within 5 days, but it took up to 13 months for the same changes to occur in arctic sediments. This lag in arctic sediments could be due to either reduced metabolic activity in the winter months or different crude oil inhibitory mechanisms in the sediments of these two areas.

Although the onset of crude oil effects was retarded in Elson Lagoon, the magnitude of the effects was essentially the same in both arctic and subarctic sediments. This plus the observation that crude oil degradation is relatively slow in arctic sediments (Haines and Atlas, 1982) suggest that the duration of crude oil effects on relative microbial activity should be much longer in arctic sediments than in temperate marine systems.

What, then, are the ecological implications of crude oil perturbation in arctic sediments? Our Kasitsna Bay study showed that the observed increase in respiration percentages in oil-treated sediments was due to a reduction in bacterial biomass production rather than an increase in respiration (Griffiths et al., 1981), which suggests that crude oil could reduce microbial productivity in the perturbed area by disrupting a vital link in the detrital food chain. The reduced microbial activity also implies a reduction in the mineralization of fixed nitrogen and phosphorus. The potential effects of crude oil perturbation on phosphorus mineralization were also documented in a more direct way. During the Kasitsna Bay study, phosphatase activity was reduced in crude-oil-treated sediments (Griffiths et al., 1982b). Inorganic nutrient regeneration from nearshore Beaufort Sea sediments is important to primary production, as has been shown in other nearshore environments (Davies, 1975; Martin and Lelong, 1981). During a recent study of microbial activity in southeastern Bering Sea sediments, a comparatively small area was discovered in which relative microbial heterotrophic activities were higher than in the surrounding sediments (Griffiths and Morita, 1981b). This was also a region in which all other microbially mediated reactions were elevated. This correlation suggests that relative microbial heterotrophic activity measurements reflect the relative activities of other microbially mediated reactions as well. The Colville River Delta is another area where elevated relative microbial activity was also observed in sediments (Fig. 3) and rapid bacterial ammonification and nitrification rates have been measured (Schell, 1974). These results undoubtedly reflect the inputs of terrestrial carbon and organic nitrogen from the Colville River drainage. In view of the results of the long-term crude oil effects studies and the role of microbial activities in primary and secondary productivity, we conclude that the Colville River Delta is potentially the most sensitive area in the southern Beaufort Sea to crude oil perturbation.

ACKNOWLEDGMENTS

This study was supported by the Bureau of Land Management through an interagency agreement with the National Oceanic and Atmospheric Administration under which a multiyear program, responding to the needs of petroleum development in the Alaskan Continental Shelf, is managed by the Outer Continental Shelf Environmental Assessment Program (OCSEAP) Office. Portions of this study were performed by A. Horowitz, T. Kaneko, J. Hauxhurst, M. I. Krichevsky, E. J. Krichevsky, and T. M. McNamara, whose contributions are gratefully acknowledged.

REFERENCES

Atlas, R. M. (1978). In "Microbial Ecology" (M. W. Loutit and J. A. R. Miles, eds.), p. 86. Springer-Verlag, Berlin.
Atlas, R. M. (1979). In "Native Aquatic Bacteria: Enumeration, Activity and Ecology" (J. W. Costerton and R. R. Colwell, eds.), p. 196.

American Society for Testing and Materials, Special Technical Publication 695.

Atlas, R. M., Horowitz, A., and Busdosh, M. (1978). *J. Fish. Res. Bd. Can.* *35*, 585.

Bulleid, N. C. (1978). *Limnol. Oceanog.* *23*, 174.

Carney, J. F. and Colwell, R. R. (1976). *Appl. environ. Microbiol.* *31*, 227.

Daley, R. J. and Hobbie, J. E. (1975). *Limnol. Oceanog.* *20*, 875.

Davies, J. M. (1975). *Mar. Biol.* *31*, 353.

Delattre, J. M., Delesmont, R., Clabaux, M., Oger, C., and Leclerc, H. (1979). *Oceanol. Acta.* *2*, 317.

Fenchel, J. A. and Jorgensen, B. B. (1977). *Adv. microb. Ecol.* *1*, 1.

Griffiths, R. P. and Morita, R. Y. (1981a). *In* "Environmental Assessment of the Alaskan Continental Shelf," p. 361. OMPA/NOAA, Juneau, AK.

Griffiths, R. P. and Morita, R. Y. (1981b). *In* "Environmental Assessment of the Alaskan Continental Shelf," p. 125. OMPA/NOAA, Juneau, AK.

Griffiths, R. P. Hayasaka, S. S., McNamara, T. M., and Morita, R. Y. (1977). *Appl. environ. Microbiol.* *34*, 801.

Griffiths, R. P., Hayasaka, S. S., McNamara, T. M., and Morita, R. Y., (1978). *Can. J. Microbiol.* *24*, 1217.

Griffiths, R. P., Caldwell, B. A., Broich, W. A., and Morita, R. Y. (1981). *Appl. environ. Microbiol.* *42*, 792.

Griffiths, R. P., Caldwell, B. A., Broich, W. A., and Morita, R. Y. (1982a). *Estuar. coast. Shelf. Sci.* *15*, 183.

Griffiths, R. P., Caldwell, B. A., Broich, W. A., and Morita, R. Y. (1982b). *Mar. Pollut. Bull.* *13*, 273.

Griffiths, R. P., Caldwell, B. A., and Morita, R. Y. (In press). *Microb. Ecol.*

Griffiths, R. P., Caldwell, B. A., and Morita, R. Y. (1982c). *Mar. Biol.*, 71, 121.

Haines, J. R., Atlas, R. M., Griffiths, R. P., and Morita, R. Y. (1981). *Appl. environ. Microbiol.* *41*, 412.

Haines, J. R. and Atlas, R. M. (1982). *Mar. environ. Res.*, 7, 91.

Hauxhurst, J. D., Kaneko, T., and Atlas, R. M. (1981). *Microb. Ecol.* 7, 167.

Hobbie, J. W. E., and Rublee, P. (1977). *In* "Aquatic Microbial Communities" (J. Cairns, Jr., ed.), p. 441. Garland Publishers, New York.

Kaneko, T., Atlas, R. M., and Krichevsky, M. (1977). *Nature 270*, 596.

Kaneko, T., Krichevsky, M. I., and Atlas, R. M. (1979). *J. gen. Microbiol.* *110*, 111.

Krichevsky, M. I. (1979). *FDA By-Line 9*, 217.

Lees, D. C. (1978). *In* "Environmental Assessment of the Alaskan Continental Shelf," p. 179. OMPA/NOAA, Juneau, AK.

Lloyd, M., Zar, J. H., and Karr, J. R. (1968). *Am. Mid. Nat. 79*, 257.

Martin, Y. P. and Lelong, P. P. (1981). *Oceanol. Acta.* *4*, 433.

Pomeroy, L. R. (1974). *Bioscience 24*, 499.

Schell, D. M. (1974). *In* "The Coast and Shelf of the Beaufort Sea" (J. C. Reed and J. E. Sater, eds.), p. 649. Arctic Inst. of N. America, Arlington, VA.

Shannon, C. E. (1948). *Bell. Syst. Technol. J. 27*, 379.

Sneath, P. H. A. and Sokal, R. R. (1973). "Numerical Taxonomy." W. H. Freeman and Co., San Francisco.

Stewart, E. D. P. (1967). *Proc. U.S. Nat. Acad. Sci. 58*, 2071.

Valdes, M. and Albright, L. J. (1981). *Mar. Biol. 64*, 231.
Walczak, C. A. (1979). *FDA By-Lines 9*, 251.
Walczak, C. A. and Krichevsky, M. I. (1980). *Int. J. Syst. Bacteriol. 30*, 622.
Zimmermann, Iturriaga, R., and Becker-Birck, J. (1978). *Appl. environ. Microbiol. 36*, 926.

TROPHIC DYNAMICS
IN AN ARCTIC LAGOON

Peter C. Craig
William B. Griffiths
Stephen R. Johnson

LGL Limited
environmental research associates
Sidney, British Columbia

Donald M. Schell

Institute of Water Resources
University of Alaska
Fairbanks, Alaska

I. INTRODUCTION

Summer in the arctic is a brief but biologically active period during which large numbers of birds and fish come to feed on an abundant supply of aquatic invertebrates in nearshore waters. This study examined major trophic pathways in Simpson Lagoon, an arctic coastal ecosystem in the central Alaskan Beaufort Sea, from the top down, that is, from consumers to producers. Thus, initial field on principal consumers (diving ducks and fish) were followed by an examination of their invertebrate prey species, then of the carbon sources which drive the nearshore food web. The interdisciplinary overview of trophic dynamics presented in this paper is based on more detailed studies of birds (Johnson and Richardson, 1981), fish (Craig and Haldorson, 1981), invertebrates (Griffiths and Dillinger, 1981),

347

and primary productivity (Schell *et al.*, 1982) conducted as part of the Beaufort Sea barrier island – lagoon ecological process studies sponsored by the Outer Continental Shelf Environmental Assessment Program.

II. STUDY AREA

Simpson Lagoon, located between Prudhoe Bay and the Colville River delta (Fig. 1), is a large and partially enclosed body of water approximately 35 km long and 3-6 km wide with an average depth of only 2 m (maximum 3 m). The lagoon floor is uniformly flat and almost featureless. In most areas, a layer of detritus covers substrates of mud and sand.

The short ice-free period in the lagoon lasts from early July to early October. The highly variable summer salinities (1-32 ppt) and water temperatures (0-14°C) fluctuate with the prevailing westward flowing Beaufort Sea current, wind, and freshwater runoff. The lagoon water is diluted by freshwater runoff and is correspondingly lower in salinity (usually 4-5 ppt) and higher in temperature (usually 2-4°C) than water beyond the barrier islands, although less so late in summer when runoff declines. Prevailing currents continually exchange lagoon water at an average rate of 10-20% d^{-1} and 100% d^{-1} when aided by exceptionally strong winds (65 km h^{-1}; Mungall, 1978). During the winter, exchange diminishes as surface ice increases in thickness to about 2 m. By late winter (April) about 90% of the lagoon volume is frozen solid. Hypersaline conditions (up to 68 ppt; Crane, 1974) develop from salt exclusion during ice formation.

Additional details about the study area appear in Craig and Haldorson (1981).

A. Biological Setting

1. Birds

Over 100 species of birds have been recorded in the Alaskan Beaufort Sea area (Johnson *et al.*, 1975; Kessel and Gibson, 1978); however, many are migrants, others are terrestrial, and a relatively small number of species of loons, gulls, terns, shorebirds, and marine waterfowl occupy nearshore coastal lagoons during the summer.

In coastal waters, only the oldsquaw duck (*Clangula hyemalis*) and two species of phalaropes (the red-necked phalarope, *Phalaropus lobatus* and the red phalarope, *P. fulicarius*) are found in large numbers for a substantial period of time (several weeks to months). Tens of thousands of oldsquaws are concentrated in coastal lagoons from mid-July to mid-August when they molt and again in late September when they feed on marine invertebrates prior to their southward migration in fall. Similarly, juvenile phalaropes concentrate during August along lagoon beaches to feed on marine invertebrates prior to their fall migration. Hence, our investigations of avian consumers focused on these key taxa, which makeup the bulk of the avian biomass during the summer open-water period.

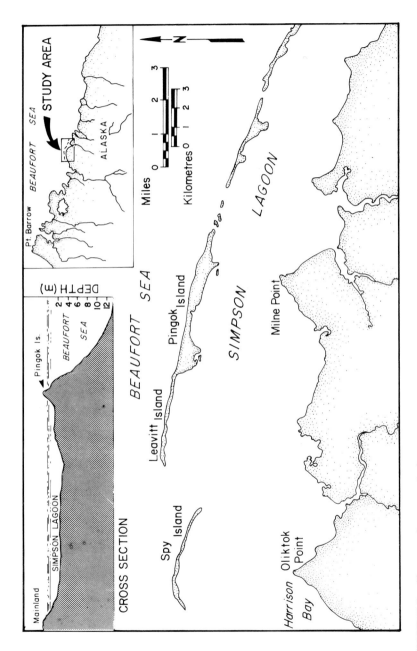

FIGURE 1. Simpson Lagoon study area.

Glaucous gulls (*Larus hyperboreus*) were also sampled in 1977 but are not considered to be a key species due to their relatively low numbers.

2. Marine mammals

Marine mammals are a negligible component of the Simpson Lagoon fauna; only a few seals were sighted during numerous aerial surveys.

3. Fish

During the brief open-water season, the relatively warm and brackish waters of Simpson Lagoon are an important feeding habitat for anadromous and marine fishes. Over 30 species have been recorded in these waters, but a few species account for most of the fish present (OCSEAP, 1978; Craig, 1983).   Key species are the three anadromous species arctic cisco (*Coregonus autumnalis*), least cisco (*C. sardinella*), and arctic char (*Salvelinus alpinus*), and the marine fourhorn sculpin (*Myoxocephalus quadricornis*) and arctic cod (*Boreogadus saida*).   Other locally abundant species are the boreal (rainbow) smelt (*Osmerus eperlanus*), broad whitefish (*Coregonus nasus*), and lake whitefish (*C. clupeaformis*).

Seasonal use of nearshore waters by these fishes differs.   The anadromous species arrive with the spring breakup, disperse along the coastline, and return to rivers or estuaries in fall to spawn or overwinter. Marine species tend to become more abundant in nearshore waters as the open-water season progresses.   In nearshore waters, both anadromous and marine fishes feed extensively on invertebrates, primarily mysids and amphipods.

In winter, with minor exceptions, only marine species are present, and even these vacate shallow areas as the ice thickens.

4. Invertebrates

The invertebrate community in Simpson Lagoon comprises infauna (animals living within bottom substrates), epibenthos (animals usually living on or near bottom substrates), and pelagic forms (animals that inhabit the water column).   Due to the shallow water and lack of a rocky substrate, no kelp community is present in the lagoon.   Infaunal organisms are restricted to the deeper (>2 m) portions of the lagoon because shallower areas (about one-third of the lagoon area) freeze solid during winter.   Abundant members of the infaunal group include polychaete worms (*Ampharete vega, Terebellides stroemi*) and bivalves (*Cyrtodaria kurriana*).   The epibenthos is dominated by amphipods (*Onisimus glacialis, Gammarus setosus*), mysids (*Mysis litoralis, M. relicta*), and isopods (*Saduria entomon*), which are found throughout the lagoon during the open-water season, generally associated with the detrital mat that covers large portions of the lagoon bottom. Pelagic forms common to Simpson Lagoon are copepods (*Calanus hyperboreus, C. glacialis*) and chaetognaths (*Sagitta elegans*).

The use of lagoon habitats by these three groups varies with season. Epibenthic and pelagic forms are generally absent in winter when much of the lagoon freezes to the bottom, recolonize the lagoon each spring, and are abundant through the summer. Infaunal organisms and some amphipods are year-round residents in deeper areas of the lagoon which do not freeze in winter.

III. METHODS

A. Vertebrate Studies

1. Birds

Birds were collected in 1977 and 1978 (Table I) with a shotgun from feeding (diving) flocks from mid-July through September. To mitigate postmortem digestion or regurgitation, the gut (proventriculus and ventriculus) and esophagus of each specimen was injected with absolute isopropyl alcohol and the esophagus plugged with a paper wad. Two-thirds of the specimens had identifiable food items in their stomachs in both years; only these birds are included in dietary analyses.

Phalaropes were collected in August in shallow areas (<1 m deep) along shorelines and in bays where they feed. Proportions of birds with identifiable food in their stomachs were 55% in 1977 and 44% in 1978. Because the two phalarope species were collected from mixed flocks feeding together, they were combined for dietary analyses. Glaucous gulls feed either singly or in loose aggregations. Of the 28 glaucous gulls collected from July to September 1977, only one, collected at a mid-lagoon location, had an empty stomach.

2. Fish

A sample of 684 specimens of 7 species was collected during summer and winter seasons 1977-79 (Table I). All summer samples were from Simpson Lagoon in the vicinity of Milne Point and Pingok Island. Winter samples were combined from a wider nearshore region: arctic cisco, least cisco, and fourhorn sculpin from Colville Delta, April-May 1978, fourhorn sculpin from Thetis Island, November 1978 and March-April 1979, boreal smelt from Simpson Lagoon and Thetis Island, November 1978, and arctic cod from Thetis Island to Narwhal Island, November 1978 and February 1979.

Fish used in diet studies were collected by gill net (87%) and fyke net (13%). Fish caught in fyke nets were used only when sample sizes from gill nets were low since those caught in fyke nets may have fed upon invertebrates attracted to or caught by the fyke net. For each species, the fish examined were generally from the most common size-class present at the time of sampling (Craig and Haldorson, 1981). However, we exercised some selection of specimens to ensure that similar size-classes were studied throughout the summer months. Fish stomachs were preserved separately in formalin and later analyzed in the laboratory.

TABLE I. Number of bird and fish stomachs (containing food) examined during summer and winter sampling periods.

Species	1977		1978		1979	Total
	Summer	Winter	Summer	Winter	Summer	
Birds						
Oldsquaw	54	—	72	—	—	126
Phalaropes	46	—	26	—	—	72
Glaucous gull	27	—	—	—	—	27
					Total	225
Fish						
Arctic cisco	55	40	52	—	—	147
Least cisco	51	23	27	—	—	101
Arctic char	60	—	17	—	—	77
Arctic cod	34	—	20	84	47	185
Fourhorn sculpin	65	9	—	45	—	119
Boreal smelt	—	—	—	39	—	39
Arctic flounder	—	—	16	—	—	16
					Total	684

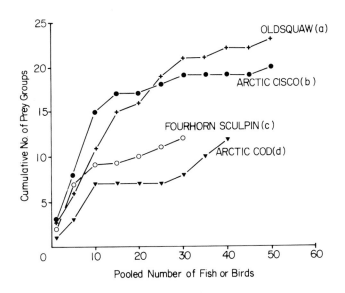

FIGURE 2. Cumulative number of prey (see Table IV) in pooled oldsquaw and fish stomachs. Sample dates: (a, b) summer 1978, (c) April-May 1979, (d) February 1979.

3. Lab and Data Analysis

Dietary analyses of bird and fish stomachs were based on wet weights of identifiable items, but the methodology differed between years. In 1977, weights were determined by directly weighing all identifiable specimens and fragments for each species or taxonomical group. In 1978 and 1979, weights of major prey species samples were determined through reconstructing their live weight by measuring a selected part of the organism and then calculating its total weight based on length and weight relations described by Griffiths and Dillinger (1981).

For detailed comparisons of fish and bird diets, food items ingested by each predator were listed as percent wet weight of identifiable contents. Then, on the assumption that most of the material not identified to the species level, such as "gammarid amphipods," was actually remains of identified species, the unidentified gammarids were allocated to identified gammarid prey species in the proportions already determined for that predator. Taxonomists who identified these samples felt that the assumption was reasonable, although a small percentage of the material identified only to taxonomic family contained uncommon species. For the present uses of these data, this procedure provides a better base for comparison than either deleting or retaining all categories of taxonomic families for mysids and amphipods. In any event, contributions of such categories were generally small (usually less than 5% of contents).

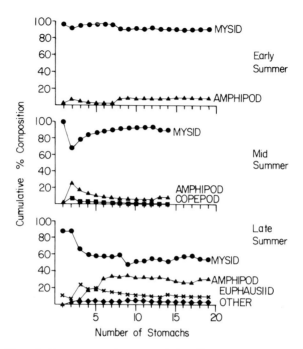

FIGURE 3. *Cumulative percent composition (by wet weight) of major food groups in the diet of arctic cisco collected during three time periods during the 1978 open-water season.*

A consumer's diet may change seasonally, so a composite estimate of a species' diet was obtained whenever possible by pooling and analyzing approximately equal numbers of samples collected during early, middle, and late periods of summer and winter. Because the variety of food items in individual fish stomachs was low (Fig. 2) and the relative proportions of major food groups did not vary greatly (Fig. 3), a sample of 10–20 stomachs appeared adequate to describe the kinds and proportions of important food items consumed by fish during any one sampling period. Actual sample sizes for fish were usually much larger than this (Table I). For oldsquaw, a sample size of about 25 was needed to adequately quantify the kinds of important food items consumed (Fig. 3), and again, the actual sample sizes were larger. We believe that the pooled data from the summer or winter periods (depending on availability of specimens) reflected the general diets of each species in nearshore waters.

The degree of dietary overlap between bird and fish species was determined by the Schoener (1968) index:

$$R_o = 1 - 0.5 \left(\Sigma \mid Pxi - Pyi \mid \right)$$

where Pxi and Pyi are the proportions of food category i in the diets of species x and y. Recent evaluations of several overlap indexes indicate that Schoener's index is a generally preferred estimate of true overlap (Linton et al., 1981; Wallace, 1981). The index ranges from 0 (diets have no food items in common) to 1, when the diets are identical in kinds and proportions of food items. A value of 0.6 or greater is arbitrarily considered to be a biologically significant overlap (Zaret and Rand, 1971).

B. Invertebrate Studies

Because our sampling program concentrated on organisms which were important foods for higher trophic levels (fish and birds), we directed sampling efforts toward epibenthic mysids, amphipods, and copepods rather than infaunal organisms.

In August 1977, three lagoon stations (Fig. 4) were sampled by a small otter trawl (4.9 m wide, 4 m long, 6.5-mm bar mesh cod end) and by SCUBA divers. Divers observed that the slow-moving trawl was not effective in capturing mobile epibenthic invertebrates. Thus, along each of 12 25-m transects, divers made five estimates of the densities of mysids and amphipods in a 10-cm-square area. The average of these estimates was extrapolated to 1 m^2 and converted to an estimated biomass by using the 1978 ash-free dry weight for the predominant size classes of mysids and amphipods observed by the diver in 1977. Because the only available data concerning epibenthos in July and September 1977 were otter-trawl samples, these were corrected by the proportion of diver-measured to trawl-measured biomass determined in August. An average biomass for the whole lagoon (a weighted mean based on the areal extent of each of stations 1-3 in the lagoon) was then determined for each sampling period (15 July, 15 August, and 15 September). The estimate for station 3 on 15 September appeared unrealistically high (68.7 g ash-free dry weight m^{-2}), possibly due to a sampling artifact (only one otter trawl sample was collected at this station and date). The exclusion of this high value introduces a conservative bias in data analyses which serves to lower estimates of invertebrate standing crop.

In 1978, a central-pursing drop net was designed for the project by modifying an epibenthic sampler developed by Clutter (1965). The net was 0.5 m in diameter and 0.75 m high, with 1.0 mm mesh. With both top and bottom of the net open, it was pushed to the bottom in shallow water; in deep water the net was weighted to free-fall. Upon reaching the sea bottom, purse lines to both net openings were immediately pulled to enclose the sample in the net. Diver observations indicated that the net effectively captured epibenthic invertebrates. Five replicate samples were collected at each of seven stations in each sampling period, weather and ice conditions permitting (Fig. 4).

C. Carbon Sources

1. Phytoplankton Primary Production

Productivity measurements were made using ^{14}C-uptake techniques (Strickland and Parsons, 1972). Light intensities were adjusted to

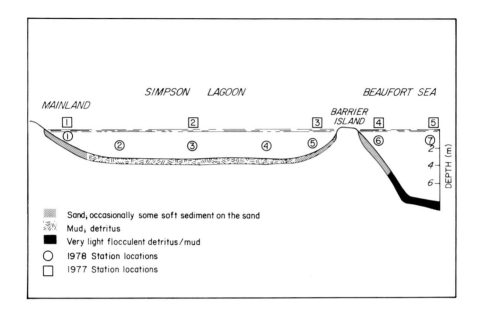

FIGURE 4. *Schematic cross section of Simpson Lagoon showing substrate types and invertebrate sampling stations for the open-water seasons of 1977 and 1978.*

approximate *in situ* intensities and incubations were typically 4-6 hours. Although techniques varied somewhat between shipboard and shore incubations, the rates observed were consistent and agreed with literature values (Alexander *et al.*, 1975). The seasonal production estimates are more uncertain since few data are available from June, early July, and September. The techniques and caveats used in estimating seasonal production are discussed in Schell *et al.* (1982).

2. *Terrestrial Carbon Inputs and Consumer Production from Allochthonous Carbon*

Coastal erosion and fluvial transport deliver large quantities of terrestrial carbon (peat) to the lagoon environment, and quantitative estimates of this input have been made by Cannon and Rawlinson (1978) and Schell *et al.* (1982). Since the role of this energy source was uncertain, we compared the natural carbon isotope abundances of the organisms within the lagoon ecosystem with the carbon isotope abundances in the energy source materials (peat vs. phytoplankton). The ^{14}C-depletions in the peat and $^{13}C/^{12}C$ ratios served as natural tracers of carbon from source materials to apical organisms (Schell, 1983; Schell et al., 1982) and enabled us to determine the critical energy sources supporting the fishes and birds.

IV. RESULTS AND DISCUSSION

A. Vertebrate Consumers

1. Trophic Spectrum

For an overview of food sources utilized by vertebrate consumers we categorized the potential food groups according to functional habitat or taxonomic units. This list included food groups known to be important to consumers in the study area and, for completeness, several basic food groups which are eaten by birds and fish in non-arctic areas. This range of foods potentially available to consumers is called a trophic spectrum, and the one used here is a slightly modified version used by Cailliet et al. (1978). Six general sources of food were recognized:

1. Water-column organisms (including zooplankton and fish).
2. Mobile epibenthos (polychaetes and crustaceans).
3. Sedentary epibenthos (crustaceans, mollusks, tunicates, and eggs from invertebrates or fish).
4. Infauna (worms such as polychaetes, and mollusks).
5. Flora (algae and vascular plants).
6. Other (detritus and miscellaneous).

The trophic spectrum indicated that the diets of vertebrate consumers were surprisingly similar (Fig. 5). Mobile epibenthic crustacea were by far the most important food group for most fishes and birds accounting for over 90% of the diet for arctic cisco, least cisco, arctic char, arctic cod, and oldsquaw. The remaining predators also fed heavily on this food category (44-64% of the diet), but additional preferences were apparent. Two supplemented their diet with water-column organisms: boreal smelt ate fish (41%), and phalaropes ate zooplankton (36%). Two predators fed on sedentary epibenthos: fourhorn sculpin and arctic flounder ate a bottom-crawling isopod (30-47%). Glaucous gulls had the most varied diet from the perspective of the trophic spectrum. These opportunistic feeders ate mobile and sedentary epibenthos, small fish and birds, and camp garbage.

Absent among vertebrate consumers in the lagoon were species that rely on infaunal organisms, sedentary epibenthos, or flora. This apparent void was only partly explained by the reduced variety of organisms inhabiting rigorous environments like Simpson Lagoon. However, a reduced variety of species does not by itself account for the observed reliance on mobile epibenthic crustaceans. Some infaunal organisms (polychaetes, bivalve mollusks) and sedentary epibenthos (stalked polychaetes, hydroids, isopods, and tunicates) were relatively abundant but little utilized. Their biomass in Simpson Lagoon was similar to the biomass of the mobile epibenthic crustaceans:

Organisms	Biomass ash-free dry weight (gm^{-2})
Infauna and sedentary epibenthos	0.5^a-2.1^b
Mobile epibenthos	0.3 -2.5^c

[a] From Griffiths and Dillinger (1981) for bivalves excluding shells.
[b] Recalculated from Crane (1974) for ash-free dry weight of worms, tunicates and bivalves excluding shells; deep lagoon stations, August 1971.
[c] From Griffiths and Dillinger (1981); deep lagoon stations, August 1977 and August 1978.

It is understandable that some infaunal organisms are out of reach to shorebirds due to water depth and lack of tidal exposure (tides are often only 10-15 cm), but they appear to be accessible to diving ducks and fish. However, oldsquaw only ate them in early summer (approximately 10% of diet). Even the arctic flounder, a fish that may eat infaunal organisms (Andriyashev, 1954), fed primarily on amphipods and isopods in Simpson Lagoon. No bivalves were found in their stomachs, and polychaetes accounted for only 3% of their diet. A slight increase in use of infauna is conceivable if the polychaetes classified as epibenthos were actually infauna when eaten, but indirect evidence (the near-absence of detrital material in fish or bird stomachs) suggests this was not the case.

If alternate sources of food such as mobile epibenthic organisms are plentiful, a consumer might not seek infaunal organisms because (1) buried organisms may be hard to find, especially in shallow lagoon waters (less than 2 m deep) where the infauna is sparse (Crane, 1974; Broad, 1978), or (2) prey size-classes vulnerable to predation may not be abundant or available at suitable depths.

2. General Food Habits and Food Chain

Mysids and amphipods were the most significant foods of birds and fishes in Simpson Lagoon. During the 1977-1979 open-water seasons, these invertebrates accounted for over 90% of all identifiable food ingested by seven of the nine species examined during at least one of the years of study (Table II). Copepods, isopods, bivalves, and smaller fish were usually of secondary importance and the remaining groups were incidental food items. For oldsquaw and the two cisco species, proportions of the food groups eaten were generally similar between years of study. However, large changes in diet were noted for the phalaropes, char, and cod. Phalaropes switched from copepods in 1977 to amphipods in 1978, and char

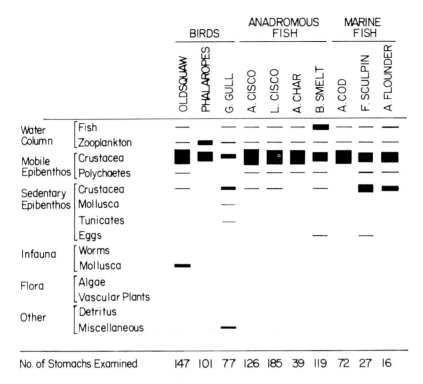

FIGURE 5. Trophic spectra of vertebrate consumers in Simpson Lagoon (combined dates and locations), 1977-1979. For each species, proportions of foods in their diet are indicated by bands; the sum of heights of all bands within each spectrum equals 100%.

switched from amphipods to mysids. Arctic cod ate varying proportions of mysids, copepods, and amphipods.

In late fall and early winter, most of the vertebrate consumers emigrate from the study area. The birds fly south and most anadromous fishes return to rivers to overwinter. Winter catches of fishes in coastal waters consisted of only three abundant species: fourhorn sculpin, arctic cod, and boreal smelt. In addition, arctic and least cisco and fourhorn sculpin were collected from the brackish waters of the Colville Delta. Stomach analyses showed that these fishes continued feeding in winter and that mysids and amphipods were again heavily utilized (Table III). Other foods were also important at this time. Fourhorn sculpin ate mostly isopods, and boreal smelt ate fish (arctic cod).

The food chain for Simpson Lagoon is very short. Fish and birds feed primarily on epibenthic invertebrates (mysids and amphipods), and these invertebrates feed directly or indirectly on phytoplankton.

TABLE II. Food groups (percent composition, wet weight) eaten by lagoon birds and fishes during the open-water period, 1977–1979.

Food item	Oldsquaw		Phalarope		Glaucous gull	Arctic cisco		Least cisco		Arctic char		Fourhorn sculpin	Arctic cod			Arctic flounder
	77	78	77	78	77	77	78	77	78	77	78	77	77	78	79	77
Mysid	68	80	8	2	7	70	87	69	66	16	89	10	88	38	59	1
Amphipod	16	12	20	96	23	25	11	21	33	78	4	81	9	18	39	58
Copepod	1	<1	65	—	3	4	<1	9	<1	1	<1	<1	2	44	1	<
Fish	3	<1	—	—	12	<1	<1	<1	—	2	6	6	—	—	<	30
Other taxa[a]	—	<1	6	—	22	<1	<1	<1	<1	3	<1	3	<1	—	<1	3
No. stomachs examined	54	72	46	46	27	55	52	51	27	60	17	65	34	20	47	16

[a] Includes hydroids, polychaetes, pteropods, cumaceans, chaetognaths, decapods, euphausiids, and birds (eaten by glaucous gulls only).

TABLE III.   Winter foods of nearshore fishes, 1977-1979 (percent composition, wet weight).

| Food item | Colville delta | | | Nearshore waters | | |
	Arctic cisco	Least cisco	Fourhorn sculpin	Fourhorn sculpin	Arctic cod	Boreal smelt
Mysids	<1	--	--	3	93	39
Amphipods	99	100	31	5	3	20
Isopods	--	--	60	78	--	<1
Fish	--	--	--	<1	2	40
Fish eggs	--	--	9	5	--	--
Polychaetes	<1	--	--	2	--	<1
Other taxa	<1	--	--	6	2	<1
No. stomachs examined	40	23	9	45	84	39

3. Principal Prey and Dietary Overlap

A list of prey for each consumer is presented in Table IV. Principal prey, arbitrarily defined as species or groups which constitute 10% or more (by wet weight) of the total diet, consisted of two mysid species, six amphipod species, and four other groups—copepods, isopods, bivalves, and fish (Fig. 6).   Several points emerge in comparing diets among the consumers.

1.   The number of principal prey categories eaten by the common vertebrate species was low (2-7) during any single sampling period, reflecting considerable dietary overlap among predators.
2.   Mysids *Mysis litoralis* and *M. relicta* and the amphipod *Onisimus glacialis* were clearly the favored prey.
3.   The common vertebrate species tended to eat similar prey during the summer but different prey in winter.
The degree of similarity between predator diets was also calculated by the Schoener overlap index. Of the 15 possible comparisons of predators during each summer, 20% (in 1977) and 27% (in 1978) showed that predators exploited similar ($C \geq 0.6$) species or groups of food organisms (Table V). These overlaps reflected the importance of the two mysid species as food sources.

TABLE IV. Foods of birds and fishes in Simpson Lagoon (% composition, net weight). Abbreviations: ARCS (arctic cisco), LSCS (least cisco), CHAR (arctic char), ARCD (arctic cod), FHSC (fourhorn sculpin), OLDS (oldsquaw), PHAL (phalarope), BORS (boreal smelt).

Food item	Summer 1977						Summer 1978						Summer 1979		Winter 1977-78		Winter 1978-79		
	ARCS	LSCS	CHAR	ARCD	FHSC	OLDS	ARCS	LSCS	CHAR	ARCD	OLDS	PHAL	ARCD	ARCS	LSCS	FHSC	FHSC	ARCD	BORS
Mysis litoralis	54.6	16.4	12.0	12.1	4.6	34.1	54.1	31.2	68.4	10.2	70.8	2.3	40.1	<0.1	—	—	3.7	81.5	34.7
Mysis relicta	15.2	52.7	3.5	75.9	5.1	33.5	32.8	35.1	20.9	28.0	8.9	—	19.3	—	—	—	0.7	11.7	4.4
Apherusa glacialis	10.8	1.0	14.6	—	<0.1	<0.1	1.3	—	—	—	—	6.7	0.6	—	—	—	—	—	—
Halirages mixtus	—	—	—	—	—	—	1.6	—	<0.1	—	—	6.7	<0.1	—	—	—	—	0.4	0.3
Onisimus glacialis	12.4	8.0	11.5	7.0	47.8	8.0	6.7	24.8	2.7	17.5	10.4	82.4	35.4	—	—	—	0.8	1.5	0.5
Gammarus setosus	1.9	2.7	45.0	0.1	16.1	1.7	<0.1	0.5	1.3	—	0.8	—	2.6	—	—	—	—	—	0.6
Parathemisto spp.	0.1	<0.1	5.0	1.8	4.2	4.9	1.7	—	<0.1	—	0.3	—	—	—	—	—	—	—	—
Pontoporeia affinis	—	0.6	0.3	0.2	0.5	—	<0.1	7.6	<0.1	0.3	0.1	—	0.5	99.5	100	30.6	3.2	0.8	0.1
Pontoporeia femorata	—	1.9	0.2	—	0.3	—	—	—	—	—	<0.1	—	<0.1	<0.1	—	—	4.2	0.4	17.6
Gammarocanthus loricatus	0.2	7.0	1.4	0.3	11.7	1.2	<0.1	0.3	0.3	—	0.7	6.7	<0.1	<0.1	—	—	<0.1	<0.1	0.6
Copepods	4.5	8.7	1.3	2.1	<0.1	1.2	0.6	—	—	44.0	<0.1	—	0.9	—	—	—	0.1	0.4	0.2
Isopods	0.3	0.1	0.2	—	6.1	2.7	<0.1	0.2	0.1	—	0.9	1.9	—	—	—	60.2	73.5	—	0.1
Cumaceans	<0.1	0.6	0.2	—	<0.1	—	<0.1	<0.1	<0.1	—	0.1	—	<0.1	—	—	—	0.1	<0.1	<0.1
Euphausiids	<0.1	—	2.8	0.5	—	—	0.7	—	—	—	0.1	—	0.4	—	—	—	—	—	—
Fish	—	0.3	1.6	—	3.3	2.7	0.2	0.2	5.8	—	0.4	—	—	—	—	—	0.7	1.5	40.5
Eggs	—	—	—	—	<0.1	—	—	—	—	—	—	—	—	—	—	9.2	3.1	<0.1	<0.1
Bivalves	—	—	—	—	—	9.6	—	—	—	—	6.2	—	—	—	—	—	—	—	—
Polychaetes	—	—	<0.1	—	<0.1	—	—	—	0.2	—	—	—	—	<0.1	—	—	9.2	—	0.2
Miscellaneous taxa	<0.1	0.2	0.3	<0.1	0.2	—	0.2	—	—	—	0.3	—	—	0.4	—	—	0.7	1.9	—
No. stomachs examined	55	51	60	34	65	54	52	27	17	20	72	26	47	40	23	9	45	84	39

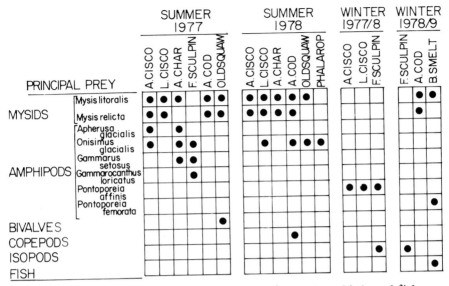

FIGURE 6. *Principal prey ($\geq$10% by weight) of nearshore birds and fish.*

Diets of some predators showed little annual variation. Arctic cisco ($C = 0.78$), least cisco (0.61) and oldsquaw (0.60) each ate essentially the same food items in the same proportions during both summers. Diets of arctic char were dissimilar between years ($C = 0.22$) and arctic cod diets varied during three summer sampling periods ($C = 0.41$ to 0.49).

In winter, fish changed their diets as indicated by very low overlap values obtained in a summer-winter comparison for arctic cisco ($C = 0.0$), least cisco (0.01), fourhorn sculpin (0.13), and arctic cod (0.26). The two ciscoes overwintering in the Colville Delta fed almost exclusively on the amphipod *Pontoporeia affinis*, accounting for an overlap of 1.00 at this time. The sculpin at this location also ate *P. affinis*, but overlap values with both cisco species ($C = 0.31$) were low because it ate isopods as well.

In nearshore coastal waters, overlap in winter diets was low:

	Arctic cod	Boreal smelt
Arctic cod	--	--
Boreal smelt	0.42	--
Fourhorn sculpin	0.08	0.10

Dietary overlap in winter was variable but seems lower at this time than in summer, suggesting a greater partitioning of food resources. This is apparent in coastal waters where the three commonly captured fishes relied in large part on different prey: arctic cod on mysids, fourhorn

TABLE V.   Dietary overlap among fishes and birds in Simpson Lagoon calculated by the Schoener overlap index.

	Summer 1977				
	Arctic cisco	Least cisco	Oldsquaw	Arctic char	Fourhorn sculpin
Arctic cisco	--	--	--	--	--
Least cisco	.47	--	--	--	--
Oldsquaw	.61	.63	--	--	--
Arctic char	.42	.31	.35	--	--
Fourhorn sculpin	.25	.29	.30	.44	--
Arctic cod	.37	.75	.56	.27	.11

	Summer 1978				
	Arctic cisco	Least cisco	Oldsquaw	Arctic char	Arctic cod
Arctic cisco	--	--	--	--	--
Least cisco	.71	--	--	--	--
Oldsquaw	.71	.52	--	--	--
Arctic char	.78	.56	.82	--	--
Arctic cod	.45	.56	.30	.34	--
Phalaropes	.11	.27	.13	.06	.22

sculpin on isopods, and boreal smelt on fish and mysids.   However, the ciscoes and sculpin in the Colville Delta all relied heavily on one prey item.

B. Invertebrate Prey

From the foregoing, it is clear that mysids and amphipods play a major role in the nearshore food web. Griffiths and Dillinger (1981) reported that the average number of mysids and amphipods on the lagoon bottom far exceeded their numbers in the water column above. Though numbers varied by species and through time, the density of key invertebrate species was usually 25 to 200 times greater in the region 0–20 cm above the lagoon bottom than in the entire 2 m of water above.

The epibenthic invertebrates in the study area appear to be trophic generalists, as shown by Schneider and Koch (1979):

Species	Principal foods
Amphipods:	
Onisimus glacialis	*Crustacean parts, diatoms*
Onisimus litoralis	*Diatoms, crustacean parts*
Gammarus setosus	*Peat, diatoms, crustacean parts*
Apherusa glacialis	*Diatoms, dinoflagellates, peat, crustacean parts*
Mysids:	
Mysis relicta	*Peat, diatoms, crustacean parts*
Mysis litoralis	*Diatoms, peat, crustacean parts*
Isopod:	
Saduria entomon	*Diatoms, polychaetes, peat*

C. Trophic Relations Between Consumers and Prey

Daily food requirements of fish and birds were compared to the availability of their major foods during the open-water season. Since the fish and birds selectively utilize the lagoon in preference to the ocean (Craig and Haldorson, 1981; Johnson and Richardson, 1981), only the feeding interactions within the lagoon were evaluated; however, food appeared to be equally abundant in the open sea beyond the barrier island. Our results are affected by the variable quality of some estimates, but the overall effect is small because the density estimates of the dominant consumer (oldsquaw ducks) and prey (mysids and amphipods) are reasonably accurate. Estimates pertaining to fish are less precise, but an error of even 100% in fish numbers or feeding rates would not greatly affect the overall results.

Daily food requirements of the key vertebrates in the study area during both 1977 and 1978 were determined using bird and fish densities of Johnson and Richardson (1981) and Craig and Haldorson (1981). Oldsquaw densities were determined by aerial surveys in both years (Table VI). Energy requirements for oldsquaws were computed using the equation (Kendeigh *et al.*, 1977)

$$\text{at } 0^{\circ}\text{C,} \quad M = 4.142 \ W^{0.544}$$

where M is the energy requirement for daily existence during the molting period (Kcal) and W is weight of bird (g). Using a digestive efficiency of approximately 70% (Owen, 1970) to convert daily existence energy requirements to intake requirements, Johnson and Richardson (1981) calculated the gross energy needs of an oldsquaw to be 240 kcal/bird-day, or 43.6 g ash-free dry weight (AFDW), assuming 5.5 Cal equals 1.0 g AFDW. Oldsquaw ducks thus consumed up to 14 mg AFDW m^{-2} day^{-1} in 1977 and 8.7 mg AFDW m^{-2} day^{-1} in 1978 (Table VI).

TABLE VI. Densities of oldsquaw ducks and their daily food requirements in Simpson Lagoon.

Survey date	1977		1978	
	Oldsquaw (no. km^{-2})	Food intake (mg AFDW m^{-2})	Oldsquaw (no. km^{-2})	Food intake (mg AFDW m^{-2})
5 June	0.0	0.0	--	--
20 June	0.2	<0.05	--	--
23 June	--	--	0.1	<0.05
5 July	6.0	0.3	15.5	0.7
15 July	--	--	183.2	8.0
25 July	--	--	79.8	3.5
28-29 July	321.1	14.0	--	--
5 August	--	--	75.4	3.3
15 August	261.0	11.4	100.7	4.4
25 August	--	--	58.2	2.5
30 August	137.1	6.0	--	--
5 September	--	--	23.2	1.0
15 September	--	--	26.3	1.2
22 September	666.3	29.1	--	--
23 September	--	--	199.2	8.7

Fish densities were determined primarily by sweeping a shoreline area of 1000 m^2 with a 91-m beach seine. Density extrapolations to central areas were made based on the 10:1 ratio of catch per unit efforts obtained using gill nets in shoreline and central lagoon waters. Fish species were combined to estimate densities of small fish (approximate average weight 15 g/fish) and large fish (approximate average weight 470 g/fish) in the lagoon in 1978 (Table VII). Fish densities for the previous summer were assumed to be similar with the exception that a brief run of millions of arctic cod like that of 11-20 August 1978 did not occur in 1977 (Craig and Haldorson, 1981). Energy requirements were assumed to be 6% of body weight per day for small fish (Craig and Haldorson, 1981) and 5% per day for large fish. Total food requirements of all fishes in Simpson Lagoon in summer were generally 1.6 mg AFDW m^{-2} day^{-1} but peaked briefly at 8.5 mg AFDW m^{-2} day^{-1} because of the arctic cod run in 1978 (Table VII).

The biomass of mysids and amphipods was determined by drop net samples. The standing crop in 1977 averaged 0.6 g AFDW m^{-2}, about three times the average standing crop in 1978 (Table VIII). Griffiths and Dillinger (1981) found that in 1978 the average standing crops of mysids and amphipods inside and outside the lagoon were similar (0.2 and 0.3 g AFDW m^{-2}, respectively), suggesting that the abundance of invertebrates in the lagoon was representative of that occurring in nearshore coastal waters.

TABLE VII. Densities and daily food requirements of small (15 g) and large (470 g) fish in Simpson Lagoon in 1978.

| Date | Fish densities (no. m^{-2} x 10^{-4}) | | | | Food intake of fish[a] |
| | Lagoon edge (7 km^2) | | Lagoon center (153 km^2) | | |
	Small fish	Large fish	Small fish	Large fish	(mg AFDW m^{-2} day^{-1})
Jul 1-10	5	5	0.5	0.5	0.2
11-20	20	20	2	2	0.7
21-31	50	40	5	4	1.6
Aug1-10	50	40	5	4	1.6
11-20	1250	40	1200	4	8.5[b]
21-31	70	40	7	4	1.7
Sep1-10	100	40	10	4	17.2
11-20	100	10	10	1	0.5
21-30	100	5	10	0.5	0.4

a Weighted mean for lagoon edge + center habitats. Ash-free dry weight calculated as 12% of wet weight of mysids and amphipods (Griffiths and Dillinger, 1981).

b Food ration for this period is based on small fish approximately 8 g in weight (the average weight of arctic cod which accounted for the run of fish at this time).

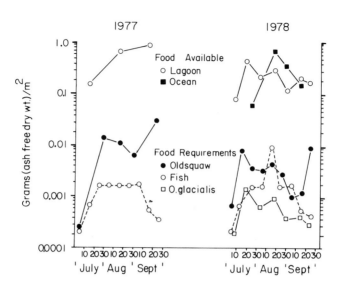

FIGURE 7. *Comparison of food available (standing crop of mysids and amphipods) and the daily food requirements of birds (oldsquaw), fish (species combined), and predatory invertebrates (the amphipod Onisimus glacialis).*

Our figures indicate that the food supply available in the lagoon exceeded consumer requirements by about 50 times during both years of study (Fig. 7). In 1977, the food supply even increased as the season progressed; thus food was not a limiting factor for consumers. In 1978, the food requirements of oldsquaw were lower than in 1977 (Table VI), but fish densities and energy needs were greater because of the run of arctic cod in mid-August (Table VII). Whether the year-to-year variations in numbers of consumers are related to differences in the energy available is not known.

In 1978 the energy requirements and food habits of the amphipod *Onisimus glacialis* were also considered since its diet consists largely of crustacean parts (Broad, 1977; Broad *et al.*, 1979). This amphipod is known to readily consume mysids. The daily ration of mysids eaten by *O. glacialis* was estimated after finding an average daily increase of 0.04 mg wet weight for first-year individuals, the dominant size class in the lagoon. Assuming that this growth represented 10% of the food consumed (Parsons *et al.*, 1977), the daily food requirement of *O. glacialis* is 0.4 mg wet weight, or approximately 0.08 mg AFDW. Of this we assumed that 10% of the diet of *O. glacialis* was mysids. From these estimates and the weighted mean density of *O. glacialis* in the lagoon, the food requirements of this species appear to have averaged 0.6 mg AFDW m^{-2} day^{-1}, which is a demand on the food resource sometimes comparable to that of the fish (Fig. 7).

1. Maintenance of the Food Supply

Did the initial early summer movement of mysids into the lagoon and their growth during the summer constitute an adequate food supply for the consumers present, or did more mysids migrate throughout the open-water season to replenish the food supply? Because biomass calculations do not distinguish between growth and immigration, we examined trends in the numbers of mysids over the summer, when no recruitment through reproduction occurs. The numbers of mysids in the lagoon thus reflected only immigration, emigration, and cropping by predators. Numbers of mysids eaten by predators were estimated by dividing the biomass of consumed mysids by the weight of the average individual as follows:

(1) For each consumer group, the biomass of mysids eaten during each interval between sampling periods (Fig. 7) was determined as

Consumed mysid biomass = $\frac{1}{2}(Y_1 + Y_2) (M) (\Delta t)$

where Y_1 = total AFDW required at first sampling date,
$\quad\quad Y_2$ = total AFDW required at second sampling date,
$\quad\quad M$ = percent mysids in diet (oldsquaw, 80%, fish, 70%, O. glacialis, 10%), and
$\quad\quad \Delta t$ = number of days between sampling dates.

(2) Because fish and presumably O. glacialis were not selective in the size of mysids they ate (Craig and Haldorson, 1981), the numbers of mysids eaten by these groups were estimated straightforwardly; however, because oldsquaw ducks consistently ate mysids larger than 8 mm (Johnson, 1984), numbers of mysids eaten by oldsquaws were calculated as biomass divided by the weight of a 10.5-mm mysid.

It is clear that consumers would quickly deplete the mysids if there was not a substantial and continual immigration of mysids into the lagoon (Fig. 8). While Simpson Lagoon provides a good feeding area for large numbers of birds and fish, these consumers effectively crop the supply of mysids entering the lagoon; they would deplete the lagoon if mysids were not continually replenished. The immigration of mysids, in turn, likely reflects the rapid exchange of lagoon water with offshore or adjacent coastal waters. A hypothesis would thus follow that arctic coastal habitats with limited water exchange are poorer feeding areas for birds and fish.

A. Carbon Sources

The energy supporting the large epifaunal populations in the lagoon could arise from three sources: in situ primary production, advection of primary production from offshore, and inputs from terrestrial sources. A small component may consist of kelp (Laminaria) detritus transported from the Stefansson Sound area to the east (Dunton et al., 1982).

Natural carbon isotope abundances show that primary production, either in situ or advected, is responsible for approximately 90% of the carbon making up amphipods and mysids in Simpson Lagoon during the summer season. Carbon derived from peat, as evidenced by radiocarbon

TABLE VIII. *Estimates of food available (mysids and amphipods) in Simpson Lagoon in 1977 and 1978*

1977 station	Habitat type (km²)	1977 biomass (g AFDW m⁻²)		
		15 Jul	15 Aug	5 Sep
1	36	0.12	0.33	0.24
2	102	0.15	0.76	1.16
3	22	0.30	0.93	_a_
Total biomass[b]		0.16	0.69	0.92

1978 station	Habitat type(km²)	1978 biomass (g AFDW m⁻²)						
		8 Jul	19 Jul	03 Aug	18 Aug	30 Aug	14 Sep	23 Sep
1	36	0.11	0.16	0.22	0.07	0.06	0.27	0.25
2	34	0.12	0.23	0.37	0.30	0.20	0.33	0.18
3	34	0.04	0.39	0.19	0.66	0.18	0.20	0.23
4	34	_c_	1.25	0.14	0.30	0.14	0.12	0.11
5	22	0.03	0.12	0.10	0.09	0.04	0.06	0.01
Total biomass[b]		0.08	0.45	0.22	0.30	0.12	0.21	0.17

_a_ Estimates for Station 3 on this date not used as they appeared unrealistically high (68.7 g AFDW m⁻²).
_b_ Weighted mean for area of represented habitat types.
_c_ Value at Station 4 assumed to be a mean of values at Stations 2 and 3.

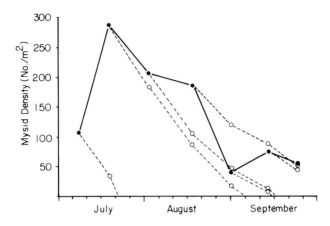

FIGURE 8. Observed density of mysids through summer 1978 (solid line) and expected rates of depletion at each date assuming cropping by predators and no new immigration (dashed lines).

depressions in consumers, is less than 10% in all invertebrate and fish samples from Simpson Lagoon. Only amphipods collected from near the mouth of the Colville River showed ^{14}C depressions equivalent to approximately 30% peat carbon (Schell *et al.*, 1982); the large quantities of peat in the river runoff and the very low primary productivity in the turbid river water probably account for the difference there.

1. Annual Primary Production Estimates

From literature estimates and our data (Table IX), we extrapolated seasonal primary production by multiplying the average effective daylight hours for June through September and summing the monthly integrated amounts. Within Simpson Lagoon we estimate annual production to be 5–7 g C m^{-2} or a total of 1.2–1.7 x 10^6 kgC yr^{-1}. This number, a small fraction of the approximately 24 x 10^6 kg C yr^{-1} derived from terrestrial inputs, dramatically illustrates the selectivity of invertebrate populations for phytoplankton carbon. Ice algae are not a major carbon source as the turbid ice and the large areas of bottom-fast ice preclude the establishment of algal populations. We estimate a seasonal ice algal production of near 0.1 g C m^{-2} yr^{-1}.

From these data we compared the carbon requirements of consumer organisms with the sources available. We included mysids, copepods, amphipods, and infauna as the major direct consumers of phytoplankton. Amphipods, especially *Onisimus* sp., are omnivorous and may be secondary rather than primary consumers of phytoplankton, but we assumed, after Griffiths and Dillinger (1981), that 10% of their diet is animal material and the rest is plant derived. Table X lists the estimated ingestion requirements based upon the following assumptions:

TABLE IX. Monthly averaged primary productivity estimates from ^{14}C incubations for Harrison Bay, Simpson Lagoon, Prudhoe Bay, and offshore.

Study area	Investigator	Primary productivity ($mg\ C\ m^{-3}\ h^{-1}$)[a]			
		May	June	July	August
Harrison Bay	Alexander et al. (1975)	--	--	--	0.61
	This study	--	--	--	0.28
Simpson Lagoon	Alexander et al. (1975)	--	0.18	1.86	0.33
	This study	--	--	--	2.27
Prudhoe Bay	Coyle (1974)	--	--	3.46	0.77
		--	--	--	0.48
Offshore	Horner and Schrader (1981)	0.07	0.09	--	--
	Schell et al. (1982)	--	0.15	1.89	--
	Alexander et al. (1975)	--	--	--	2.39

[a] No data available for September from nearshore waters. Average productivity rates from August (1.0 mg C m^{-3} h^{-1}) were used. Offshore values are from stations less than 10 km north of the barrier islands of Simpson Lagoon and Stefansson Sound.

Copepods:

(1) Average summer standing stocks are equivalent to 25 mg C m^{-2} (Griffiths and Dillinger, 1981).
(2) Summer data are representative of the period July–November, and winter populations are approximately 5% of summer (Tarbox et al., 1979).
(3) Copepods ingest 40% of their body weight per day (Parsons et al., 1977), and the same percentage of body carbon is ingested.

Mysids and amphipods:

(1) Total population growth is calculated from seasonal densities and growth equations for first-year class Mysis litoralis and Onisimus glacialis, the dominant species (Griffiths and Dillinger, 1981).
(2) Growth is 10% of ingestion (Parsons et al., 1977).

Infauna:

(1) Infaunal biomass being about the same as epifauna (Crane and Cooney, 1975; and this paper), infauna ingest the same amount of carbon each year as mysids and amphipods. Food is sinking phytoplankton or fecal pellets derived from phytoplankton.
(2) Summer ingestion rates are twice the winter rates due to temperature increases.

From the results in Table X, we calculated the total amount of carbon ingested each year by secondary consumers to be approximately 6.9 g C m^{-2}. We estimate annual primary productivity in Simpson Lagoon to be 5–7 g C m^{-2} yr^{-1} from the data of Alexander et al. (1975). The primary and secondary productivity data appear to balance.

2. Transport of Primary Production into the Lagoon

The finding that terrestrial carbon does not appreciably support the epibenthic fauna of Simpson Lagoon underscores the close links between the summer populations of mysids and amphipods in the lagoon with those offshore. Given the uncertainties in estimating production, the standing stocks of primary producers and consumers are in rough balance with energy inputs. Immigration of mysids and amphipods into the lagoon compensates for predation by birds and fish, and the primary production is rapidly grazed. Nevertheless, growth rates of phytoplankton are low, as evidenced by low primary productivity and a long (3–4 day) turnover time for standing stocks, which argues for a large advective input of phytoplankton. Evidence that this process is important was observed by Alexander et al. (1975), who found higher rates of primary production in offshore deeper (> 2 m) water. Further, the observed rapid mixture of river inflow with seawater during the open-water season requires advection of sea water into the lagoon. Mungall (1978) estimated exchange times as short as 24 h under strong wind conditions.

TABLE X. Carbon ingested by secondary consumers, Simpson Lagoon

	Ingestion (g C m^{-2})			
Group	July–Sept	Oct–mid-Feb	Mid-Feb–June	Total
Copepods	0.9	0.6	0.1	1.6
Mysids	1.1	0.2	0.1	1.4
Amphipods	0.5	0.6	0.2	1.3
Infauna	1.0	0.8	0.8	2.6
Total	3.5	2.2	1.2	6.9

We can conclude that *in situ* and advected primary production drive the lagoon ecosystem. Primary productivity in the lagoon is probably too low to sustain the grazer biomass, and advection of offshore deeper water provides the major supplement to consumer food requirements. The low salinity of the lagoon and the stresses imposed on phytoplankton by the euryhaline water may result in more rapid sinking rates. The accumulation of phytoplankton at the seafloor may account for the high densities of mysids and amphipods observed there.

V. DISCUSSION

A superabundance of food is available to fish and birds in Simpson Lagoon during the summer: invertebrate studies show that roughly 50 times more food (mysids and amphipods) is available than is required daily by vertebrate consumers. The epibenthic prey are probably easy to catch on the smooth lagoon floor, although some may escape into the detrital mat on the lagoon bottom.

The high diet overlap among consumers also indicates that food is not a limiting factor. Overlap occurs not simply from the reduced variety of species in the arctic, for there are several sources of food in the lagoon (hydroids, polychaetes, mollusks, isopods, and tunicates) which predators do not appreciably exploit. Instead, consumers feed primarily on the abundant epibenthic invertebrates and their diets overlap considerably. This implies a lack of competition for food because the available food supply is not finely partitioned by the predators. Limited food resources and strong competition for food would give rise to specialized and non–overlapping feeding habits among predators, such as have been documented in other species-poor ecosystems in non-lagoon waters of the Arctic (Tyler, 1978) and Antarctic (Targett, 1981). Some exceptions to the foregoing pattern occur—fourhorn sculpin, for example, eat more isopods than other consumers, boreal smelt eat more fish, and differences in the mouths of different fish species indicate a degree of feeding specialization—nevertheless the epibenthic food resource remains the principal diet of most fishes.

Some consumers prefer different feeding habitats; however, this partial partitioning of lagoon habitat does not obscure the point that the lagoon's supply of epibenthos (especially mysids) is highly dependent on immigration or dispersal from outside areas. Both oldsquaw ducks (which feed mainly in open lagoon waters) and fish (which probably feed near shorelines) are feeding on the same food supply, but at different points along its path through the lagoon.

Can the premise that food is superabundant be extrapolated to other nearshore areas along the Beaufort Sea coastline? The available data indicate that the kinds of birds and fish in Simpson Lagoon are generally similar to those all along the Beaufort Sea coastline, and at all locations studied the birds and fish rely on epibenthic invertebrates as their principal food resource (e.g., Kendel et al., 1975; Furniss, 1975; Griffiths et al., 1975, 1977; Bendock, 1977; Johnson, 1983). But in August 1978 an event in Simpson Lagoon suggested that fish may on occasion reduce the food supply to low levels. During this month a large school of arctic cod entered the lagoon and consumed a large quantity of mysids during their nine-day visit, which may have contributed to a decline in mysid densities at that time. This decline did not substantially affect the food base, but it is conceivable that the school of cod, given its estimated size and rate of food consumption, could have seriously depleted the available food supply if it had remained in the lagoon an extra two weeks or so. A second note of caution is that a high incidence of empty fish stomachs has been recorded at some coastal locations (summarized in Craig and Haldorson, 1981). While the occurrence of empty stomachs may reflect a number of factors (e.g., periodicity in feeding, regurgitation or digestion of food after capture, or reduced feeding by anadromous fish during migration to fresh water), the data also are consistent with either of two very different interpretations: (1) fish may not have to feed continuously in order to satisfy their nutritive requirements, or (2) fish are not getting enough to eat at some locations or in some years.

A. Factors Contributing to Food Abundance

1. Immigration

Despite the large standing stock of epibenthos in Simpson Lagoon in summer, birds and fish could theoretically consume it all within 2-6 weeks. But the food supply remains relatively high throughout the summer despite an increase in consumer demand. Potential reasons are growth of the invertebrates present in the lagoon and immigration of new invertebrates into the lagoon. (Reproduction does not occur during summer.) Griffiths and Dillinger (1981) showed that immigration is by far the more important. Mysis litoralis, the major prey species, moves into the lagoon in early summer; this and the immigrations that follow are critical events for the fish and birds. Shallow nearshore habitats become attractive to feeders only after being repopulated each year by key invertebrates and remain so only by continued immigration. If these immigrations are obstructed by natural events or artificial alterations in the nearshore environment, the lagoon might remain a poor feeding area for that particular summer.

Productivity in Simpson Lagoon is linked to offshore areas in a more fundamental way. Isotopic studies of organisms collected in the lagoon indicate that fish ultimately obtain 90% of the carbon in their tissues from modern marine primary production (Schell et al., 1982). Relatively little of the large input of terrestrially derived organic carbon appears to enter trophic pathways leading to birds and fish in the marine ecosystem (but note Schell's (1983) discussion about the role of peat in the freshwater ecosystem).

2. Seasonally Limited Availability

A factor that contributes to the abundance of epibenthos is that these invertebrates are protected from most vertebrate consumers for almost nine months of the year. Birds and anadromous fishes can eat them only during the short open-water period, although marine fish have almost year-round access to the epibenthos.

3. Habitat Disturbance

Ecological succession, the orderly process of community change, is maintained at an early stage in Simpson Lagoon. The lagoon is a pioneer community because periodic physical disturbance allows few invertebrates to be permanent residents. Several features of the lagoon make life difficult. The lagoon is shallow, wind-churned, and turbid. Its nearly featureless bottom is covered with an unstable mud-sand substrate. Summer water temperatures and salinities are highly variable, and a rapid flushing rate ensures that the system is influenced by events outside its borders. In winter almost the whole lagoon freezes solid and little free water remains under the ice elsewhere; at this time water may become hypersaline. In spring there may be some ice-gouging of substrates as well as rapid and extreme fluctuations in salinity.

Species diversity in early successional stages is typically low, but the species present are typically very numerous. In Simpson Lagoon, it appears that the numerical success of the epibenthic invertebrates is due to their mobile life style and tolerance of a wide range of physical conditions. The mobile mysids and amphipods are widespread and seem to utilize virtually the entire nearshore zone, being tolerant of the wide ranges of water temperature and salinity (Busdosh and Atlas, 1975; Broad et al., 1979) that occur there. As a generalization, the epibenthic invertebrates are colonizers, which Williams (1969) described as "versatile species—creatures of the ecotone, physiologically and ecologically tolerant of many conditions and requiring of few."

The contention that Simpson Lagoon is a disrupted environment is reflected in the high degree of food overlap among consumers. As disturbance to an ecosystem increases, the amount of dietary overlap also increases (Tyler, 1978). Tyler found that overlap is low in systems that are physically constant. In Dease Strait, a deep-water arctic area with year-round constant temperatures and almost year-round ice cover, the assemblage of fishes has a very strong partitioning of food resources (low overlap), quite unlike the physically disturbed habitat of Simpson Lagoon (high overlap). This relationship is shown in Fig. 9. Tyler suggested that

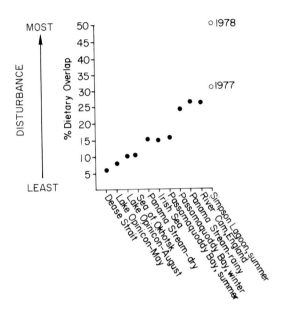

FIGURE 9. Relationship between dietary overlap among vertebrate consumers and the degree of habitat disturbance in the system (Tyler, 1978). Data are ranked by percentage overlap; solid dots indicate overlap among fish communities at various locations, and open circles indicate overlap among major consumers (fish and birds) in Simpson Lagoon, summers 1977 and 1978. Redrawn from Tyler (1978).

regularly repeated perturbations to a system would allow the persistence of high-turnover r-type species (Gadgil and Solbrig, 1972), accompanied by weakening of food resource partitioning and coexistence of predators that would otherwise not be possible.

To compare the Simpson Lagoon data with Tyler's findings, the percent overlap was calculated from principal prey diagrams (Fig. 6), the method used by Tyler (1978). For the summer of 1977, there are seven entries of principal prey and six consumers. The possible number of recurrences of principal prey in consumer diets is thus 35. Since there are 11 actual recurrences, the overlap was 11/35 (31%). There was a 50% overlap the following summer. Analyzed in this fashion, Simpson Lagoon is a "most disturbed" habitat during the summer months (Fig. 9).

In winter, one might predict that overlap would be low since ice cover would dampen physical fluctuations; however, the widely separated overlap values obtained at different places and times during this period make interpretation difficult. Overlap was low in coastal waters (10%, 1978-79 winter) but high in the Colville Delta (50%, 1977-78 winter). Reasons for this difference may reflect several factors (*e.g.*, different fish species, different habitats, small sample sizes).

In summary, the bistic community of Simpson Lagoon is maintained at an early successional stage by harsh and periodic physical disturbances. During the brief open-water season, the lagoon is invaded by pioneer species which exploit an apparently abundant food resource.

ACKNOWLEDGMENTS

We thank Gunter Weller and Dave Norton for their continued interest and assistance in our research. We especially acknowledge Joe Truett, Carl Walters, John Richardson, Alan Birdsall, and Aaron Sekerak for their discussions and overall improvements to our programs. We greatly appreciate the contributions of the many field biologists, laboratory technicians, office staff, and administrators who have participated in this project over the years.

REFERENCES

Alexander, V., Coulon, C., and Chang, J. (1975). *In* "Environmental Studies of an Arctic Estuarine System - Final Report" (V. Alexander *et al.*, eds.), EPA-660/3-75-026, Environmental Research Laboratory, U.S.E.P.A. Corvallis, OR.

Andriyashev, A.P. (1954). "Fishes of the northern seas of the U.S.S.R." Izdatel'stvo. Akad. Nauk. SSSR, Moskva-Leningrad. Translation from Russian, Israel Program for Scientific Translations. Jerusalem, 1964.

Bendock, T. (1977). *In* "Environmental Assessment of the Alaskan Continental Shelf, Final Report of Principal Investigators, March 1979," Vol. 4, p. 670. NOAA, Boulder.

Broad, A.C. (1977). *In* "Environmental Assessment of the Alaskan Continental Shelf, Quarterly Report," Vol. 9, p. 109, NOAA, Boulder.

Broad, C. (1978). *In* "Environmental Assessment of the Alaskan Continental Shelf, Annuual Report of Principal Investigators, March 1978," Vol. 5, p. 1. NOAA, Boulder.

Broad, C., Benedict, A., Dunton, K., Koch, H., Mason, D., Schneider, D., and Schonberg, S. (1979). *In* "Environmental Assessment of the Alaskan Continental Shelf, Annual Report of Principal Investigators," Vol. 3, p. 361. NOAA, Boulder.

Busdosh, M., and Atlas, R. (1975). *J. Fish. Res. Board Can. 32,* 2564.

Cailliet, G.M., Antrim B.S., and Ambrose, D.A. (1978). *In* "Fish Food Habits Studies: Gut Shop 78" (S. Lipovsky and C. Simenstad, eds.), p. 118. Sea Grant Pub., Washington, D.C.

Cannon, P.J., and Rawlinson, S.E. (1978). *In* "Annual Report of Principal Investigators," Vol. 10, p. 687. NOAA, Boulder.

Clutter, R.I. (1965). *Limnol. Oceanogr. 10(2),* 293.

Coyle, K.O. (1974). M.S. Thesis. Univ. of Alaska, Fairbanks.

Craig, P.C. (1983). "Fish use of coastal waters of the Alaskan Beaufort Sea. Environmental assessment of the Alaskan Continental Shelf, Sale 87 Synthesis." NOAA, Juneau.

Craig, P.C. and Haldorson, L. (1981). In "Environmental Assessment of the Alaskan Continental Shelf, OCS Final Report of Principal Investigators," Vol. 7, p. 384. NOAA, Boulder.

Crane, J. (1974). M.S. Thesis, Univ. Alaska, Fairbanks.

Crane, J.J. and Cooney, R.T. (1975). In "Environmental Studies of an Arctic Estuarine System - Final Report" (V. Alexander, ed., p. 299. EPA-660/3-75-026, Environmental Research Laboratory, U.S.E.P.A., Corvallis, OR.

Dunton, K., Reimnitz, E., and Schonberg, S. (1982). Arctic 35, 465.

Furniss, R. (1975). Alaska Dept. Fish and Game Annual Rept. 16.

Gadgil, M. and Solbrig, O. (1972). Am. Nat. 106, 14.

Griffiths, W., Craig, P.C., Walder, G., and Mann, G. (1975). Arctic Gas Biol. Rep. Ser. 34(2).

Griffiths, W., DenBeste, J., and Craig, P. (1977). Arctic Gas Biol. Rep. Ser. 40(2).

Griffiths, W. and Dillinger, R. (1981). In "Environmental Assessment of the Alaskan Continental Shelf, Final Report of Principal Investigators," Vol. 8, p. 1. NOAA, Boulder.

Horner, R., and Schrader, G.C. (1981). In "Environmental Assessment of the Alaskan Continental Shelf, Final Report of Principal Investigators," Arctic 35, p. 485. NOAA, Boulder.

Johnson, S. R. (1983). In "Environmental Characterization and Biological Use of Lagoons in Eastern Beaufort Sea." Rept. by LGL Ecological Res. Assoc., Inc. to NOAA, Off. Mar. Poll. Assess., Juneau.

Johnson, S. R. (1984) Can. Wldl. Serv. Spec. Pub. Ottawa.

Johnson, S.R., Adams, W.J., and Morrell, M.R. (1975). "Birds of the Beaufort Sea." Dept. Envir. (Canada), Victoria, B.C.

Johnson, S. R. and Richardson, W.J. (1981). In "Environmental Assessment of the Alaskan Continental Shelf, Final Report of Principal Investigators," Vol. 7, p. 190. NOAA, Boulder.

Kendeigh, S.C., Dol'nik, V.R., and Gavrilov, V. (1977). In "Granivorous Birds in Ecosystems, Their Evolution, Populations, Energetics, Adaptations, Impact and Control" (J. Pinkowski and S.C. Kendeigh, eds.), p. 127. Internat. Biol. Prog. 12. Cambridge Univ. Press, Cambridge.

Kendel, R., Johnston, R., Lobsiger, V., and Kozak, M. (1975). Beaufort Sea Tech. Rep. 6.

Kessel, B. and Gibson, D. (1978). "Status and Distribution of Alaskan Birds." Cooper Ornith. Soc., Los Angeles, CA.

Linton, L., Davies, R., and Wrona, F. (1981). J. Anim. Ecol. 50, 283.

Mungall, C. (1978). In "Environmental Assessment of the Alaskan Continental Shelf, Annual Report of Principal Investigators," Vol. 10, p. 732. NOAA, Boulder.

OCSEAP (Outer Continental Shelf Environmental Assessment Program) (1978). "Environmental Assessment of the Alaskan Continental Shelf: Interim synthesis: Beaufort/Chukchi." NOAA, Boulder.

Owen, R.B., Jr. (1970). Condor 72, 153.

Parsons, T.R., Takahashi, M., and Hargrave, B.   (1977).   "Biological oceanographic processes."   Second edition.   Pergamon Press, New York.

Schell, D.M. (1983). *Science 219*, 1068.

Schell, D.M., Ziemann, P., Parrish, D., and Brown, E. (1982). "Foodweb and Nutrient Dynamics in Nearshore Alaskan Beaufort Sea Waters." Cumul. Summ. Rept. for NOAA, Boulder.

Schneider, D. and Koch, H.   (1979).   *In* "Environmental Assessment of Selected Habitats in the Beaufort and Chukchi Littoral System." Vol. 3, p. 503. NOAA, Boulder.

Schoener, T. (1968). *Ecol. 49*, 704.

Strickland, J.D.H. and Parsons, T.R., (1972). "A Practical Handbook of Seawater Analysis." Bulletin 167, Fish. Res. Bd. Canada.

Tarbox, K., Busdosh, M., LaVigne, D., and Robbilliard, G.   (1979).   *In* "Environmental Studies of the Beaufort Sea - Winter 1979," part II. Woodward-Clyde Consultants, Anchorage, AK.

Targett, T. (1981). *Mar. Ecol. Prog. Ser. 4*, 243.

Tyler, A.V. (1978). Tech. Paper No. 5050, Oregon Agric. Expt. Station, Corvallis.

Wallace, R. (1981). *Trans. Am. Fish. Soc. 110*, 72.

Williams, E. (1969). *Quart. Rev. Biol. 44*, 345.

Zaret, T.H. and Rand, A.S., (1971). *Ecology, 52*, 336.

TROPHIC RELATIONSHIPS OF VERTEBRATE CONSUMERS IN THE ALASKAN BEAUFORT SEA

Kathryn J. Frost
Lloyd F. Lowry

Alaska Department of Fish and Game
Fairbanks, Alaska

I. INTRODUCTION

The Beaufort Sea (Fig. 1) is an important feeding area for several species of vertebrate consumers. Although estimates of annual primary production by ice-associated algae and phytoplankton are low, amounting in combination to 16-36 g C m^{-2} yr^{-1} (Schell *et al.*, 1982), species such as the bowhead whale (*Balaena mysticetus*) leave the highly productive Bering Sea, where annual phytoplankton production is estimated at 120-400 g C m^{-2} yr^{-1} (Iverson and Goering, 1979), and migrate 3000 km or more to the Beaufort Sea to feed. The biological and oceanographic processes which produce favorable feeding conditions in the Beaufort Sea are very poorly known.

Recent studies in the United States and Canada have greatly increased the available information on distribution, abundance, and food habits of vertebrate consumers in the Arctic. Information obtained has allowed the construction of food webs which delineate trophic relationships among major components of the arctic fauna (Lowry and Burns, 1980; Bradstreet and Cross, 1982; Schell *et al.*, 1982). Most studies conducted, however, have not considered the quantities of food consumed on an annual basis, although such information is important when considering the interrelations of species feeding on similar prey. In this chapter we compile and review the available information on biomass and food habits and use it to produce estimates of the quantities of prey consumed by major vertebrate species on the continental shelf of the Alaskan Beaufort Sea, an area of approximately 50,000 km^2. Such estimates can be of great value in evaluating the ecological role of various consumers. We consider only those species which feed primarily on organisms connected to the pelagic/planktonic food web.

THE ALASKAN BEAUFORT SEA:
ECOSYSTEMS AND ENVIRONMENTS

381

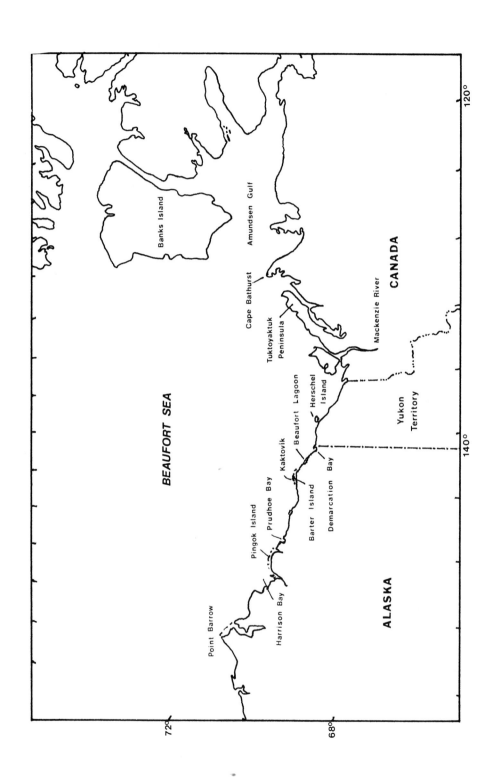

II. POPULATION SIZE, BIOMASS, AND FOODS OF MAJOR VERTEBRATE CONSUMERS

A. Marine Mammals

Of the nine species of marine mammals regularly found in the Beaufort Sea (Eley and Lowry, 1978), three species, bowhead whales, white whales, and ringed seals, are abundant and are directly linked to the pelagic food web (Table I). To determine the food consumption of these species during their stay in the Alaskan Beaufort Sea, it is necessary to know the residence time, population size and biomass, and diet composition of each.

1. Bowhead Whales.

During their spring northward migration, bowhead whales travel through the shore lead which extends north from Bering Strait to Point Hope then northeastward near the Alaskan coast to Point Barrow (Marquette, 1977). Whales pass Barrow from late April to at least early June (Durham, 1979). After passing Barrow, they move east through a series of leads which run from Point Barrow to the northwestern portion of Banks Island (Braham et al., 1980; Ljungblad, 1981). Bowheads have been observed near Banks Island as early as May (Braham et al., 1980) and by June are seen in the polynya which forms at the mouth of Amundsen Gulf between Cape Bathurst and Banks Island (Fraker et al., 1978).

In July and August, bowheads feed in the southeastern Beaufort Sea and outer Amundsen Gulf west of long. 122°W (Fraker and Bockstoce, 1980). Few bowheads summer in the Alaskan part of the Beaufort Sea (Frost and Lowry, 1981b).

During the fall, most bowheads migrate westward near shore in relatively shallow water. Sightings of whales in recent years were most common off the Mackenzie estuary and Tuktoyaktuk Peninsula in August and west of there along the Yukon Territory coast in September (Fraker and Bockstoce, 1980). Ljungblad et al. (1980) reported sightings of groups of bowheads along the Alaskan coast east of Barter Island near Demarcation Bay in late September. Groups of bowheads have been sighted near and to the east of Barrow in the same time period (Braham and Krogman, 1977; Fraker et al., 1978). The latest sightings of bowheads in the Alaskan Beaufort Sea have occurred in October and November (Ljungblad et al., 1980; Lowry and Burns, 1980; Ljungblad, 1981).

From the available information we conclude that most bowheads spend June through early September feeding in Canadian waters. The bowhead population migrates westward along the continental shelf during September and October. The timing of the migration appears to be greatly affected by freeze-up but in most years extends for about a month beginning in mid-September. Since bowheads also appear in the Chukchi Sea in early October, the average bowhead may spend only about 25 days foraging in the Alaskan Beaufort Sea.

FIGURE 1. Map of the Beaufort Sea showing locations mentioned in text.

TABLE I. Seasonal Abundance and Primary Foods of Mammals which
Regularly Occur in the Alaskan Beaufort Sea.

| Common name | Abundance | | Primary foods |
Scientific name	Winter/ spring	Summer/ fall	
Bowhead whale Balaena mysticetus	Absent	Common	Small to medium-sized zooplankton and nekton
Gray whale Eschrichtius robustus	Absent	Rare	Epibenthos
White whale Delphinapterus leucas	Absent	Common	Fish, crustaceans, and cephalopods
Bearded seal Erignathus barbatus	Uncommon	Uncommon	Epibenthos and infauna
Ringed seal Phoca hispida	Common	Common	Fish, medium-sized nekton, and epibenthos
Spotted seal Phoca largha	Absent	Uncommon	Fish, crustaceans, and cephalopods
Walrus Odobenus rosmarus	Absent	Uncommon	Infauna and epibenthos
Polar bear Ursus maritimus	Common	Common	Ringed seals, bearded seals, and carrion
Arctic fox Alopex lagopus	Common	Rare	Carrion and ringed seal pups

The best estimate of present abundance, based on counts of animals made near Point Barrow during the spring migration, is 3125-3987 (median 3556) bowheads (Dronenburg et al., 1982). Based on satellite photos and aerial surveys, it appears that all animals passing Point Barrow head eastward. However, surveys conducted in the Alaskan and Canadian sectors of the Beaufort Sea have accounted for fewer than the total estimated number of whales, suggesting that some animals may spend part or all of the summer on feeding grounds in the Chukchi Sea or elsewhere in the Canadian Arctic. For our purposes, we consider that 3556 bowheads feed in the Canadian Beaufort Sea in summer and subsequently migrate through and feed in the Alaskan Beaufort Sea.

The average length of bowheads harvested by Eskimos in 1973-77 was 10.2 m (Marquette, 1977). By applying the length-weight relationship that Lockyer (1976) developed for black right whales, Balaena glacialis, (weight (t) = 0.0132 X length (m)$^{3.06}$), Draper et al. (1979) calculated that the average bowhead harvested weighed approximately 15.9 metric tons. Griffiths and Buchanan (1982) stated that the average length of whales observed in the eastern Beaufort Sea was 13.5 m, from which they calculated an average weight of 45 metric tons. Although the sample of whales taken by Eskimo hunters is probably biased toward small animals, it seems unlikely that the average length of whales in the population is 13.5 m since sexual maturity occurs at a length of 11-12 m and physical maturity is reached at 14-15 m (Marquette, 1977). For our purposes, we assume that the average whale is 12 m long. Based on this assumption, the average bowhead weighs 26 t and the biomass of 3556 bowheads is about 92,500 metric tons.

Foods eaten by bowhead whales are poorly known in comparison to other baleen whales. Commercial whalers removed only the baleen and blubber of bowheads and thus seldom observed their stomach contents. Most published reports of bowhead foods (MacGinitie, 1955; Tomilin, 1957; Johnson et al., 1966; Durham, 1972; Mitchell, 1975) present only general information, often based on indirect observations, which indicates that bowheads feed on small and medium-sized zooplankton and sometimes also benthic organisms such as amphipods and mysids.

Since 1976 we have examined samples of prey items from the gastrointestinal tracts of 14 whales taken in the Beaufort Sea (Lowry et al., 1978; Lowry and Burns, 1980; Frost and Lowry, 1981b). One whale had small clams in the colon; among the others, euphausiids were the major food in seven and copepods in six. Other prey groups, although they comprised many species and occurred quite frequently, were never a major component of the samples. Either *Thysanoessa raschii* or *Calanus hyperboreus* was usually the dominant prey species.

Quantitative data on the stomach contents of bowhead whales taken during the fall (Lowry et al., 1978; Lowry and Burns, 1980) indicate that copepods composed approximately 60% of the contents at Kaktovik, while at Barrow copepods did not occur. Euphausiids composed about 37% of stomach contents at Kaktovik and 92% at Barrow (Table II). Giving equal weight to both the Barrow and Kaktovik samples, the overall proportions of prey in the diet in the Alaskan Beaufort Sea would be approximately euphausiids (*Thysanoessa* spp.) 65%; copepods (*Calanus* spp.) 30%; hyperiid amphipods (*Parathemisto* spp.) 1%; and other organisms, including primarily gammarid amphipods, 4%.

Published estimates of daily food consumption rates for large whales range from 1 to 4% of total body weight (Sergeant, 1969; Brodie, 1975, 1980). Brodie (1975) estimated that an average antarctic fin whale (*Balaenoptera physalus*) that fed for 120 days and fasted for the remainder of the year would consume about 2.1% of its total body weight per day during the feeding season. Since the average fin whale considered by Brodie was much larger (48.0 t) than what we are assuming for bowheads (26.0 t), and since, generally speaking, there is an inverse relationship between size and feeding rate (Sergeant, 1969), we will assume the daily food consumption rate of bowheads in the Alaskan Beaufort to be 3% of the total body weight per day.

2. White Whales.

Many characteristics of the distribution and movements of white whales are similar to those described for bowheads. White or belukha whales migrate along the Alaskan coast in spring, generally in association with bowheads (Braham and Krogman, 1977), arriving in the eastern Beaufort Sea in late May and early June (Fraker et al., 1978). They appear in the Mackenzie estuary in late June or early July. Many whales remain in

TABLE II. *Quantitative Composition (%) of Stomach Contents from Bowhead Whales.*

Prey item	Kaktovik, autumn 1979 Specimen number						Barrow, autumn 1976 Specimen number		
	1	2	3	4	5	Mean[a]	6	7	Mean[b]
Copepod	100	99	23	88	<1	60	--	--	--
Euphausiid	--	<1	68	5	98	37	97	87	92
Mysid	--	<1	7	--	1	<1	--	--	--
Hyperiid amphipod	<1	<1	<1	<1	--	<1	2	3	3
Gammarid amphipod	<1	<1	<1	2	<1	<1	1	10	5
Other invertebrate	<1	<1	--	2	1	1	--	<1	<1
Fish	<1	<1	1	2	--	<1	--	--	--
Sample volume (mL)	2406	545	400	131	358		18	33	
Estimated total contents (L)	44	18	22	18	36		unk.	109	

[a]*Calculated from the volume and percent composition of each sample and the estimated total contents of stomachs from which samples were taken.*
[b]*Calculated as the simple average of the two sample percentages.*

the estuary until early to mid–August, while others occur in the eastern Beaufort Sea and western Amundsen Gulf (Fraker *et al.*, 1978). Few white whales occur in the Alaskan Beaufort Sea during July and August (Frost and Lowry, 1981b).

White whales move westward during the latter half of August and September. Most sightings have occurred in deep water north of the continental shelf (Frost and Lowry, 1981b). To calculate food consumption, we assume that whales pass through the Alaskan Beaufort Sea during a period of 30 days in late August and September.

Fraker *et al.* (1978) summarized estimates of white whale abundance in the Mackenzie estuary. The maximum in 1976 was 5500–6000, while estimates for 1977 and 1978 were 5500 and 6600, respectively (Fraker, 1978). We will use 6000 as an estimate of the number of whales which summer in the Mackenzie estuary and migrate through the Beaufort Sea.

Since there are no direct observations of weights of white whales in the Beaufort Sea, we derive a biomass estimate using the length-weight relationship determined for white whales in the St. Lawrence estuary (log weight (kg) = 2.605 X log length (cm) - 3.807; Sergeant and Brodie, 1969). Reported mean lengths of whales taken in the Mackenzie region are 4.1 and 4.3 m for males and 3.6 and 3.9 m for females (Sergeant and Brodie, 1969; Fraker *et al.*, 1978). Since most of the animals taken are adults (Fraker *et al.*, 1978), these measurements are greater than the length of an average individual in the population. We use 4.0 m and 3.5 m as the average length of male and female white whales, respectively. Based on the Sergeant-Brodie length-weight relationship, the average weight for males and females is 940 and 660 kg. Assuming a 50:50 sex ratio, an average white whale then weighs about 800 kg, and the total biomass of 6000 whales is 4800 metric tons.

There are few direct observations of white whale stomach contents in the Beaufort Sea. Seaman *et al.* (1982) reported arctic cod (*Boreogadus saida*) as a major food at Barrow and Point Hope during the spring migration, while other fishes such as saffron cod (*Eleginus gracilis*), herring (*Clupea harengus*), smelt (*Osmerus mordax*), and sculpins (Family Cottidae) were eaten during summer at more southern locations. Fraker *et al.* (1978) speculated that arctic cod is a major food offshore but found little feeding inshore near the Mackenzie estuary. Considering the importance of arctic cod in the summer diet of white whales in other areas of the Arctic (*e.g.*, Kleinenberg *et al.*, 1964), we assume that 80% of the diet in the Alaskan Beaufort Sea is composed of arctic cod. The remaining 20% may consist of organisms such as shrimps, cephalopods, and other fishes.

Sergeant (1969) estimated the food consumption of white whales as 5.1% of the total body weight per day, which is the value we use for our calculations.

3. Ringed Seals.

Ringed seals are present in the Beaufort Sea throughout the year. Although they can and do occur in all sea ice types, the seasonal cycle of sea ice has a great effect on their distribution and regional abundance (Burns, 1970; Frost and Lowry, 1981a). During the summer ringed seals range freely. With the onset of winter freeze-up, their movements become increasingly restricted. It is generally considered that many seals which have summered in the Beaufort Sea move west and south with the advancing ice and disperse throughout the Chukchi and Bering seas. Others remain in the Beaufort, probably concentrating in areas of abundant prey (Lowry *et al.*, 1980). As ice forms, the seals make and maintain breathing holes in the fast ice or in thin first-year ice in the pack (Smith *et al.*, 1978). Lairs in which to haul out, give birth, and nurse young are excavated in the snow which covers the ice (Smith and Stirling, 1975).

As the snow cover melts in May and June, birth and haul-out lairs collapse, and increasing numbers of seals appear near holes and leads (Burns and Harbo, 1972; Smith, 1973). From this time until July, seals haul out for long periods of time while the annual molt occurs (Finley, 1979). The number of hauled-out seals diminishes in July, and ringed seals are rarely seen on the ice for the remainder of the year.

Distribution of ringed seals during open water is poorly understood as seals spend virtually all of their time in the water feeding. Their distribution is believed to be greatly influenced by that of their prey (Frost and Lowry, 1981b; unpubl. data).

Direct estimates of ringed seal abundance are available only from surveys conducted in late spring, which presumably reflect abundance of seals during the previous winter. In order to estimate the total number of ringed seals, it is necessary to know what proportion of the population is on the ice during surveys. Smith (1973, 1975) considered that 50% of ringed seals were usually in the water at the times of his aerial counts. Detailed surface observations by Finley (1979) indicated that under ideal circumstances 70% or more of the seals may be hauled out at one time. Since all surveys do not occur under ideal conditions, we will use Smith's estimate that observed densities are 50% of actual densities. The observed spring density of ringed seals in the Beaufort Sea (including the Yukon coast), derived from 7 years of aerial surveys, is 0.40 seal km^{-2} (Frost and Lowry, 1981b). Using the above ratio, the actual density is 0.80 seal km^{-2}, resulting in an estimate of 40,000 seals in the study area in winter and spring.

Freeze-up in the Beaufort Sea is usually well underway by November, at which time ringed seals appear in large numbers at coastal locations in the Bering Sea, presumably having moved south from the Chukchi and Beaufort seas. In the latter part of June, there is a mass influx of ringed seals back into some areas of the Beaufort Sea. Thus, the winter-spring density derived above is applicable from about 1 November to 1 July. Although it is apparent that many seals that have wintered to the west and south move into the Beaufort in late June or early July, the actual magnitude of the summer increase in ringed seal abundance is poorly known (Frost and Lowry, 1981b). For these calculations, we assume that the number of ringed seals in the Alaskan Beaufort Sea doubles during summer to about 80,000 individuals.

The average weight of 929 ringed seals taken in the Bering, Chukchi, and Beaufort Seas was 34.3 kg (Burns, Frost, and Lowry, unpubl.). Therefore, the estimated winter and summer ringed seal biomasses in the Alaskan Beaufort Sea are 1372 t and 2744 t, respectively.

Stomach contents of ringed seals collected in the Alaskan Beaufort Sea from 1972 to 1980 show that arctic cod, nektonic crustaceans (hyperiid amphipods and euphausiids), and benthic crustaceans (gammarid amphipods, mysids, shrimps, and isopods) are the major foods eaten (Lowry et al., 1978, 1979, 1980; Frost and Lowry, 1981b). The primary prey type varies seasonally as follows: benthic crustaceans in April-June; nektonic crustaceans in August-September; and arctic cod in November-March. Largest amounts of food are found in seals which feed on nektonic crustaceans or arctic cod.

Between November and early April, prey consisted of 75-98% arctic cod and 0-17% hyperiid amphipods, except for one collection made just outside of the barrier islands east of Prudhoe Bay in November, where the principal prey were mysids and gammarid amphipods (Table III). The mean stomach contents of the 131 November-April seals is 195 ml, composed of arctic cod (83%), hyperiid amphipods (5%), and other organisms (12%). We use these values as representative of the diet in November through March.

The two May samples reported in Table III contained widely different prey organisms. An additional 21 stomachs collected at various locations near Barrow between late March and June 1972-77 contained a mean volume of 48.1 ml, made up of 34% gammarid amphipods, 20% isopods, 23% mysids, 12% euphausiids, 4% shrimps, and 6% arctic cod. Considering May data from Table III and these additional 21 specimens, the overall food composition of our specimens collected in spring is arctic cod, 6%;

TABLE III. Ringed Seal Stomach Contents (% volume) from the Alaskan Beaufort Sea.

Prey type	Autumn (Nov.)				Winter (Feb.-Apr.)			Spring (May)		Summer (Aug., Sep.)			
	Barrow 1977	Barrow 1978	Prudhoe 1977	Prudhoe 1978	Barrow 1978	Barrow 1979	Prudhoe 1979	Barrow 1976	Prudhoe 1979	Barrow 1976	Pingok 1980	Prudhoe 1977	Beaufort Lagoon 1980
Euphausiid	—	—	—	—	—	—	2	90	—	99	<1	—	44
Mysid	—	—	—	45	1	1	—	—	3	—	<1	4	<1
Hyperiid amphipod	5	4	12	7	17	—	—	—	—	—	<1	92	<1
Gammarid amphipod	—	—	—	30	1	—	—	5	44	—	<1	1	4
Shrimp	1	—	—	—	3	—	—	—	13	—	1	—	<1
Other invertebrate	—	3	—	12	—	—	—	5	29	1	—	—	7
Arctic cod	90	93	86	6	75	98	96	—	9	1	98	2	45
Other fishes	4	—	1	<1	2	—	1	—	2	—	<1	<1	<1
Mean volume of contents (mL)	369	172	168	148	51	224	248	47	22	362	150	216	67
Depth range (m)	10-100	10-60	30-60	4-10	30-90	20-60	20-40	10-15	20-40	5-10	14-21	20-30	3-40
Sample size	14	18	19	22	16	18	24	3	5	2	8	13	16

euphausiids, 20%; mysids, 19%; isopods, 16%; gammarid amphipods, 32%;and other invertebrates, 7%. Mean volume of contents was 43.5 ml. We use these values to represent the composition of the ringed seal diet in April through June.

Samples collected prior to 1980 suggested that euphausiids and hyperiid amphipods were the major food of ringed seals in August and September in the Alaskan Beaufort Sea (Lowry et al., 1979, 1980). However, collections made in 1980 indicate that in some circumstances arctic cod are the major prey (Table III), perhaps at times or in locations where euphausiids or hyperiids are not adequately abundant. Averaging the four August–September samples shown in Table III gives an average volume of contents of 149 ml consisting of 30% arctic cod, 21% euphausiids, 44% hyperiid amphipods, and 5% other organisms.

There are no data on the foods of ringed seals in the Alaskan Beaufort Sea during July and October. Values for stomach contents volumes and food composition for those months have been assigned as the average of values for the preceding and following months.

Ringed seals exhibit a seasonal cycle in feeding intensity which can be measured by changes in volume of stomach contents, the proportion of seals with empty stomachs, and the average fatness of seals in different seasons (McLaren, 1958; Johnson et al., 1966). In addition, daily caloric requirements vary greatly with age of the seal (Parsons, 1977). The average daily food requirement for prey of average caloric value ranges from about 9% of body weight in pups to 3% in adults (see Lowry et al., 1980). We use 6% of the body weight per day as the average daily requirement. This value can be adjusted to reflect monthly changes in food consumption by considering the ratio of observed stomach contents volume in a given month to the average volume for all months combined, which results in the following feeding rates (expressed as percent of total body weight per day): November to March, 8.4%; April to June, 1.9%; July, 4.1%; August and September, 5.6%; and October, 7.4%.

B. Seabirds

Although many species of birds occur along the Beaufort Sea coast, few are common in marine habitats (Divoky, this volume). Several of the most abundant species (e.g., oldsquaw (Clangula hymenalis) and eiders (Somateria sp.) feed almost entirely in nearshore marine waters and lagoons on benthic and near-bottom organisms and thus are not considered as part of this study. Eleven species (or species groups) of seabirds commonly forage in marine waters of the Beaufort Sea on organisms connected to the pelagic food web. Most species are migrants, entering the area in early June when open-water areas form. Some individuals breed in coastal and barrier island areas, and their feeding activities are somewhat restricted to the vicinity of the nest site. Nonbreeders and adults after fledging of young are more free to move throughout the area and forage in marine waters. Migration west and south from the Beaufort Sea is affected by freeze-up but generally occurs in September. During migration large numbers of birds transit the area, but many are not feeding. To determine food consumption, we assume that birds feed in the study area for an average of 90 days per year (about mid-June to mid-September). Ross' and ivory gulls, which are not known to breed in the Alaskan Beaufort Sea but migrate through the area in fall, are considered to reside there for 30 days per year.

Data on bird weights as well as estimates of the number of individuals of each species in the study area (Divoky, pers. commun.) were used to estimate the biomass values for each species (Table IV). Species with either a small population size or a population of small individuals had small biomasses; large, abundant seabirds had large biomasses.

The diet of seabirds in the study area was studied during 1976–79 by Divoky (1979; this volume) and summarized qualitatively by Schamel (1978). We have used those sources, supplemented with other relevant data

TABLE IV. *Summary of Abundance and Biomass of Marine Birds in the Alaskan Beaufort Sea (Divoky, pers. commun.).*[a]

Species	Estimated number of individuals	Average individual size (kg)	Estimated total biomass (tons)
Black-legged kittiwake (Rissa tridactyla)	5,000	0.40	2.0
Glaucous gull (Larus hyperboreus)	7,000	1.20	8.4
Ivory gull (Pagophila eburnea)	1,000	0.40	0.4
Ross' gull (Rhodostethia rosea)	10,000	0.20	2.0
Sabine's gull (Xema sabini)	30,000	0.20	6.0
Arctic tern (Sterna paradisaea)	100,000	0.12	12.0
Jaegers (Stercorarius spp.)	30,000	0.50	15.0
Black guillemot (Cepphus grylla)	1,000	0.40	0.4
Thick-billed murre (Uria lomvia)	1,000	1.0	1.0
Loons (Gavia spp.)	50,000	2.0	100.0
Phalaropes (Phalaropes fulicarius and Lobipes lobatus)	200,000	0.06	12.0

[a] *Abundances and biomasses are preliminary approximations used for calculating food requirements.*

from arctic and subarctic localities (Uspenskiy, 1959; Swartz, 1966; Divoky, 1976; Bradstreet, 1980; Hunt et al., 1981), to derive the diet composition summarized in Table V. Arctic cod, composing up to 90% of the diet, are a major food of most species. Euphausiids are eaten by several species, and when they wash up in abundance on beaches they can dominate the diet of terns and gulls (Divoky, 1980). Copepods are the major food of phalaropes and are also eaten by Sabine's gulls.

TABLE V. Estimated Composition (%) of the Diet of Marine Birds in the Alaskan Beaufort Sea.

Bird species or group	Copepod	Euphausiid	Hyperiid amphipod	Arctic cod	Other
Black-legged kittiwake	--	2	1	90	7
Glaucous gull	--	9	1	50	40
Ivory gull	--	10	--	80	10
Ross' gull	--	40	--	40	20
Sabine's gull	13	10	--	10	67
Arctic tern	--	18	2	40	40
Jaegers	--	--	--	40	60
Black guillemot	--	--	--	80	20
Thick-billed murre	--	2	2	90	6
Loons	--	--	--	50	50
Phalaropes	90	--	--	--	10

Estimates of daily food consumption of seabirds range from 15% to 40% of total body weight (Swartz, 1966; Livingston, 1980; Hunt et al., 1981) and vary with individual size, activity, time of year, and availability of food. We use 25% of the total body weight as an estimate of daily food consumption for all species.

C. Fishes

Frost and Lowry (1983) indicated that at least 19 species of fish occur in Beaufort Sea marine waters less than 400 m deep. Of those, only two, arctic cod and leatherfin lumpsucker (Eumicrotremus derjugini), feed predominantly on planktonic organisms. Arctic cod is by far the most abundant species (Frost and Lowry, 1981b, 1983). Although plankton-eating fishes such as capelin (Mallotus villosus) and herring are occasionally present in the Beaufort Sea (*e.g.*, McAllister, 1962), little is known about their distribution and abundance.

Available data do not allow a direct estimate of arctic cod biomass in the study area, but an indirect assessment of stock size can be made based on data presented here. The estimated amount of arctic cod consumed annually by predators in the study area is approximately 28,630 metric tons. Although arctic cod are sometimes cannibalistic (Baranenkova *et al.*, 1966; Frost and Lowry, 1981b), we do not consider consumption of cod by cod in our calculations since there are few data on this relationship. Such consumption, even at a very low rate, would greatly influence estimates of total cod stock size. A conservative estimate of stock size, excluding larvae and fry, can be derived by assuming that the annual consumption by predators equals the maximum sustainable yield of the arctic cod stock. Data on the relationship between total stock size and sustainable yield are sparse, particularly for arctic species. In temperate regions, sustained yield for fish stocks should be 1/4 to 1/2 of standing stock, or conversely stock size should be two to four times the sustained yield (Sheldon *et al.*, 1977). For our calculations we assume the arctic cod stock size to be three times the estimated amount consumed annually by predators, or 85,890 metric tons.

Arctic cod eat a variety of prey, including benthic organisms, planktonic organisms, and species associated with the undersurface of ice. In 157 arctic cod stomachs collected in the northeastern Chukchi and Beaufort seas in August and September 1977 in waters 40–400 m deep, calanoid copepods (primarily *Calanus hyperboreus* and *C. glacialis*) and the gammarid amphipod *Apherusa glacialis* were the major foods, followed by hyperiid amphipods, mysids, chaetognaths, euphausiids, and shrimps (Lowry and Frost, 1981). Similar summer foods have been reported from arctic Canada (Bohn and McElroy, 1976; Bradstreet and Cross, 1982), the Barents Sea (Hognestad, 1968), and the Siberian Arctic (Moskalenko, 1964).

More recent studies (Lowry *et al.*, 1981; Craig *et al.*, 1982) indicate substantial dietary variability in relation to years, season, water depth, ice conditions, and fish size (Table VI). Mysids and gammarid amphipod are most important in shallow nearshore waters, with mysids predominating during winter. Copepods are of greatest significance in deeper ice–covered waters. Fishes (mostly young–of–the–year arctic cod) were eaten principally by large cod in shallow water. They composed 59% by weight of the winter offshore sample but occurred in only 14% of the stomachs examined (Craig *et al.*, 1982). Such variability makes it difficult to estimate an average annual diet, but based on the data in Table VI we use the following as a preliminary approximation: copepod, 50%; euphausiid, 5%; mysid, 20%; hyperiid amphipod, 1%; gammarid amphipod, 12%; arctic cod, 5%; and other organisms, 7%.

Craig and Haldorson (1981) estimated that arctic cod in summer consume about 6% of their body weight per day. In our arctic cod samples collected at Beaufort Lagoon, the weight of stomach contents exceeded 5% of the total weight of the fish in only 5 of 91 stomachs examined. The maximum quantity of food in a single stomach (2.03 g) was 9.8% of the weight of the fish. We consider 6% of the body weight per day to be a reasonable estimate of the average daily food consumption of arctic cod.

TABLE VI. Quantitative Composition (% Wet Weight) of Stomach Contents of Arctic Cod Collected in the Beaufort Sea.

Prey	Simpson Lagoon Summer 1977-79[a]	Pingok Island August 1980 <10 cm[b]	Pingok Island August 1980 >10 cm[b]	All[c]	Beaufort Lagoon Sept. 1980[d]	Central Beaufort Winter 1978-80[a] Near-shore	Central Beaufort Winter 1978-80[a] Off-shore
Copepod	1-44	1	3	55	56	<1	22
Euphausiid	0-1	<1	8	8	8	0	3
Mysid	39-88	56	17	6	6	93	0
Hyperiid amphipod	0-2	<1	<1	<1	1	0	4
Gammarid amphipod	8-41	22	20	12	20	2	13
Other invertebrate	<1	17	17	18	5	<1	1
Fish	<1	3	34	<1	5	2	59

[a]*Craig et al., 1982.*
[b]*Lowry et al., 1981; water depth 5 and 10 m; open water.*
[c]*Lowry et al., 1981; water depth 14-19 m; in ice.*
[d]*Frost and Lowry, 1981b; water depth 5-30 m; in ice.*

III. TROPHIC INTERACTIONS AMONG MAJOR VERTEBRATE CONSUMERS

From the data and assumptions presented in the previous sections, we estimated the amount of each major prey category eaten annually by each major vertebrate consumer. Results of calculations for seabirds (Table VII) indicate that phalaropes are the major consumers of copepods; gulls and terns are the major consumers of euphausiids; and loons, jaegers, terns, and gulls are the major consumers of arctic cod. In total the seabirds we considered are estimated to consume about 3546 t of food annually, comprising about 7% copepods, 2% euphausiids, 44% arctic cod, and 46% other organisms.

For ringed seals, calculations indicate that large quantities of arctic cod are consumed during the months of August through March, while consumption of euphausiids and hyperiid amphipods is greatest in July to October (Table VIII). Overall, the food consumed by ringed seals annually is composed of about 55% arctic cod, 10% euphausiids, 18% hyperiid amphipods, and 17% other organisms.

In aggregate, bowhead and white whales, ringed seals, seabirds, and arctic cod consume approximately 2 million metric tons of food annually in the study area, comprising approximately 48% copepods, 7% euphausiids, 1% hyperiid amphipods, 6% arctic cod, and 37% other organisms (Table IX). Arctic cod, the major consumers of copepods, are estimated to eat almost 1 million metric tons annually. Bowhead whales are estimated to eat only about 2% of the total amount of copepod biomass consumed

TABLE VII. Estimated Quantities of Food (tons) Consumed by Marine birds in the Alaskan Beaufort Sea.

Bird species or group	Copepod	Euphausiid	Hyperiid amphipod	Arctic cod	Other
Black-legged kittiwake	--	1	<1	40	3
Glaucous gull	--	17	2	94	76
Ivory gull	--	<1	--	2	<1
Ross' gull	--	6	--	6	3
Sabine's gull	18	14	--	14	90
Arctic tern	--	49	5	108	108
Jaegers	--	--	--	135	202
Black guillemot	--	--	--	7	2
Thick-billed murre	--	<1	<1	20	1
Loons	--	--	--	1125	1125
Phalaropes	243	--	--	--	27
Total	261	87	8	1552	1638

annually by these species of predators. Arctic cod and bowheads are the major consumers of euphausiids, with cod estimated to consume more than double the quantity eaten by bowheads. Largest amounts of hyperiid amphipods are eaten by arctic cod and ringed seals. Major consumers of arctic cod, in decreasing order of estimated amounts eaten annually, are arctic cod, ringed seals, white whales, and seabirds.

IV. DISCUSSION AND CONCLUSIONS

Our estimates of quantities of prey consumed are imprecise. Errors in estimation of biomass, residence times, food consumption rate, and diet composition will cause errors in determining the quantity of prey consumed. Although refinements in the data base will modify the values calculated for each predator species, we consider major changes unlikely. Arctic cod are the most important secondary consumer in the region and provide the bulk of the diet of ringed seals, several species of seabirds, and probably also white whales (Tables VII, VIII, IX). Arctic cod probably play a similar pivotal role in nearshore marine ecosystems throughout the Arctic (Klumov, 1937; Bradstreet and Cross, 1982; Craig et al., 1982). It is difficult at present to assess whether populations of vertebrate consumers experience food limitation in the Alaskan Beaufort Sea. In the eastern Beaufort Sea, the average total zooplankton biomass has been estimated at 28 g m^{-2} (Griffiths and Buchanan, 1982). Our calculations indicate that vertebrate consumers in the Alaskan Beaufort consume 22.6 g

TABLE VIII. Estimated Quantities of Food (tons) Consumed by Ringed Seals in the Alaskan Beaufort Sea.

Month	Arctic cod	Euphausiid	Hyperiid amphipod	Other
November	2800	--	138	484
December	2894	--	143	500
January	2894	--	143	500
February	2614	--	129	452
March	2894	--	143	500
April	47	156	--	579
May	48	162	--	598
June	47	156	--	579
July	628	698	767	1395
August	1429	1000	2096	238
September	1383	968	2028	230
October	3525	629	1511	629
Total	21,203	3,769	7,098	6,684

m^{-2} of zooplankton annually, which agrees favorably with standing stock estimates from the eastern Beaufort. Note, however, that zooplankton biomass estimates (Griffiths and Buchanan, 1982) include a substantial quantity of hydrozoans, which have never occurred in our samples of stomach contents and which we do not consider potentially significant foods for vertebrate consumers due to their morphology and low caloric value. Preliminary calculations based on estimates of primary production (Frost and Lowry, 1981b) indicate that zooplankton stocks are probably adequate in years of high primary productivity, but may be as low as one-fourth of our estimated quantities consumed in years of low primary production. In years of low zooplankton availability, euryphagous species such as ringed seals, arctic cod, and glaucous gulls may be able to switch to other types of prey such as mysids, isopods, and gammarid amphipods. Comparatively stenophagous species such as phalaropes and bowhead whales may feed at sub-optimal levels or may seek food elsewhere. Food limitation causes decreased fledging and nesting success in seabirds (Divoky, 1980; Drury et al., 1981). Stirling et al. (1977) documented a substantial decline in numbers and productivity of ringed seals in the eastern Beaufort Sea, which they correlated with heavy ice conditions that might have reduced primary productivity.

If food is limiting in the Beaufort Sea, then competition among consumers must occur. Effects of competition may be lessened somewhat by specific aspects of the feeding ecology of consumer species. Based on our investigations, copepods consumed by bowhead whales are almost all larger than about 2.5 mm cephalothorax length, which includes primarily the genera Calanus and Metridea. In contrast, arctic cod frequently

TABLE IX. Total Quantities of Food (tons) Consumed Annually by the Major Vertebrates in the Alaskan Beaufort Sea. The Percent of Prey Eaten by the Consumer is Given in Parentheses.

			Prey category			
Consumer	Copepod	Euphausiid	Hyperiid amphipod	Arctic cod	Other	Total
Bowhead whale	20,812 (2.2)	45,094 (31.5)	694 (2.6)	--	2,775 (0.4)	69,375
White whale	--	--	--	5,875 (4.8)	1,469 (0.2)	7,344
Ringed seal	--	3,769 (2.6)	7,098 (26.7)	21,203 (17.3)	6,684 (0.9)	38,754
Marine birds	261 (<0.1)	87 (<0.1)	8 (<0.1)	1,552 (1.3)	1,638 (0.2)	3,546
Arctic cod	940,495 (97.8)	94,050 (65.8)	18,810 (70.7)	94,050 (76.7)	733,586 (98.3)	1,880,991
Total	961,568	143,000	26,610	122,680	746,152	2,000,010

consume large numbers of the much smaller (0.9–1.6 mm cephalothorax length) genera *Derjuginia, Pseudocalanus,* and *Acartia* (Frost and Lowry, 1981b; unpubl. data). Since small copepods prefer to eat small phytoplankton such as dinoflagellates, and larger copepods forage most efficiently on larger diatoms (Parsons and LeBrasseur, 1970), the trophic subsystems exploited by arctic cod and bowheads may be somewhat isolated. It seems likely that bowhead whales may be able to efficiently filter euphausiids when they occur in low densities (Nemoto, 1970), while ringed seals probably require dense concentrations of euphausiids for efficient feeding.

Competition for food is of greatest concern with respect to bowhead whales. The western arctic population of bowhead whales was greatly reduced by commercial whalers during the late 1800s and early 1900s. The stock, which probably originally numbered 14,000–26,000 (Breiwick *et al.*, 1981), is now estimated at approximately 3500 (Dronenburg *et al.*, 1982). Populations of other vertebrate consumers such as ringed seals and arctic cod may have increased subsequent to the decimation of the bowheads, and their foraging activities may presently be retarding the recovery of the whale stock (Lowry *et al.*, 1978). Laws (1977) suggested that the reduction of baleen whale stocks in the Antarctic has resulted in increased productivity and population size of other consumers due to an increase in food availability. Kawamura (1978) reviewed the interspecific feeding relationships of southern hemisphere baleen whales and concluded that the abundance of sei whales (*Balaenoptera borealis*) increased with the reduction in abundance of right whales (*Eubalaena australis*). He further suggested that the present relationship in abundance of the two species may be causing competition for food, which would explain the lack of recovery in size of the right whale population in spite of protection from hunting for over 40 years. It is probably noteworthy that the feeding apparatus of right whales and bowhead whales is very similar, and, based on the morphology of the baleen, they appear to be specialized primarily for feeding on copepods (Nemoto, 1970). However, in the Beaufort Sea, euphausiids appear to be eaten more commonly than copepods. Euphausiids are also exploited by arctic cod and to a lesser extent by ringed seals. Furthermore, the abundance of seals should be enhanced by a large, productive arctic cod population. The question posed by Laws (1977) for the Antarctic is also most relevant to the Beaufort Sea: can the original balance in numbers of animals be regained with appropriate management, which means, in essence, will the bowhead whale stock recover to its former abundance? Further studies of foods, trophic relationships, and population dynamics of vertebrate consumers will be of great value in addressing this question.

REFERENCES

Baranenkova, A. S., Ponomarenko, V. P., and Khokhlina, N. S. (1966). *Vopr. Ikhtiol. 6*, 498–518. (Transl. from Russian, 1977, available as *Fish. mar. Serv. Transl. Ser. No. 4025*, 37 p.)

Bohn, A., and McElroy, R. O. (1976). *J. Fish. Res. Board Can. 33*, 2836–2840.

Bradstreet, M. S. W. (1980). *Can. J. Zool. 58*, 2120–2140.

Bradstreet, M. S. W., and W. E. Cross (1982). *Arctic 35*, 1–12.

Braham, H. W., and Krogman, B. D. (1977). "Population biology of the bowhead (*Balaena mysticetus*) and belukha (*Delphinapterus leucas*) in the Bering, Chukchi, and Beaufort seas." Processed rep., U.S. Dept. Commerce, NOAA, Natl. Marine Mammal Lab., NMFS, Seattle, 29 p.

Braham, H. W., Fraker, M. A., and Krogman, B. D. (1980). *Mar. Fish. Rev. 42(9–10)*, 36–46.

Breiwick, J. M., Mitchell, E. D., and Chapman, D. G. (1981). *Fish. Bull. 78(4)*, 843–853.

Brodie, P. F. (1975). *Ecology 56*, 152–161.

Brodie, P. F. (1980). *Rept. Int. Whaling Comm.*, SC/32/SP 18.

Burns, J. J. (1970). *J. Mammal. 51*, 445–454.

Burns, J. J., and Harbo, S. J., Jr. (1972). *Arctic 25*, 279–290.

Craig, P. C., and Haldorson, L. (1981). *In* "Environmental Assessment Alaskan Continental Shelf, Final Reports Principal Investigators," Vol. 7, Biological Studies, p. 384–678. OCSEAP, Boulder.

Craig, P. C., Griffiths, W., Haldorson, L., and McElderry, H. (1982). *Can. J. Fish. Aquat. Sci. 39*, 395–406.

Divoky, G. J. (1976). *Condor 78*, 85–90.

Divoky, G. J. (1979). *In* "Environmental Assessment Alaskan Continental Shelf, Annual Reports Principal Investigators," Vol. 1, p. 330–599. OCSEAP, Boulder.

Divoky, G. J. (1980). In "Environmental Assessment Alaskan Continental Shelf, Annual Reports Principal Investigators," Vol. 1, p. 110–141. OCSEAP, Boulder.

Draper, H. H., Milan, F. A., Osborn, W., and Schaefer, O. (1979). "Report of the nutrition panel for the aboriginal/subsistence whaling panel." Unpubl. rep. to Int. Whaling Comm. 40 p.

Dronenburg, R. B., Carroll, G. M., Rugh, D. J., and Marquette, W. M. (1982). Rept. Int. Whaling Comm. SC/34/ PS 9. 25 p.

Drury, W. H., Ramsdell, C., and French, J. B., Jr. (1981). In "Environmental Assessment Alaskan Continental Shelf, Final Reports Principal Investigators," Vol. 11, Biological Studies, p. 175–487. OCSEAP, Boulder.

Durham, F. E. (1972). Can. J. Res. D. 20, 33–46.

Durham, F. E. (1979). Nat. Hist. Mus. Los Angeles Cty. Contrib. Sci. 314. 14 p.

Eley, T., and Lowry, L. F. (1978). In "Environmental Assessment Alaskan Continental Shelf. Interim Synthesis: Beaufort/Chukchi," p. 134–151. OCSEAP, Boulder.

Finley, K. J. (1979). Can. J. Zool. 57, 1985–1997.

Fraker, M. A. (1978). "The 1978 whale monitoring program, Mackenzie estuary, N.W.T." Rep. prepared for Esso Resources Canada Ltd. by F. F. Slaney and Co. Ltd., Vancouver, Canada. 28 p.

Fraker, M. A., and Bockstoce, J. R. (1980). Mar. Fish. Rev. 42(9–10), 57–64.

Fraker, M. A., Sergeant, D. E., and Hoek, W. (1978). "Beaufort Sea Project Tech. Rep. No. 4." 114 p.

Frost, K. J., and Lowry, L. F. (1981a). In "Handbook of Marine Mammals," Vol. 2 (R. J. Harrison and S. H. Ridgway, eds.), p. 29–53. Academic Press, London and New York.

Frost, K. J., and Lowry, L. F. (1981b). "Feeding and trophic relationships of bowhead whales and other vertebrate consumers in the Beaufort Sea." Final Report Contract No. 80–ABC–00160 submitted to NMML/NMFS/NOAA, Seattle. 142 p.

Frost, K. J., and Lowry, L. F. (1983). NOAA Tech. Rep., NMFS Spec. Sci. Rep., Fish. Ser. 764, 22 p.

Griffiths, W. B., and Buchanan, R. A. (1982). In "Behavior, disturbance responses and feeding of bowhead whales Balaena mysticetus in the Beaufort Sea, 1980–81," (W. J. Richardson, ed.), p. 347–455. Unpubl. rep. by LGL Ecological Research Associates, Inc., Bryan, TX for U.S. Bureau of Land Management, Washington, D.C.

Hognestad, P. T. (1968). In "Symposium on the Ecology of Pelagic Fish Species in Arctic Waters and Adjacent Seas" (R. W. Blacker, ed.). Int. Counc. Explor. Sea. Rep., Vol. 158.

Hunt, G. L., Jr., Burgeson, B., and Sanger, G. A. (1981). In "The Eastern Bering Sea Shelf: Oceanography and Resources," Vol. 2 (D. W. Hood and J. A. Calder, eds.), p. 629–647. Univ. Washington Press, Seattle.

Iverson, R., and Goering, J. (1979). PROBES Progress Rep. 1, 145–161.

Johnson, M. L., Fiscus, C. H., Ostenson, B. T., and Barbour, M. L. (1966). In "Environment of the Cape Thompson region, Alaska" (N. J. Wilimovsky and J. N. Wolfe, eds.), p. 897–924. U.S. Atomic Energy Comm., Oak Ridge, TN.

Kawamura, A. (1978). *Rept. Int. Whaling Comm. 28*, 411–420.

Kleinenberg, S. E., Yablokov, A. V., Bel'kovich, B. M., and Tarasevich, M. N. (1964). "Belukha (*Delphinapteras leucas*) – investigation of the species." (Transl. from Russian by Israel Program for Scientific Translations, Jerusalem, 376 p.)

Klumov, S. K. (1937). *Izv. Akad. Nauk SSSR, Ser. Biol., No. 1.* In Russian.

Laws, R. M. (1977). *Philos. Trans. R. Soc. Lond. B 279*, 81–96.

Livingston, P. (1980). "Marine bird information synthesis." U.S. Dept. Commerce, Nat. Mar. Fish. Serv., Northwest and Alaska Fisheries Center, Seattle. 25 p.

Ljungblad, D. K. (1981). "Aerial surveys of endangered whales in the Beaufort Sea, Chukchi Sea, and northern Bering Sea. Final Report: Fall 1980." Naval Oceans Systems Center Tech. Doc. 449. 211 p.

Ljungblad, D. K., Platter-Reiger, M. F., and Shipp, F. S., Jr. (1980). "Aerial surveys of bowhead whales, North Slope, Alaska. Final Report: Fall 1979." Naval Oceans Systems Center Tech. Doc. 314. 181 p.

Lockyer, C. (1976). *J. Cons. Int. Explor. Mer 36*, 259–273.

Lowry, L. F., and Burns, J. J. (1980). *Mar. Fish. Rev. 42(9-10)*, 88–91.

Lowry, L. F., and Frost, K. J. (1981). *Can. Fld-Naturalist 95*, 186–191.

Lowry, L. F., Frost, K. J., and Burns, J. J. (1978). *Can. Fld-Naturalist 92*, 67–70.

Lowry, L. F., Frost, K. J., and Burns, J. J. (1979). In "Environmental Assessment Alaskan Continental Shelf, Final Reports Principal Investigators," Vol. 6, Biological Studies, p. 573–630. OCSEAP, Boulder.

Lowry, L. F., Frost, K. J., and Burns, J. J. (1980). *Can. J. Fish. aquat. Sci. 37*, 2254–2261.

Lowry, L.F., Frost, K.J., and Burns, J.J. (1981). In "Environmental Assessment Alaskan Continental Shelf, Annual Reports Principal Investigators," Vol. 1, p. 149–189. OCSEAP, Boulder.

MacGinitie, G. E. (1955). *Smithson. Misc. Coll. 128(9).* 201 p.

Marquette, W. M. (1977). "The 1976 catch of bowhead whales (*Balaena mysticetus*) by Alaskan Eskimos, with a review of the fishery, 1973–1976, and a biological summary of the species." U.S. Dept. Commerce, Nat. Mar. Fish. Serv., Seattle. 80 p.

McAllister, D. E. (1962). *Natl. Mus. Can. Bull. 185*, 17–39.

McLaren, I. A. (1958). *Fish. Res. Board Can. Bull. 118.* 97 p.

Mitchell, E. (1975). In "Proceedings of the Canadian Zoologists Annual Meeting, June 2-5, 1974" (D. B. Burt, ed.), p. 123–133.

Moskalenko, B. K. (1964). *Vopr. Ikhtiol. 4(3):32*, 433–443. In Russian.

Nemoto, T. (1970). In "Marine food chains" (J. H. Steele, ed.), p. 241–252. Oliver and Boyd, Edinburgh.

Parsons, J. L. (1977). M.S. Thesis, Univ. Guelph, Ontario. 82 p.

Parsons, T. R., and LeBrasseur, R. J. (1970). In "Marine food chains" (J. H. Steele, ed.), p. 325–343. Oliver and Boyd, Edinburgh.

Schamel, D. (Ed.) (1978). In "Environmental Assessment Alaskan Continental Shelf. Interim Synthesis: Beaufort/Chukchi," p. 152–173. OCSEAP, Boulder.

Schell, D. M., Ziemann, P. J., Parrish, D. M., and Brown, E. J. (1982). "Food web and nutrient dynamics in nearshore Alaskan Beaufort Sea waters." Cumulative summary rep. for RU 537 to OCSEAP. 135 p.

Seaman, G. A., Lowry, L. F., and Frost, K. J. (1982). *Cetology 4 *, 1–19.

Sergeant, D. E. 1969. *Fisk Dir. Skr., Ser. Havunders. 15(3)*, 246-258.
Sergeant, D. E., and Brodie, P. F. (1969). *J. Fish. Res. Board Can. 26*, 2561-2580.
Sheldon, R.W., Sutcliffe, W.H. Jr., and Paranjape, M.A. (1977). *J. Fish. Res. Board Can. 34*, 2344-2353.
Smith, T. G. (1973). *Fish. Res. Board Can. Tech. Rep. No. 427*. 18 p.
Smith, T. G. (1975). *Arctic 28*, 170-182.
Smith, T. G., and Stirling, I. (1975). *Can. J. Zool. 53*, 1297-1305.
Smith, T. G., Hay, K., Taylor, D., and Greendale, R. (1978). *Inst. N. Affairs Publ. No. Q5-8160-022-EE-A1*, Ottawa, Canada. 85 p.
Stirling, I., Archibald, W. R., and DeMaster, D. (1977). *J. Fish. Res. Board Can. 34*, 976-988.
Swartz, L. G. (1966). *In* "Environment of the Cape Thompson Region" (N. J. Wilimovsky and J. N. Wolfe, eds.), p. 611-678. U.S. Atomic Energy Comm., Washington, D.C.
Tomilin, A. G. (1957). *In* "Mammals of the USSR and Adjacent Countries" (V. G. Heptner, ed.). (Transl. from Russian by Israel Program for Scientific Translations, 1967.)
Uspenskiy, S. M. (1959). *Bull. Mosc. Soc. Nat. Res. Biol. 64*, 39-51.

ECOLOGY OF SHOREBIRDS
IN THE ALASKAN BEAUFORT LITTORAL ZONE

Peter G. Connors

Bodega Marine Laboratory
University of California
Bodega Bay, California

I. INTRODUCTION

Along the Beaufort Sea coast of Alaska, as in many arctic and subarctic regions, shorebirds (Charadriiformes: Charadrii: sandpipers, plovers and their close relatives) constitute a prominent segment of the avifauna (Bailey, 1948; Gabrielson and Lincoln, 1959; Pitelka, 1974). The 20 species listed in Table I are all migratory, many traveling annually between arctic breeding grounds and wintering areas as distant as southern South America. Most species nest on the tundra, where plant cover and insect food are plentiful for a brief period in June and July. Littoral habitats along the coast, including salt marsh, brackish sloughs, and arctic beaches, are all but ignored during this period, but they play an increasingly important role as the season progresses. During August many species forage in the littoral zone in densities much higher than tundra breeding densities, apparently storing energy for the southward migration. The extent of this habitat shift varies widely among species, as does the relative use of different habitats within the littoral zone (Table II), but it is an important and consistent feature of Beaufort coast shorebird communities.

Most of the information presented in this paper derives from ground-based censuses of fixed transects in a variety of tundra and littoral habitats, repeated throughout the summer season at three Beaufort sites: Barrow, Fish Creek Delta, and Prudhoe Bay (Connors et al., 1979, 1984; Myers and Pitelka, 1980). Data collected in this way provide a quantitative and consistent description of seasonal changes in habitat use at different sites, and multi-year studies at one site (Barrow) permit discussion of annual variation in movements, population densities, and habitat use. Information drawn from other published studies extends the discussion to other sites along the Alaska Beaufort Sea coast.

TABLE I.   Shorebirds occurring regularly along the Alaskan Beaufort Sea coast.[a]

Number	Species name	Number	Species name
1	Semipalmated plover Charadrius semipalmatus	11	Dunlin Calidris alpina
2	Lesser golden plover Pluvialis dominica	12	Semipalmated sandpiper Calidris pusilla
3	Black-bellied plover Pluvialis squatarola	13	Western sandpiper Calidris mauri
4	Ruddy turnstone Arenaria interpres	14	Sanderling Calidris alba
5	Common snipe Gallinago gallinago	15	Stilt sandpiper Calidris himantopus
6	Whimbrel Numenius phaeopus	16	Buff-breasted sandpiper Tryngites subruficollis
7	Red knot Calidris canutus	17	Long-billed dowitcher Limnodromus scolopaceus
8	Pectoral sandpiper Calidris melanotos	18	Bar-tailed godwit Limosa lapponica
9	White-rumped sandpiper Calidris fuscicollis	19	Red phalarope Phalaropus fulicaria
10	Baird's sandpiper Calidris bairdii	20	Red-necked phalarope Phalaropus lobatus

[a]From Bailey (1948), Pitelka (1974), Johnson et al. (1975), Kessel and Gibson (1978), and Martin and Moiteret (1981). These sources also name 19 shorebird species as visitors.

II. HABITAT DESCRIPTIONS

Quantitative analyses of six measured habitat variables provided objective reasons for grouping the census transects in several habitat types (Connors et al., 1984). Brief descriptions of these habitats follow.

A. Gravel beach

Shorelines of most arctic spits and barrier islands and parts of the mainland shore consist of gravel beaches; these are ice-scoured and subject to gravel movement during open-water storms.   There is no benthic infauna of major importance to shorebirds, but shallow-water zooplankton are sometimes abundant. Upper parts of beaches are sparsely vegetated with salt-tolerant plants such as Honckenya peploides and Elymus arenarius.

TABLE II. Habitat use of Beaufort Sea shorebirds in Alaska

Species	Breeding status[a]	Breeding habitat[b]	Migrant status[a]	Migrant habitat[b]	Littoral Habitat[c]
1	Rare	Both	Rare	Littoral	Flats
2	Common	Tundra	Common	Tundra	
3	Uncommon	Tundra	Common, east, central	Littoral	Flats
4	Uncommon, coastal	Both	Common	Littoral	Both
5	Rare, inland	Tundra	Rare		
6	Rare, central, inland	Tundra	Rare		
7	Rare, west	Tundra	Rare, west	Littoral	Beaches
8	Common	Tundra	Common	Tundra	
9	Uncommon	Tundra	Uncommon, east	Littoral	Flats
10	Common	Both	Uncommon	Both	Both
11	Common, west; Rare, east	Tundra	Common	Both	Both
12	Common	Both	Common	Littoral	Flats
13	Uncommon, west	Both	Uncommon, west	Littoral	Flats
14	Rare, west	Tundra	Common	Littoral	Beaches
15	Uncommon, east, central	Tundra	Common, east, central	Littoral	Flats
16	Uncommon	Tundra	Uncommon	Tundra	
17	Uncommon	Tundra	Common	Both	Flats
18	Uncommon, central, inland	Tundra	Rare		
19	Common	Tundra	Common	Littoral	Both
20	Common, east; Uncommon, west	Tundra	Common	Littoral	Both

[a]From Bailey (1948), Pitelka (1974), Johnson et al. (1975), Kessel and Gibson (1978), and Martin and Moiteret (1981).

[b]Both indicates tundra and littoral. From Connors et al. (1979), Myers and Pitelka (1980), Martin and Moiteret (1981), and Connors et al. (1984).

[c]Flats includes salt marsh, mudflats, and slough edges. From Connors et al. (1984).

B. Littoral flats and salt marsh

These habitats, differing principally in degree of vegetative cover, are usually almost flat, slightly above mean sea level, and protected from normal wave action. They are maintained by periodic flooding with salt water during high storm tides. Substrates are mud or sand, sometimes mixed with gravel, and they harbor populations of invertebrates important as shorebird prey. Characteristic salt-tolerant plants include *Puccinellia phryganodes*, *Carex subspathacea*, *Carex ursina*, *Stellaria humifusa*, *Cochlearia officinalis*, and *Dupontia fischeri*.

C. Slough and small lagoon edges

These brackish and estuarine areas vary from small streams to lagoons of about 1 km diameter with openings to salt water at least during storm conditions. Border substrates vary from gravel to fine mud, and borders may be narrow or wide, including broad areas of mudflat or salt marsh. In these cases they merge with the previous category, and bird use is similar.

D. Mainland shores

Beaches along the mainland may be exposed or partially protected by barrier islands and vary from gravel to fine sand. They are sometimes broad and relatively flat but more frequently are narrow and backed by a tundra bank. Although some mainland beaches support moderate densities of shorebirds, especially near the mouths of sloughs and lagoons, bird densities are usually lower in these habitats than in any other littoral habitat type. Mainland beach transects censused at Barrow were broad gravel shores. They are combined with other gravel beach transects in the analyses reported here.

E. Tundra

Non-littoral habitats, classed as tundra, vary from well-drained uplands to very wet lowlands. Coastal lowland tundra is distinguished from salt marsh by the absence of saltwater influence. Littoral habitats are at least occasionally inundated by salt water and differ from tundra in the absence of tundra vegetation and the presence of salt-tolerant plants.

III. SEASONAL HABITAT USE

The transect census data yield a quantitative picture of shorebird habitat use. Considering the entire shorebird community, a general pattern emerges at all Beaufort study sites, demonstrated in Fig. 1. Contrasting the tundra with the littoral zone, we find a marked shift in bird use toward the littoral zone in late summer. Birds arrive at breeding areas in late May and early June, while most shoreline areas are frozen and inaccessible. As

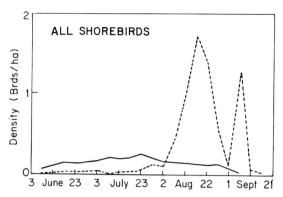

FIGURE 1. Seasonal densities of shorebirds on tundra (solid line) and littoral transects (dotted line) near Barrow, 1976 (Connors et al., 1979).

snow melt progresses, birds establish territories and nest on newly exposed tundra. After about 3 weeks of incubation, eggs hatch from late June through mid-July. Shorebird prey during this period are mainly freshwater zooplankton, spiders, and insect larvae and adults (Holmes and Pitelka, 1968; Johnsgard, 1981).

For some species (red and red-necked phalaropes, pectoral sandpiper) nesting participation by one sex ends before young have hatched. Together with non-breeding or failed-breeding adults of other species, these released adults occur in increasing numbers on littoral mudflats, in salt marshes, and along lagoon edges. Meltoff proceeds along ocean shorelines during July, and adults of some species move to these habitats in late July and early August as they are freed from nesting duties. For most species, the heaviest use of littoral habitats occurs as recently fledged juveniles move there to accumulate fat for southward migration, foraging on oligochaetes and insect larvae on mudflats and on a wide variety of marine zooplankton along ocean shorelines. By mid-August the littoral zone has become a major foraging area for shorebirds of many species. By early September, few birds remain.

The marked shift from tundra to littoral habitats indicated in Fig. 1 is heavily influenced by the high densities of juvenile red phalaropes that occur near Barrow in August. Figure 1 is a composite of many different species patterns that exhibit this shift to different degrees and that differ in timing of population movements as well as in relative use of different habitats within the littoral zone. Variation in use of littoral habitats is demonstrated by Figs. 2 and 3, contrasting 5-year mean densities on tundra transects (Myers and Pitelka, 1980) with 4-year mean densities on nearby littoral transects (Connors et al., 1984). Golden plovers are almost restricted to tundra habitats, even after the young fledge. Densities in August are slightly lower than nesting-season densities. In contrast, dunlin densities increase in both habitats in August, but the shift to littoral habitats is pronounced. Some species, most notably red phalaropes, move to the littoral zone even more emphatically, with peak densities in August approximately 70 times higher in littoral habitats than on tundra (Connors et al., 1984). A few species such as semipalmated sandpiper use littoral

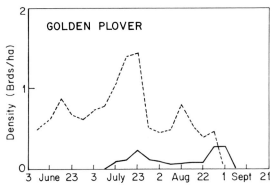

FIGURE 2. *Seasonal densities of golden plovers on tundra (dotted line) and littoral transects (solid line) near Barrow, 1975–1978 (Myers and Pitelka, 1980; Connors et al., 1984).*

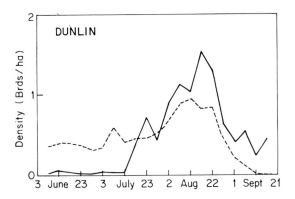

FIGURE 3. *Seasonal densities of dunlins on tundra (dotted line) and littoral transects (solid line) near Barrow, 1975–1978 (Myers and Pitelka, 1980; Connors et al., 1984).*

habitats even during the breeding season, foraging along slough edges and on mudflats, where these occur near nesting sites. Connors et al. (1979) grouped the common Barrow shorebirds in four categories reflecting these different seasonal patterns of tundra and littoral use, suggesting that species vulnerability to oil spills and coastal development differs according to these patterns.

IV. LITTORAL HABITAT PREFERENCES

Quantitative habitat descriptions of 50 by 50 m plots on all littoral transects at three sites show that the subjective distinctions made among gravel beach, littoral flat, and slough edge transects reflect the major habitat distinctions, and that birds recognize these differences; bird use for most species varies according to habitat type. The differences in habitat use within the littoral zone (Fig. 4) are revealing in terms of foraging ecology of each species and also in the susceptibility to environmental perturbations, which may affect habitats differently. Ruddy turnstones, sanderlings, and red phalaropes are birds of gravel beaches in late summer. Densities are as high as or higher than in more protected habitats. The red phalarope graph reflects an unusually high density in 1977 on one lagoon-edge transect. In three other years gravel-beach densities averaged higher than lagoon-edge densities. Baird's sandpipers and dunlins also forage commonly along gravel shores, which are much more extensive than littoral flats and edges at Barrow. All other species show much less use of gravel shores and vary only in the proportion of foraging time spent on littoral flats or along edges of sloughs and lagoons. To a great extent, types of shorebird prey in these latter habitats are similar (principally larvae and adults of chironomid flies, as well as oligochaete worms), so this distinction may be less important than the contrast with gravel beach use,

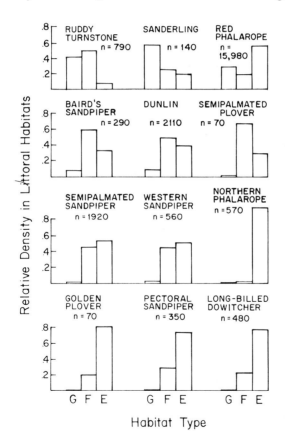

FIGURE 4. Relative use of littoral habitats by common shorebirds near Barrow, 1975-1978: gravel beach (G), littoral flat (F), slough edge (E) (Connors et al., 1984).

where the prey base is a wide variety of small crustaceans and marine zooplankton.

V. GEOGRAPHIC VARIATION

The habitat use patterns derived principally from studies at Barrow apply generally to most coastal areas of the Alaska Beaufort Sea, modified to reflect local habitat availability and regional trends in shorebird distribution. There is no clear geographic cline in littoral habitat types across the Beaufort coast, but many areas lack the extensive system of gravel beaches on spits or barrier islands found near Barrow. Since these habitat forms are often concentration areas for phalaropes, probably because of effects involving prey availability and protected foraging conditions during onshore winds (Connors and Risebrough, 1977; Johnson and Richardson, 1981), the distribution of phalaropes and other gravel-beach shorebirds is widespread but not uniform along the coast. Similarly, salt-marsh habitats occur in small patches in almost all areas, but are extensive at only a few sites, usually in river delta areas. At Fish Creek delta west of the Colville delta, transect characteristics reflected this contrast of more extensive salt marsh and less gravel beach compared to transects at Prudhoe Bay and Barrow (Connors et al., 1984). Bird use within habitat types was similar at all three sites, but the overall pattern of shorebird use varied among sites because of the different mix of habitats available. At the Canning River delta, all littoral habitat types are well represented and many species occur in the littoral zone (Martin and Moiteret, 1981). Habitats and shorebird numbers on the Simpson Lagoon barrier islands were similar to those found on Barrow Spit (Johnson and Richardson, 1981).

Local densities of shorebirds also reflect primarily longitudinal gradients in species abundance. Many species are somewhat uniformly distributed as breeders or migrants across the entire Alaska Beaufort Sea coast; others are significantly more common in eastern or western regions (Table II). For example, the western sandpiper is a regular breeder and common migrant at Barrow, near the eastern limit of its breeding range, but is much less common as a migrant to the east. Black-bellied plovers and stilt sandpipers are common migrants in littoral habitats along the eastern and central Alaska Beaufort Sea coast, but are infrequent near Barrow. Both phalarope species occur as breeders and migrants throughout the region, but reds are more common to the west while red-necks are more common to the east. Ratios of reds to red-necks in migration varied from 30:1 at Barrow (Connors et al., 1984) to 1:40 at Herschel Island, 70 km east of the Alaska border (Vermeer and Anweiler, 1975). At intermediate points, red-necked phalaropes outnumber reds at mainland sites, with reds more common on barrier islands, but densities are closer to equal (Johnson and Richardson, 1981; Martin and Moiteret, 1981; Connors et al., 1984).

At several points along the coast, physiographic features such as river deltas, points, bays, and island groups, coupled with prevailing wind and current conditions, combine to create concentration areas of shorebirds during migration, as at Barrow and the Plover Islands (Pitelka, 1974; Weller et al., 1978; Norton and Sackinger, 1981). Migrant shorebirds occur at all sites along the Alaska Beaufort Sea, however, and the littoral-zone community of shorebirds present at each site reflects primarily the geographic distribution and habitat preferences of each species, as well as the local availability of different littoral habitat types.

VI. SHOREBIRD DIETS

Littoral-zone diets of common Beaufort shorebirds have been reported in Connors and Risebrough (1976-1979) and in Johnson and Richardson (1981), primarily from analysis of stomach contents of collected birds. The central point emerging from the limited data is that diets of most species correspond to the habitats in which species forage rather than to strong species differences in diet preference within habitats. Diets of many species overlap broadly within habitat type, while prey taken by a single species in different habitats or at different times within a habitat may differ markedly, reflecting differences in prey availability. Of course species do exhibit differences in foraging methods, and some distinctions are imposed by size and bill morphology, but in general, diet differences correspond to differences in habitat use. On littoral flats, in salt marshes, and along small lagoon and slough shorelines, shorebirds forage principally on chironomid fly larvae in the substrate, but in several locations small oligochaete worms are also taken, and adult chironomids are important prey when present early in the post-breeding period. Along gravel beaches on marine shores, most species take a variety of marine zooplankton and a few species of amphipods associated with the substrate or the underside of ice. Prey species identified from Barrow varied widely at different locations and at different times within and between seasons, but differences in prey between shorebird species at one time and place were slight (Table III). This similarity of diet extends from red (and probably red-necked) phalaropes, which forage while swimming in shallow water near gravel beaches, to ruddy turnstones, sanderlings, dunlins, and occasionally other shorebirds that forage while walking at the water's edge. All these species tend to select the larger items from the available mix of zooplankton (Connors and Risebrough, 1977), and the species favored at Barrow included amphipods of the genera *Apherusa* and *Onisimus*, euphausiids (*Thysanoessa*), copepods (*Calanus*), and decapod zoea. These apparent diet preferences probably vary widely, however, depending on availability of species within the local zooplankton community.

Table III, together with Fig. 4 summarizing littoral zone habitat use for the entire season, conveys a general idea of the expected diets of shorebirds in the Barrow littoral. There is, however, a strong seasonal component to shorebird diets in the arctic. Diets of most species during the breeding season on tundra center on larvae and adults of tipulid and chironomid flies, but also include other insects, spiders, berries, and freshwater zooplankton (Holmes and Pitelka, 1968; Johnsgard, 1981). As shorebirds shift increasingly to littoral habitats in late July, many species take adult chironomid flies. By mid-August these adult flies are no longer available, and the same birds have shifted to other prey. Along shorelines, the mix of zooplankton changes continually (Redburn, 1974; Connors·and Risebrough, 1977). Thus the diets of shorebirds change seasonally as

TABLE III.   Groups of species with overlapping diets in littoral habitats near Barrow

Habitat	Diet	Species
Marine shores, gravel beaches	Marine zooplankton, including copepods, euphausiids, decapod zoea	4, 11, 14, 19
	Amphipods	10, 19
Small lagoons	Copepods	19, 20
Mudflats, salt marsh, lagoon and slough edges	Adult chironomid flies	4, 11, 13, 19
	Chironomid larvae	4, 11, 12, 13, 17, 19
	Oligochaetes	4, 11

species change foraging habitats and also as prey availability within habitats changes.

VII. PREMIGRATORY FAT DEPOSITION

Rates of fat accumulation and amounts of subcutaneous fat on juvenile shorebirds in August, prior to southward migration, vary among species (Connors et al., 1984). Collected specimens were assigned scores from 1 to 5 corresponding to museum fat designations (1 = no fat; 2 = little fat; 3 = moderate fat; 4 = heavy fat; 5 = excessive fat). Two species showed a significant increase in fat score with date (red phalarope mean score $\bar{x}$ 2.6, Spearman correlation coefficient r_s = .40, p < .01; dunlin $\bar{x}$ 2.5, r_s = .41, p < .05). Both species remain late in the arctic and do not arrive at the latitude of California until October. Two other species had much higher mean fat scores which did not increase significantly with date (ruddy turnstone $\bar{x}$ = 3.3; sanderling $\bar{x}$ = 3.8). Both species occur in California at least one month earlier. These data suggest that the late summer period spent foraging in the littoral zone is important for accumulation of energy reserves for southward migration, but the nature of the fat accumulation process in relation to speed and timing of migration differs among species. Semipalmated sandpipers leave the Arctic sooner than these four species, and with only moderate fat ($\bar{x}$ = 2.6); they may need to replenish fat reserves more frequently during migration.

Johnson (1978) also found an increase in fat levels of red phalaropes during August at Simpson Lagoon, and his data and ours show that even closely related species differ in fat accumulation schedules. At both sites red-necked phalaropes had consistently higher fat levels than red phalaropes. Differences in foraging ecology or metabolism that might account for this result remain a puzzle.

Differences even occur within a single species. Juvenile red phalaropes are common along Beaufort Sea shorelines for three or four weeks in August and early September, but adults leave the Arctic soon after nesting duties are completed. If juveniles require the long foraging period to accumulate energy reserves necessary for migration, perhaps adults have already achieved higher fat levels when they leave the tundra. Fat levels measured in 14 adult male phalaropes collected along shorelines 15 July-3 August were significantly higher than levels in 20 juveniles taken 8-12 August ($p < .02$, Mann-Whitney test). This further suggests that juveniles must remain beyond mid-August to acquire energy stores necessary for successful migration.

VIII. ANNUAL VARIATION

Although high variation in environmental and biological parameters is commonly accepted as a prominent attribute of arctic ecosystems, data to examine annual variation are scarce. The frequent censuses on fixed littoral-zone transects during four post-breeding seasons at Barrow, 1975-1979 (Connors et al., 1984), together with a similar set of tundra transects censused during the same years (Myers and Pitelka, 1980) provide a unique opportunity to address this topic.

Littoral post-breeding densities at five-day intervals are presented for two species in Fig. 5. Despite a six-fold difference in peak densities among years, the correspondence in annual variation between species is remarkably close in both timing and magnitude. It is evident that migrant densities of western and semipalmated sandpipers do not fluctuate randomly or independently. Considering all common shorebird species at Barrow, annual variation in post-breeding migrant densities is not closely related to variation in local breeding densities or to temperatures during the breeding season. It is correlated, however, with post-breeding temperatures. Furthermore, groups of species (such as the two sandpipers discussed here) fluctuate similarly from year to year, and these groups often share the same littoral zone post-breeding habitat, but not the same breeding habitat. This all suggests that late summer conditions in the littoral zone are important in determining post-breeding migrant densities. Since this effect occurs after the breeding season, annual variation in breeding productivity is apparently not involved, but post-fledging survival once birds reach the littoral zone, geographic movements of birds over local or large areas, or the length of time individuals remain at one site during migration may be important factors. These factors are potentially affected by habitat changes, storm-water levels, weather stress, or changes in prey densities and foraging profitability.

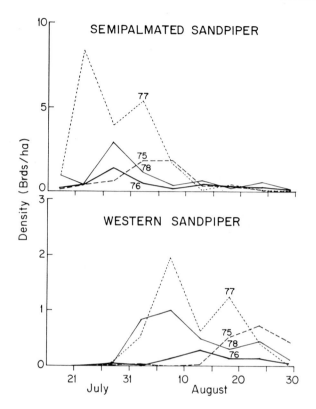

FIGURE 5. Annual variation in sandpiper post-breeding densities on littoral transects near Barrow, 1975-1978.

IX. CONCLUSION

All these results point to the importance of arctic littoral habitats during the post-breeding period in the life cycles of most arctic-breeding shorebirds. Species differ in degree of dependence on the littoral zone, in habitat preference within littoral areas, in foraging ecology, timing of movements, fat accumulation schedules, and several other factors. For most species, however, the post-breeding period in arctic littoral habitats, along with the more obvious events of breeding, migration and winter periods, must be noted in any discussions of population management or of evolution of species morphology, ecology, and life history.

ACKNOWLEDGMENTS

For guidance, suggestions and criticism I thank F.A. Pitelka, J.P. Myers, D.W. Norton, and K.G. Smith. For field assistance and suggestions I thank C.S. Connors, J.T. Carlton, B.S. Bowen, S. Gellman, R. Greenburg, F. Gress, K. Hirsch, and C. Hohenberger.

REFERENCES

Bailey, A.M. (1948). "Birds of Arctic Alaska." Colorado Mus. Nat. Hist., Denver.

Connors, P.G. and Risebrough, R.W. (1976). In "Environmental Assessment of the Alaskan Continental Shelf," Ann. Repts., Vol. 2, p. 401. NOAA, Boulder.

Connors, P.G. and Risebrough, R.W. (1977). In "Environmental Assessment of the Alaskan Continental Shelf," Ann. Repts., Vol. 3, p. 402. NOAA, Boulder.

Connors, P.G. and Risebrough, R.W. (1978). In "Environmental Assessment of the Alaskan Continental Shelf," Ann. Repts., Vol. 2, p. 84. NOAA, Boulder.

Connors, P.G. and Risebrough, R.W. (1979). In "Environmental Assessment of the Alaskan Continental Shelf," Ann. Repts., Vol. 1, p. 271. NOAA, Boulder.

Connors, P.G., Myers, J.P. and Pitelka, F.A. (1979). In "Shorebirds in Marine Environments" (F.A. Pitelka, ed.), p. 100. Studies in Avian Biology No. 2, Cooper Ornithological Society, Los Angeles.

Connors, P.G., Connors, C.S. and Smith, K.G. (1984). In "Environmental Assessment of the Alaskan Continental Shelf," Final Repts. In press. NOAA, Boulder.

Gabrielson, I.N., and Lincoln, F.C. (1959). "The Birds of Alaska." Stackpole, Harrisburg, PA.

Holmes, R.T. and Pitelka, F.A. (1968). Syst. Zool. 17, 305.

Johnsgard, P.A. (1981). "The Plovers, Sandpipers and Snipes of the World." University of Nebraska Press, Lincoln.

Johnson, S.R. (1978). In "Beaufort Sea Barrier Island Lagoon Ecological Process Studies, 1976–1977," p. 1. NOAA, Boulder.

Johnson, S.R., Adams, W.J. and Morrell, M.R. (1975). "The Birds of the Beaufort Sea." Report to Canadian Wildlife Service, Edmonton.

Johnson, S.R. and Richardson, W.J. (1981). In "Environmental Assessment of the Alaskan Continental Shelf," Final Repts. Vol. 7, p. 109. NOAA, Boulder.

Kessel, B. and Gibson, D.D. (1978). "Status and Distribution of Alaska Birds." Studies in Avian Biology No. 1, Cooper Ornithological Society, Los Angeles.

Martin, P.D. and Moiteret, C.S. (1981). "Bird Populations and Habitat Use, Canning River Delta, Alaska." U.S. Fish and Wildlife Service, Fairbanks.

Myers, J.P. and Pitelka, F.A. (1980). "Effect of Habitat Conditions on Spatial Parameters of Shorebird Populations." Report to U.S. Dept. of Energy, Washington, D.C.

Norton, D.W. and Sackinger, W.M. (eds.). (1981). "Beaufort Sea Synthesis - Sale 71." Office of Marine Pollution Assessment, Juneau.

Pitelka, F.A. (1974). Arctic and Alpine Res. 6, 161.

Redburn, D.R. (1974). "The Ecology of the Inshore Marine Zooplankton of the Chukchi Sea near Point Barrow, Alaska." Unpub. Thesis, Univ. of Alaska, Fairbanks.

Vermeer, K. and Anweiler, G.G. (1975). *Wilson Bull. 87*, 467.
Weller, G., Norton, D. and Johnson, T. (eds.). (1978).   "Environmental
    Assessment of the Alaskan Continental Shelf, Interim Synthesis:
    Beaufort/Chukchi." NOAA, Boulder.

THE PELAGIC AND NEARSHORE BIRDS OF THE ALASKAN BEAUFORT SEA: BIOMASS AND TROPHICS

George J. Divoky

College of the Atlantic[1]
Bar Harbor, Maine

I. INTRODUCTION

The Alaskan Beaufort Sea differs from other Alaskan seas both in being high arctic and in lacking rock cliffs and talus that can support major seabird breeding colonies. Instead the Alaskan Beaufort coast is important as a summering area for non-breeding birds, post-breeding staging area, and migratory pathway. The North Slope and Arctic Canada to the east support large numbers of tundra-nesting loons, waterfowl, phalaropes, gulls, jaegers, and terns that move to the Beaufort Sea after breeding. While some species use the region primarily as a migratory pathway, others molt there or accumulate fat reserves prior to migration. For the latter species, the distribution and availability of prey can be expected to be a major factor in determining their distribution and abundance.

This report discusses the feeding habits and large-scale distributions of the more numerous bird species that occupy the Alaskan Beaufort Sea from early August through mid-September, comparing pelagic and nearshore regimes and regions within these regimes. Seabirds can be useful indicators of the location and relative amounts of secondary and tertiary productivity (Brown, 1979; 1980). Whenever possible I have tried to correlate bird densities with appropriate physical and biological oceanographic data.

Pitelka (1974) reviewed all Alaskan Beaufort Sea birds and their status. Specific data gathered on the cruises discussed here that deal with small-scale distribution, habitat use, migration, and uncommon species are presented elsewhere (Divoky, 1983).

A. Study Area

The Alaskan Beaufort Sea has a narrow continental shelf (average of 55 km between the 20-m and 200-m contours) and is ice covered for eight months of the year. Observations presented here were conducted during the time of least ice cover. Sea-surface temperatures in August typically are below 5°C; the extreme western area regularly has warmer sea-surface

---

1. Current address: University of Alaska, Fairbanks

THE ALASKAN BEAUFORT SEA:
ECOSYSTEMS AND ENVIRONMENTS

417

temperatures than further east due to Bering Sea water that has passed through the Chukchi Sea and entered the Beaufort (Johnson, 1956). The water column is highly stratified with low nutrient levels near the surface (Hufford, 1974; Horner, 1981), resulting in low biological productivity (Schell *et al.*, 1982). Pelagic waters on the shelf move to the east, while movement north of the shelf is to the west (Aagaard, this volume).

The nearshore regime (inside the 20-m contour) averages 40 km in width near Barrow and decreases to 10 km in the east (Fig. 1). It contains three major barrier island systems and associated lagoons separated by unprotected coasts and bays. Ice scour of benthic habitats in the nearshore can be severe (Barnes *et al.*, this volume). Nearshore waters in August are brackish and water temperatures are regularly between 5° and 10°C (Craig and Haldorson, 1981). Currents are wind-driven and primarily to the west.

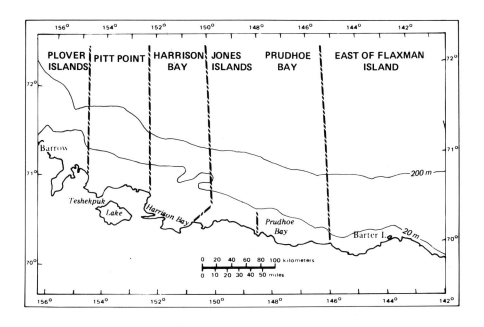

FIGURE 1.   The five pelagic regions and six nearshore regions of the Alaskan Beaufort Sea.   The pelagic and nearshore regimes are separated by the 20-m contour.

Two principal regimes are compared in this paper, the pelagic and nearshore. The two regimes, separated by the 20-m contour, are distinct ecological units (Truett, 1981; Carey *et al.*, 1982) (Fig. 1). Most of our pelagic observations were made on the shelf (inside 200 m) with fewer observations on the continental slope (Fig. 2). Because nearshore observations were rarely as close as 300 m to land, the littoral zone is not included in the data presented here; bird densities there are discussed by Connors (this volume).

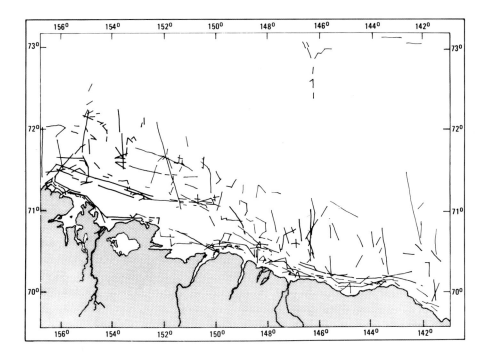

FIGURE 2. Locations of bird observations in the Alaskan Beaufort Sea.

I divided the Alaskan Beaufort Sea into longitudinal regions: five for the pelagic regime and six for the nearshore, based on coastal morphology and oceanographic parameters. These divisions are shown in Fig. 1 and listed below. A prominent coastal feature is used to name the region:

> *Plover Islands*—Point Barrow to Cape Simpson
> ($156°30'$W to $154°30'$W)
> *Pitt Point*—Cape Simpson to Cape Halkett
> ($154°30'$W to $152°10'$W)
> *Harrison Bay*—Cape Halkett to Thetis Island
> ($152°10'$W to $150°10'$W)
> *Jones Islands*—Thetis Island to Egg Island
> ($150°10'$W to $148°30'$W)
> *Prudhoe Bay*—Egg Island to Brownlow Point
> ($148°30'$W to $145°50'$W)
> *East of Flaxman Island*—Brownlow Point to
> Demarcation Point
> ($145°50'$W to $141°00'$W)

For pelagic data the Jones Island and Prudhoe Bay regions are treated as a single unit. For nearshore data they are divided.

II. METHODS

We censused pelagic areas from icebreakers and used smaller vessels in nearshore waters. Other than the difference in the height of the observer above the sea (18 m vs 6 m), methods were the same for both regimes. Cruises were conducted in pelagic waters in 1971, 1972, and 1976-78, nearshore cruises in 1976-78. Observations were made from 2 August to 18 September in pelagic waters and from 2 to 30 August in nearshore waters. The location of cruise tracks is shown in Fig. 2.

Twenty-minute census intervals were used in 1971 and 1972 and 15-minute intervals in 1976-78. The number of intervals (or transects) in the two regimes in each region is given in Table I. Observations were conducted only when the vessel was going at least 3 knots. All birds observed out to 300 m to one side of the ship were identified to species or species group and counted, with notes taken on age, sex, and activities whenever possible. The area censused in each transect was computed and the density (birds km^{-2}) obtained for each species. Biomass density (kg km^{-2}) was derived from the density figure and the average weight of the species. Average weights used, presented in Table II, have been obtained from specimens collected for stomach contents analysis or from the literature. For both regimes in each region, an average biomass density for each species was obtained by pooling data from all stations in that area.

TABLE I. Number of 15- or 20-minute Transects of Bird Observations in the Five Regions of the Alaskan Beaufort Sea.

Region	Pelagic	Nearshore
Plover Islands	199	108
Pitt Point	206	76
Harrison Bay	229	82
Jones Islands	--	132
Prudhoe Bay	364	128
East of Flaxman Island	313	65

I attempted to include only those species and individuals that feed or potentially feed in the area. For example, the majority of the eiders seen in the Alaskan Beaufort are flying migrants, and only sitting eiders were used in density calculations. Loons and oldsquaw are commonly seen both flying and sitting in the nearshore, and all sightings in that regime were included in density calculations. They do not occur in pelagic waters until the last days of August, when migration begins (Timson, 1976; Divoky, 1983). Because few loons and oldsquaw sit on the water in the pelagic regime, they are not included in biomass calculations. Surface-feeding species are always included in density calculations, as even when they are migrating through the Beaufort they are low enough to be searching for prey.

TABLE II.   *Average Weights of Species Regularly Encountered in the Pelagic and Nearshore Alaskan Beaufort Sea.*

Species	Weight (kg)
Surface-feeding	
Red phalarope (*Phalaropus fulicarius*)	0.05
Red-necked phalarope (*Phalaropus lobatus*)	0.05
Pomarine jaeger (*Stercorarius pomarinus*)	0.8
Parasite jaeger (*S. parasiticus*)	0.5
Long-tailed jaeger (*S. longicaudus*)	0.4
Glaucous gull (*Larus hyperboreus*)	1.2
Black-legged kittiwake (*Rissa tridactyla*)	0.4
Ross' Gull (*Rhodostethia rosea*)	0.2
Sabine's gull (*Xema sabini*)	0.2
Arctic tern (*Sterna paradisaea*)	0.1
Diving	
Arctic loon (*Gavia arctica*)	2.2
Red-throated loon (*G. stellata*)	1.8
Short-tailed shearwater (*Puffinus tenuirostris*)	0.6
Oldsquaw (*Clangula hyemalis*)	0.8
King eider (*Somateria spectabilis*)	1.5
Common eider (*S. mollissima*)	2.5
Thick-billed murre (*Uria lomvia*)	1.1
Black guillemot (*Cepphus grylle*)	0.5

Note: Jaeger biomass computed using 0.6 kg for all sightings; loon biomass computed using 2 kg for all sightings; eider biomass computed using 2 kg for all sightings.

An icebreaker can attract large flocks of ship followers when arctic cod (*Boreogadus saida*) are washed onto the ice by propwash and the shifting of ice during icebreaking. At such times it can be hard to census certain species accurately, primarily glaucous gulls, black-legged kittiwakes, and jaegers. In addition, the garbage thrown overboard from icebreakers can attract large flocks of scavenging glaucous gulls. While every attempt was made to count only those individuals that would be present regardless of the ship, the error in censusing these species in pelagic waters is larger than for others. Such problems do not occur in the nearshore.

Stomach contents were obtained by collecting birds with a shotgun and removing the esophagi and stomachs. Prey items were identified to the lowest possible taxonomic unit, and prey species or species groups were weighed to the nearest 0.1 g. Contents of stomachs collected in the pelagic and nearshore regimes were pooled separately for each bird species, and the percent of the total weight and the frequency of occurrence (percent stomachs with the prey item) were computed for each prey species, or group. Prey species or groups found in small amounts in bird stomachs are not listed. Summaries of the stomach contents are presented here, detailed information on all prey species is in Divoky (1978a). Terms used to describe feeding methods are from Ashmole (1971).

A. Species Present in the Alaska Beaufort Sea

1. Surface Feeders. Of the surface-feeding species in the Alaskan Beaufort (Table II), the phalaropes, jaegers, and Sabine's gulls breed on tundra and winter in pelagic waters in the tropics or temperate southern hemisphere (Ashmole, 1971). The arctic tern breeds on both marine islands and tundra and winters at the edge of the ice in the Antarctic (Salomonsen, 1967). All of the above species are most abundant in the Beaufort Sea during August.

The glaucous gull breeds on coastal islands as well as inland tundra. Unlike the previously mentioned species, it has a large nonbreeding population, and birds seen in the pelagic Beaufort in August are primarily nonbreeders. Most glaucous gulls leave the Beaufort in September (Bailey, 1948) and winter in subarctic areas south of the pack ice. The black-legged kittiwake breeds on cliffs as far north as Cape Lisburne, 480 km southwest of Barrow (Springer *et al.*, 1982), and all birds seen in August can be assumed to be nonbreeders or possibly failed breeders. Kittiwakes winter in the subarctic. Ross' gull breeds on wet tundra in northern Siberia (Dementiev and Gladkov, 1970) and moves eastward to the Point Barrow region in September. Its wintering range is thought to be in the pack ice either in the Arctic Basin or the western Bering Sea.

Of the surface-feeding species only the phalaropes obtain prey exclusively by surface seizing, sitting on the water and grasping individual prey items. Arctic terns feed exclusively by aerial methods of dipping (picking prey from the water's surface while remaining in the air) and surface plunging (dropping to the water's surface with usually little if any submersion of the body). The remaining species use both surface and aerial techniques. The two large jaegers also feed by kleptoparasitism (stealing prey from other birds). Both jaegers and glaucous gulls will kill and eat phalaropes. Surface plunging, the deepest feeding method employed, probably allows prey to be taken to a maximum depth of 0.5 m.

2. Diving Species. The diving species present in the Beaufort Sea are a diverse group. Loons nest on freshwater ponds and lakes and use both marine and freshwater habitats until the young fledge, when all move to marine waters. The oldsquaw and king eider breed primarily on tundra, the common eider primarily on marine islands and at coastal locations. In all three waterfowl species the males leave the females shortly after egg-laying and move to or migrate over marine habitats, as do females and young when nesting is completed (Johnson and Richardson, 1981, 1982). The thick-billed murre, a cliff-nesting alcid, breeds in small numbers at Cape Parry, 600 km east of the Alaskan border in the Canadian Beaufort (Höhn, 1955). The nearest Alaskan colony is at Cape Lisburne (Springer *et al.*, 1982), and most birds seen in summer can be assumed to be nonbreeders. Black guillemots breed in coastal debris at scattered localities along the Beaufort coast; the most nests are in the Plover Islands (Divoky et al., 1974; Divoky, 1978b). The short-tailed shearwater breeds in the southern hemisphere and spends the austral winter primarily in subarctic waters in the Bering Sea (Sanger and Baird, 1977). A few move north into the Chukchi Sea and fewer still into the Beaufort. Although diving is its primary feeding method, it also uses surface-feeding methods (Ainley and Sanger, 1979).

Unlike the surface-feeding species, most diving species do not undertake major migrations after leaving the Beaufort. The short-tailed shearwater migrates to the southern hemisphere to breed, but all other species move to subarctic and north temperate waters and winter in coastal and neritic habitats, chiefly north of 50°N.

Diving species are able to exploit prey in the water column and benthos in shallow areas. Information on depths of dives and diving behavior for loons and waterfowl can be found in Palmer (1962, 1976). All of the diving species in the Beaufort Sea can dive to 20 m.

III. RESULTS

A. Pelagic

The pelagic regime is characterized by an almost complete absence of diving species in most regions and a marked east-west gradient, with the extreme western region having over 10 times the biomass of the eastern extreme (Table III). Although diving species made up 68% of the biomass in the pelagic waters north of the Plover Islands, diving bird biomass was extremely low in all other regions. The high density of diving species in the Plover Islands is due primarily to one dispersed flock of short-tailed

TABLE III.   Average Biomass (kg km^{-2}) of Surface-feeding and Diving Species in the Pelagic and Nearshore Alaskan Beaufort Sea.

Species	Plover Is.	Pitt Point	Harrison Bay	Jones Island	Prudhoe Bay	East of Flaxman Is.
Pelagic[a]						
Surface	3.0	3.1	1.0		1.0	0.6
Diving	6.4	0.1	0.01		0.01	0.01
Total	9.4	3.2	1.1		1.1	0.61
Nearshore						
Surface	6.2	1.3	1.0	0.8	2.0	0.5
Diving	43.7	56.6	12.2	48.8	63.0	33.7
Total	49.9	57.9	13.2	49.6	65.0	34.2

[a] Jones Island and Prudhoe Bay regions are combined for pelagic species.

shearwaters seen on 15 September 1978. Short-tailed shearwaters have
been seen on only 9 percent of all transects conducted in pelagic waters of
the Plover Islands region, and most of the sightings are from 1978. For all
other years, the density of diving species in the Plover Islands region was
similar to the rest of the Beaufort Sea. Biomass of surface-feeding species
was similar for the two westernmost regions ($\sim$3 kg km^{-2}) and low in the
central and eastern regions ($\leq$1 kg km^{-2}).

Three species of diving birds are regularly in pelagic waters, excluding
those species that are present primarily as migrants. The short-tailed
shearwater is sporadic but sometimes abundant (Fig. 3). The thick-billed
murre and black guillemot have been seen in all pelagic regions but at low
densities ($<$.05 kg km^{-2}). No east-west gradient in pelagic densities was
observed for these two alcids.

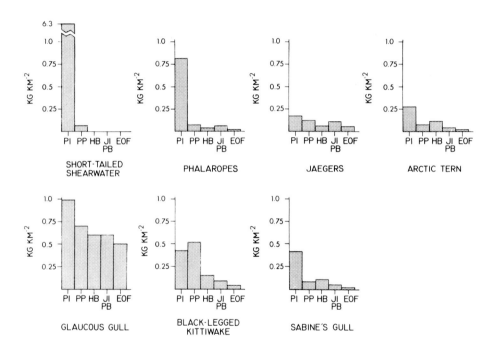

FIGURE 3. Biomass densities of common seabird species in the five regions
of the Alaskan Beaufort Sea pelagic regime.

The low numbers of guillemots and murres found in the Beaufort Sea is
reflected in our small sample size of birds collected (Table IV). Arctic cod
are almost certainly the main prey items. My observations at a black
guillemot colony on the Plover Islands have shown that chicks are fed
almost exclusively on arctic cod, with four-horned sculpin (*Myoxocephalus
quadricornis*) and snailfish (*Liparis* spp.) being taken when the pack ice is
well offshore.

TABLE IV. *Percent of Total Weight and Frequency of Occurrence of Prey Items in Thick-billed Murred and Black Guillemot Stomachs Collected in the Pelagic Beaufort Sea.*

Prey	Thick-billed Murre (3 stomachs sampled with total contents of 18 g)		Black Guillemot (1 stomach sampled with total contents of 1 g)	
	% wt.	% freq.	% wt.	% freq.
Arctic cod	99	100	100	100
Amphipods	1	33	--	--

All sightings of shearwaters in the Beaufort Sea have been in areas where Bering Sea water was present near the surface. Shearwaters penetrate the Beaufort Sea only as far east as the presence of Bering Sea waters, and their presence could be expected to be intermittent. They are not common or regular in adjacent areas of the Chukchi Sea. While we have not made sightings east of 153°40'W, large flocks have been reported north of Pingok Island (in the Jones Islands) and Flaxman Island (L. Lowry, pers. comm.; Bodfish, 1936). No information on local prey consumed is available, but they take both fish and zooplankton in the Bering Sea (Sanger and Baird, 1977).

The glaucous gull accounted for over 50% of the biomass of surface-feeding birds in pelagic waters and while most common in the Plover Islands region (Fig. 3), its decrease to the east was not extreme. This probably reflects the opportunistic feeding habits of the species (Ingolfsson, 1967). Glaucous gulls collected in pelagic waters were primarily preying on small birds and arctic cod (Table V). Because small birds (phalaropes) are uncommon over most of the pelagic waters it is likely that arctic cod is the primary prey item in most areas.

TABLE V. *Percent of Total Weight and Frequency of Occurrence of Prey Items in Glaucous Gull Stomachs Collected in the Pelagic and Nearshore Beaufort Sea.*

Prey	Pelagic (9 stomachs sampled with total contents of 20 g)		Nearshore (9 stomachs sampled with total contents of 98 g)	
	% wt.	% freq.	% wt.	% freq.
Arctic cod	17	56	60	33
Small birds	75	22	--	--
Amphipods	1	22	1	22
Thysanoessa sp.	--	--	13	33
Saduria entomon	--	--	12	11

The black-legged kittiwake was common in the Plover Islands and Pitt Point regions but uncommon in the central and eastern Beaufort Sea. In pelagic waters, kittiwakes took primarily arctic cod and a small amount of zooplankton (Table VI).

TABLE VI. Percent of Total Weight and Frequency of Occurrence of Prey Items in Black-legged Kittiwakes' Stomachs Collected in the Pelagic and Nearshore Beaufort Sea.

Prey	Pelagic (3 stomachs sampled with total contents of 134 g)		Nearshore (1 stomach sampled with total contents of 11.2 g)	
	% wt.	% freq.	% wt.	% freq.
Arctic cod	95	92	5	27
Amphipods	1	12	67	7
Shrimp	1	4	--	--
Apherusa glacialis	--	--	14	28
Mysis sp.	--	--	11	14

Jaegers were slightly more common in the Plover Islands region than regions to the east. We observed the two larger jaeger species obtaining arctic cod from glaucous gulls and kittiwakes in feeding flocks during icebreaking. Long-tailed jaegers could be expected to take arctic cod and zooplankton. Phalaropes are occasionally preyed upon by parasitic jaegers in the nearshore and are probably also taken in pelagic waters.

Phalaropes, Ross' gull, Sabine's gull, and arctic tern are found in their highest densities in the Plover Islands region and decrease sharply to the east to extremely low densities. Ross' gull was encountered in numbers only in the Plover Islands in mid-September 1976 when it averaged .1 kg km^{-2}, with an average of $<.05$ kg km^{-2} for all years combined. All phalaropes collected for stomach contents were red phalaropes, and their stomachs frequently contained less than 0.1 g of prey, frequently in the form of crustacean chyme. Percent weight was therefore difficult to compute and only frequency of occurrence is presented (Table VII). The average length of whole prey items is also given. The 8 stomachs collected in the pelagic regime did not differ from the 68 from the nearshore so the data were pooled. Gammarid amphipods, especially *Apherusa glacialis,* appear to be the major prey items. Connors and Risebrough (1977) found copepods, *A. glacialis,* and decapod zoea to be the major prey items in the littoral zone at Point Barrow, and at Simpson Lagoon (Jones Islands region) phalaropes took primarily copepods, amphipods, and mysids (Johnson and Richardson, 1981).

Arctic cod was the principal prey of arctic terns in the pelagic regime and was frequently found in Sabine's gull stomachs (Tables VIII and IX). Both species consume pelagic zooplankton, however. No Ross' gulls were collected in the pelagic Beaufort, but it is likely that arctic cod and gammarid amphipods are the principal prey items as they are in the Chukchi Sea and nearshore of the Plover Islands region (Divoky, 1976).

TABLE VII. Frequency of Occurrence and Average Length of Prey Items in 76 Red Phalarope Stomachs Collected in the Pelagic and Nearshore Beaufort Sea.

Prey	Percent frequency	Size of prey (mm)
Unidentified gammarid amphipods	31	5
Unidentified crustaceans	30	--
Apherusa glacialis	18	8
Mysis sp.	13	14
Copepods	11	2.5
Thysanoessa sp.	5	11
Larval fish	1	15

TABLE VIII. Percent of Total Weight and Frequency of Occurrence of Prey Items in Sabine's Gull Stomachs Collected in the Pelagic and Nearshore Beaufort Sea.

Prey	Pelagic (6 stomachs sampled with total contents of 15 g)		Nearshore (32 stomachs sampled with total contents of 60.6 g)	
	% wt.	% freq.	% wt.	% freq.
Arctic cod	13	67	4	3
Thysanoessa sp.	13	17	4	3
Parathemisto sp.	53	17	--	--
Shrimp	3	17	--	--
Apherusa glacialis	--	--	49	56
Mysis	--	--	24	28
Unident. amphipods	1	17	6	19

B. Nearshore

Total avian biomass supported in the nearshore Alaska Beaufort Sea is over 10 times the amount present in the pelagic waters (Table III). Except for Harrison Bay and the area east of Flaxman Island, the nearshore had rather consistent values between 49 and 65 kg km^{-2}. The low values for Harrison Bay are probably valid as that area was extensively sampled, and Johnson and Richardson (1981) also found low bird densities there. Our sampling east of Barter Island in the easternmost region has not been as thorough, and the data on biomass of diving birds are partially based on Spindler's (1981) data, which are not directly comparable to ours. The densities for this area should be considered minimum estimates. While no east-west trend is evident in diving species in the nearshore, surface-

TABLE IX. *Percent of Total Weight and Frequency of Occurrence of Prey Items in Arctic Tern Stomachs Collected in the Pelagic and Nearshore Beaufort Sea.*

Prey	Pelagic (6 stomachs sampled with total contents of 40.4 g)		Nearshore (48 stomachs sampled with total contents of 78.6 g)	
	% wt.	% freq.	% wt.	% freq.
Arctic cod	64	78	20	21
Thysanoessa sp.	35	22	23	23
Amphipods	1	4	22	38
Mysis sp.	--	--	13	29
Apherusa glacialis	--	--	9	17
Sand lance	--	--	12	2

feeding species had their highest total density in the Plover Islands area (6.2 kg km^{-2}) with low values (0.5 to 2 kg km^{-2}) for areas to the east.

The diving species numerous in the nearshore regime are absent from pelagic waters except as migrants (Fig. 4). Oldsquaw constitute 80% of the biomass of diving birds in the nearshore, with loons and eiders making up the remainder. While loons had their highest densities in the western part of the Alaskan Beaufort, oldsquaw and eiders showed no such trend. Eiders were most abundant in the Prudhoe Bay region, where they are common as breeders on barrier islands (Divoky, 1978b). Murres were common in the Plover Islands in 1978, giving them a relatively high average biomass for that region (Fig. 4). Guillemot biomass in the Plover Islands was 0.2 kg km^{-2}, with densities east of there <0.01 kg km^{-2}.

Oldsquaw were found to be feeding on a wide range of invertebrates (Table X). Mysids and amphipods are the primary prey of oldsquaws in Simpson Lagoon (Johnson and Richardson, 1981) and in the Beaufort Lagoon (east of Flaxman region) (S.R. Johnson, pers. comm.). The common and king eiders appear to take mainly large isopods in nearshore waters (Table XI). Two arctic loons collected in the nearshore Beaufort Sea in early July both contained amphipods (Divoky, 1978a). Fish, as well as amphipods, are important to loons; adults are frequently seen carrying arctic cod from nearshore waters to inland nesting areas.

Surface-feeding species present in the nearshore are the same as those in the pelagic regime. All reach their highest densities in the Plover Islands region (Fig. 5). The decrease to the east in the total biomass of surface-feeding species is similar to that found in pelagic waters with the exception that the Prudhoe Bay region has the highest density east of the Plover Islands.

As in pelagic waters, glaucous gulls constituted more than 50% of the surface-feeding biomass. Arctic cod are an important prey item, but amphipods and euphausiids are also frequent in stomachs (Table V). The remaining surface-feeding species all showed a sharp decline east of the Plover Islands. In nearshore waters black-legged kittiwakes, Sabine's gulls,

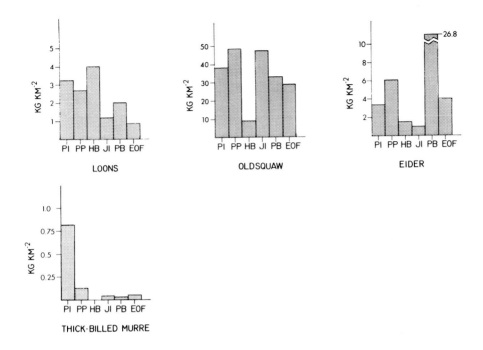

FIGURE 4. Biomass densities of the common diving species in the six regions of the Alaskan Beaufort Sea nearshore regime.

TABLE X. Percent of Total Weight and Frequency of Occurrence of Prey Items in Oldsquaw Collected in the Nearshore Beaufort Sea.

| Prey | Oldsquaw (93 stomachs sampled total contents of 317.8) | |
	% wt.	% freq.
Amphipods	23	38
Mollusks	22	59
Mysis sp.	20	16
Thysanoessa sp	17	13

and arctic terns all feed to a large extent on zooplankton, with amphipods, mysids, and euphausiids the most important groups (Tables VI-IX). Arctic cod are of less importance than in the pelagic regime. Phalaropes feed on zooplankton in both regimes. Probably because of their small size, copepods are apparently taken regularly only by phalaropes.

TABLE XI. *Percent of Total Weight and Frequency of Occurrence of Prey Items in Common and King Eider Stomachs Collected in the Nearshore Beaufort Sea.*

Prey	Common Eider (3 stomachs sampled with total contents of 42 g)		King Eider (16 stomachs sampled with total contents of 178.6 g)	
	% wt.	% freq.	% wt.	%freq.
Saduria entomon	83	67	89	63
Mysis sp.	15	33	--	--
Mollusks	1	66	2	31
Amphipods	1	33	2	19

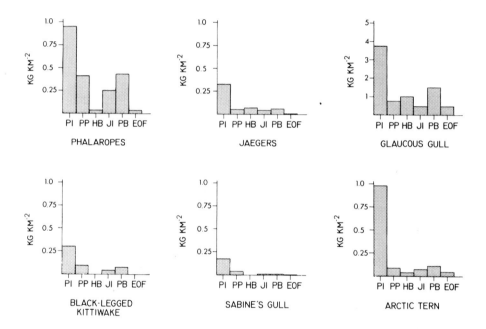

FIGURE 5. *Biomass densities of the common surface-feeding species in the six regions of the Alaskan Beaufort Sea nearshore regime.*

IV. DISCUSSION

A. *Seasonality of Bird Use*

The observations presented here were made during the season of maximum abundance of birds in the study area. The Alaskan Beaufort Sea has nearly complete ice cover from November through May, and birds can be expected to be absent or rare. Black guillemots winter in the lead systems in the northern Chukchi Sea (Bailey, 1948; Nelson, 1969) and can be expected to occur where the leads extend into the extreme western Beaufort Sea.

When birds return to the arctic in May, ice cover is still nearly complete and many species (phalaropes, jaegers, small gulls, and terns) complete their migration over land (Richardson and Johnson, 1981; Woodby and Divoky, 1982). Those species that enter the Beaufort Sea from the leads in the Chukchi Sea (eiders, oldsquaw, glaucous gulls, and murres) (Woodby and Divoky, 1982; Bailey, 1948) can be expected to occur offshore as they have been observed in the Canadian Beaufort (Searing et al., 1975). Unlike the Chukchi and the eastern Beaufort Sea, however, the Alaskan Beaufort Sea lacks well-developed leads and polynyas in spring, and fewer birds would be expected to land offshore. While a nearshore spring migration occurs (Richardson and Johnson, 1982), it is not major, and the lack of open water prevents birds from feeding or roosting except in areas affected by river runoff (Schamel, 1978).

From early June through July the amount of open water increases, allowing birds access to prey throughout most of the Beaufort Sea. As birds finish breeding, the numbers of birds in the nearshore increase slowly until late July, when numbers increase rapidly as many young and adults leave the tundra (Schamel, 1978; Johnson and Richardson, 1982). Limited shipboard and aerial censusing (Divoky, unpubl.; Harrison, 1977) has shown few birds to be present offshore during this period.

By mid-September, when our observations cease, many species have left. Those that remain are actively migrating, and densities of feeding birds can be expected to be less. Ross' gull is the one species whose densities might be expected to increase since they do not enter the Beaufort Sea until mid-September.

B. Pelagic Regime

The bird densities in the pelagic waters of the Beaufort Sea are the lowest of any of the seas adjacent to Alaska, with an average density of 2 kg km^{-2}. The low numbers of birds supported by the pelagic waters can be at least partially attributed to the low annual primary productivity of the Beaufort, which averages between 15 and 30 g C m^{-2} (Schell et al., 1982), and resulting low prey densities (Horner, 1981). The lack of diving species in most areas of the pelagic western Beaufort Sea is another indication of low prey density. Surface feeders are able to feed where prey densities are low because they can search for food while flying, whereas diving species require more abundant and reliable food sources. In general, diving species in pelagic waters are found over the continental shelf, and the narrow shelf of the Alaskan Beaufort Sea provides little suitable habitat.

Arctic cod is the primary prey of birds in the pelagic regime, but seabirds take a very small amount of the total arctic cod biomass consumed by vertebrates. Frost and Lowry (1981), using bird densities provided by the current study, estimated that seabirds consume 3% of the total arctic cod biomass consumed annually by vertebrates: ringed seals (*Phoca hispida*), belukha whales (*Delphinapterus leucas*), and cannibalism by arctic cod are the major consumers. The low percentage consumed by birds is due to their low densities and their short residency time. Estimates of the total annual production of smaller fish consumed by large predators, including seabirds, was estimated to be 10% in the North Sea (Evans, 1971) and 8% in the North Pacific (Sanger, 1972).

Aside from low densities and a paucity of diving species, the most striking aspect of the western Beaufort Sea pelagic bird community is the difference between the relatively high densities found west of Cape Halkett (Plover Islands and Pitt Point regions) and the low densities present to the east. During the period of observations there was a post-breeding movement of most surface-feeding species west to Point Barrow and then southwest to the Pacific. Many of these species undertake long migrations after leaving the arctic and require abundant prey to build up fat reserves before beginning these migrations. The results presented here, as well as other information on bird movements in pelagic waters, show that birds move westward through the Beaufort Sea until they reach the area west of Cape Halkett, where they begin to actively feed and search for food. Birds accumulate in this area, resulting in high densities, before continuing their post-breeding migration.

Three species, the short-tailed shearwater, glaucous gull, and black-legged kittiwake, have pelagic populations that are not involved in breeding and are able to respond to prey abundance during the entire summer and early fall. These species have a similar distributional pattern, the highest densities occurring west of Cape Halkett.

It appears that the higher bird densities in the western Beaufort Sea are due to the input of warm subarctic water resulting in higher prey densities. The advection of Bering Sea water into the Beaufort Sea (the Bering Sea Intrusion) consists of two water masses, Alaska Coastal Water and Bering Sea Water (Aagaard, this volume). The former water mass mixes with arctic surface water soon after it passes Point Barrow and is not clearly identifiable east of 148°W. The latter moves east to at least 143°W. Satellite images from periods when ice is decomposing indicate that when the intrusion meets the water moving westward in the Beaufort Gyre and the nearshore regime, eddies are formed in the Plover Islands and Pitt Point regions (Solomon and Ahlnas, 1980).

While the importance of the Bering Sea Intrusion to Beaufort Sea biological oceanography was recognized by Johnson (1956), the studies related to oil development in the central Alaskan Beaufort Sea were usually too far from the major effects of the intrusion to provide further information on it. Schell et al. (1982), showed a tongue of high primary productivity (35–40 g C m^{-2} yr^{-1}) from Point Barrow east to 152°W along the 20-m contour and speculated that eddies north of the Plover Islands have annual production rates approaching 50 g C m^{-2}. Areas east of 151°W have annual productivity rates between 15 and 27.5 g C m^{-2}.

While zooplankton sampling has demonstrated the presence of Bering Sea species and higher densities of other species in the intrusion, no clear evidence of higher total zooplankton abundance in the intrusion is available. Johnson (1956) found the area from Point Barrow to 150°W to have copepods characteristic of the Bering Sea. The arctic copepod *Arcatia longiremis* was most common west of 150°W, and barnacle larvae reached their highest densities directly north of Point Barrow. McConnell (1977) in an analysis of zooplankton collected in 1971 east of 155°W found only one species, a pteropod (*Clione limacina*), with a distinctly western distribution, but did find an area of high zooplankton diversity and abundance at 148°W which she attributed to a pulse of Bering Sea water. Data collected by Horner (1981) in 1973 and 1976–78 showed barnacle larvae to be most abundant in the intrusion but failed to find any other effect of the intrusion on zooplankton densities or larval fish. Lowry and Frost (1981) showed that arctic cod were most common west of $148^{\circ}50'$W, although their data, from bottom tows, are of use in explaining the densities of surface feeders only in showing a general trend. Marine mammal distribution reflects the influences of the intrusion. Bowhead whales (*Balaena mysticetus*) concentrate in the fall in the area north of the Plover Islands (Braham and Krogman, 1977) as do belukha whales (J. Burns, pers. comm.). Seals present on the shorefast ice have been shown to be most common west of 153°W (Burns and Harbo, 1972; Burns and Eley, 1978).

The importance of Bering Sea water, and its associated higher productivity, to seabirds in the Chukchi Sea was recognized by Springer *et al.* (1982) for cliff-nesting species as well as for pelagic populations (Divoky, unpub. data).

If the Bering Sea Intrusion is the primary factor causing high prey abundance in the extreme western Beaufort Sea, the distribution and abundance of prey could be expected to be patchy in space and time and poorly sampled by opportunistic zooplankton tows. Bering Sea water moves into the Beaufort Sea in pulses, and the strength of the intrusion shows yearly variability (Johnson, 1956; Hufford *et al.*, 1974). Physical oceanographic sampling in the Plover Islands and Pitt Point regions demonstrates that certain stations show no effects of the intrusion and that the temperature and depth of the Bering Sea water varies greatly among stations where the intrusion is observed (Horner, 1981). In addition, the presence of eddies would increase the patchiness of zooplankton. For these reasons it is not surprising that the available zooplankton data fail to coincide with data for marine bird and mammal densities.

The patchiness of bird densities in the Plover Islands region is demonstrated by observations made on 4 September 1977. Transects began at $72^{\circ}10'$N, $155^{\circ}00'$W and went south to $71^{\circ}33'$N. All transects were in ice-free waters. In the first 4.25 hours of observation the average biomass was 0.4 kg km^{-2}. Approaching the 20-m contour, a feeding flock of 300 arctic terns, 20 black-legged kittiwakes, 10 Sabine's gulls, 5 red phalaropes, and 2 pomarine jaegers was observed. The flock was over a convergence line that caused enough wave action to be clearly visible on the ship's radar as sea clutter. The east side of the convergence had a Secchi depth of 9 m while the west side had a Secchi depth of 6 m, with both sides having 8°C water in the entire water column (27 m) (Horner, 1981). Stomach samples showed that arctic terns were taking primarily euphausiids while Sabine's gulls were taking hyperiid amphipods. Bongo tows on both sides of the convergence showed only moderate densities of either prey species. Bird censusing conducted next to and over the convergence produced an average of 6.9 kg km^{-2} in 1.75 hours of observation with a high of 30.7 kg km^{-2} directly over the convergence. Densities to the east and west of the convergence were low with 0.5 kg km^{-2} in 4.5 hours of observations.

The processes by which the intrusion increases prey abundance are not known, but it could

(1) bring water with higher nutrient or phytoplankton levels into the Beaufort Sea, thus increasing primary productivity or providing food for zooplankton, or

(2) bring water with high zooplankton standing stocks into the Beaufort Sea, thus directly providing prey.

Major washups of dead and dying *Thysanoessa raschii* that occur regularly on the Plover Islands (Boekelheide, 1980) may be composed of Pacific expatriates that perish as the Bering Sea water loses its integrity in the western Beaufort Sea. An integrated study of the physical and biological components of the extreme western Beaufort Sea is needed in order to delineate the processes involved.

C. Nearshore Regime

The nearshore regime is characterized by high densities of diving species that prey primarily on epibenthos. Surface feeders are most common in the Plover Islands region with lesser numbers present to the east. Their primary prey is zooplankton. The rather consistent biomass levels of diving species indicate that the epibenthic system that supports these species (Johnson and Richardson, 1982) varies little between regions. This agrees with the results of benthic sampling (Carey et al., 1982). An exception is Harrison Bay, where the low avian biomass levels are probably related to the shallowness of the bay (averaging <10 m) and the major input of turbid fresh water from the Colville River. Epibenthic-feeding diving species are probably limited to the nearshore due to the energy required in diving more than 20 m. However, there is a change in the benthic populations of amphipods and mollusks at 20 m (Carey et al., 1982) that may be related to the lack of epibenthic feeders in deeper waters.

The abundance of surface feeders in the vicinity of the Plover Islands indicates that these waters are substantially more productive than nearshore waters to the east. Shoreward transport of marine production may occur at depth (Truett, 1981), and the higher productivity in the pelagic Plover Islands region may be transported to the adjacent nearshore. Mixing of the waters of the two regimes may occur near the Plover Islands since the principal movement of the nearshore waters is to the west and as it passes Point Barrow it meets the northeastward moving offshore water.

V. SUMMARY

The Alaskan Beaufort Sea has two distinct bird communities in August and early September. A low-biomass pelagic community of primarily surface feeders preys principally on arctic cod. The higher biomass nearshore community is composed of diving species feeding on epibenthos with a small surface-feeding component preying principally on zooplankton. Prey abundance and availability for surface feeders are apparently enhanced by the intrusion of Bering Sea water as biomass densities of these species are highest in the extreme western Beaufort in both pelagic and nearshore regimes. With the exception of low densities in Harrison Bay, the biomass of diving birds is rather consistent throughout the nearshore and is supported largely by epibenthic crustaceans.

The data presented here are the first to provide information on the large-scale seabird distribution patterns in the Alaskan Beaufort Sea and to compare densities and feeding habits in the pelagic and nearshore regimes. This overview provides a framework that facilitates the interpretation of the many site- and region-specific ornithological studies that have recently been conducted in the area.

As pointed out by Ashmole (1971) and Brown (1980), seabirds are rarely given their due significance as components of marine ecosystems. However, as field ornithologists know (see for instance Pitelka, 1974), the relative ease and efficiency with which birds can be censused makes them ideal subjects for biogeographic analysis. The data presented showing an area of high avian biomass in the extreme western Beaufort, and the differences in species composition and biomass between the pelagic and the nearshore regimes, suggest a number of areas and phenomena that deserve further research by biological oceanographers.

ACKNOWLEDGMENTS

Data from 1971 and 1972 were gathered as part of the Smithsonian Institution's participation in the U.S. Coast Guard's Western Beaufort Sea Ecological Cruises. Later cruises were conducted through the Arctic Project Office of the Outer Continental Shelf Environmental Assessment Program, sponsored by the National Oceanic and Atmospheric Administration and the Bureau of Land Management.

The assistance of the captains and crews of the U.S. Coast Guard icebreakers Glacier and Northwind, and the Naval Arctic Research Laboratory research vessels Alumiak and Natchik is gratefully acknowledged.

A. Edward Good, George E. Watson, Doug A. Woodby, Kenneth L. Wilson, Doug J. Forsell, Tom Scharffenburg, and the author conducted shipboard observations and collected specimens for stomach contents analysis. Frank A. Pitelka, William H. Drury, and Steven R. Johnson provided useful suggestions on an earlier version of this manuscript.

REFERENCES

Ainley, D.G., and Sanger, G.A. (1979). In "Conservation of Marine Birds of Northern North America" (J.C. Bartonek and D.N. Nettleship, eds.), p. 95. U.S. Dept. Int. Fish Wildl. Serv., Wildl. Res. Rep. 11.
Ashmole, N.P. (1971). In "Avian Biology" (D.S. Farmer and J.R. King, eds.), Vol. 1, p. 223. Academic Press, New York.
Bailey, A.M. (1948). Pop. Ser. No. 8., Colorado Mus. Nat. History, Denver.
Bodfish, H.H. (1936). "Chasing the Bowhead." Harvard Univ. Press, Cambridge, MA.
Boekelheide, R.J. (1980). M.S. Thesis. Univ. California, Davis.

Braham, H.W., and Krogman, B.D. (1977). "Population Biology of the Bowhead (*Balaena mysticetus*) and Beluga (*Delphinapterus leucas*) Whale in the Bering, Chukchi and Beaufort Seas." Nat. Mar. Fish. Serv. Rept., Seattle.

Brown, R.G.B. (1979). *Ibis 121*, 283-92.

Brown, R.G.B. (1980). *In* "Behavior of marine animals" (J. Burger, B. Olla and H.E. Winn, eds.), Vol. 4, p.1. Plenum Press, New York.

Burns, J.J., and Eley, T.J. (1978). *In* "Environmental Assessment of the Alaskan Continental Shelf, Annual Report of Principal Investigators," Vol. 1, p. 99. OCSEAP, Boulder.

Burns, J.J., and Harbo, S.J. (1972). *Arctic 25*, 279.

Carey, A.G., Ruff, R.E., Scott, P.H., Walters, K.R., and Kern, J.C. (1982). *In* "Environmental Assessment of the Alaskan Continental Shelf," Vol. 2, p. 27. OCSEAP, Boulder.

Connors, P., and Risebrough, R.W. (1977). *In* "Environmental Assessment of the Alaskan Continental Shelf, Annual Report of Principal Investigators," Vol. 2, p. 402. OCSEAP, Boulder.

Craig, P.C., and Haldorson, L. (1981). *In* "Environmental Assessment of the Alaskan Continental Shelf, Final Report of Principal Investigators," Vol. 8, p. 384. OCSEAP, Boulder.

Dementiev, G.P., and Gladkov, N.A. (1970). "Birds of the Soviet Union." Vol. 3. Israeli Program for Scientific Translation, Jerusalem.

Divoky, G.J. (1976). *Condor 78*, 85.

Divoky, G.J. (1978a). *In* "Environmental Assessment of the Alaskan Continental Shelf, Annual Report of Principal Investigators," Vol. 1, p. 549. OCSEAP, Boulder.

Divoky, G.J. (1978b). *In* "Environmental Assessment of the Alaskan Continental Shelf, Annual Report of Principal Investigators," Vol. 1, p. 482. OCSEAP, Boulder.

Divoky, G.J. (1982). "The pelagic and nearshore birds of the Alaskan Beaufort Sea." Final Rept. to OCSEAP.

Divoky, G.J., Watson, G.E. and Bartonek, J.C. (1974). *Condor 76*, 339.

Evans, P.R. (1971). *In* "North Sea Science" (E.D. Goldberg, ed.), p. 400. MIT Press, Cambridge, MA.

Frame, G.W. (1973). *Auk 90*, 552.

Frost, K.J., and Lowry, L.F. (1981). "Feeding and trophic relationships of bowhead whales and other vertebrate consumers in the Beaufort Sea." Rept. to Natl. Mar. Fish. Serv., Seattle.

Gabrielson, I.N., and Lincoln, F.C. (1959). "The birds of Alaska." The Stackpole Company and Wildlife Management Institute, Harrisburg, PA.

Harrison, C.S. (1977). *In* "Environmental Assessment of the Alaskan Continental Shelf, Annual Report of Principal Investigators," Vol. 3, p. 285. OCSEAP, Boulder.

Höhn, E.O. (1955). *Can. Field-Nat. 69*, 41.

Horner, R.A. (1981). *In* "Environmental Assessment of the Alaskan Continental Shelf, Annual Report of Principal Investigators," Vol. 13, p. 65. OCSEAP, Boulder.

Hufford, G.L. (1974). *J. Geophys. Res. 78*, 274.

Hufford, G.L., Fortier, S.H., Wolfe, D.E., Poster, J.F., and Noble, D.L. (1974). "Physical oceanography of the western Beaufort Sea." U.S. Coast Guard Oceanogr. Rept. CG-373.

Ingolfsson, A. (1967). Ph.D. dissertation, Univ. Michigan, Ann Arbor.
Johnson, M.W. (1956). "The plankton of the Beaufort and Chukchi Sea areas of the Arctic and its relation to hydrography." Arct. Inst. of N. Am. Tech. Pap. No. 1.
Johnson, S.R., and Richardson, W.J. (1981). In "Environmental Assessment of the Alaskan Continental Shelf, Annual Report of Principal Investigators," Vol. 7, p. 180. OCSEAP, Boulder.
Johnson, S.R., and Richardson, W.J. (1982). Arctic 35, 291.
Lowry, L.F., and Frost, K.J. (1981). Can. Field-Nat. 95, 186.
McConnell, M. (1977). M.S. Thesis. Univ. Rhode Island, Kingston.
Nelson, R.K. (1969). "Hunters of the Northern Ice." Univ. of Chicago Press, Chicago.
Palmer, R.S. (ed.) (1962). "Handbook of North American Birds. Vol. 1. Loons through Flamingos." Yale Univ. Press, New Haven, CT.
Palmer, R.S. (ed.) (1976). "Handbook of North American Birds. Vol. 3. Waterfowl." Yale Univ. Press, New Haven, CT.
Pitelka, F.A. (1974). Arctic and Alpine Res. 6(2), 161.
Richardson, W.J., and Johnson, S.R. (1982). Arctic 34, 108.
Salomonsen, F. (1967). Biol. Medd. Kobenhaven 24, No. 1, 1.
Sanger, G.A. (1972). In "Biological oceanography of the northern North Pacific Ocean" (A.Y. Takenouti, ed.), p. 589. Idenitsu shoten, Tokyo.
Sanger, G.A., and Baird, P.A. (1977). In "Environmental Assessment of the Alaskan Continental Shelf, Annual Report of Principal Investigators," Vol. 12, p. 372. OCSEAP, Boulder.
Schamel, D.L. (1978). Can. Field-Nat. 92, 35.
Schell, D.M., Ziemann, P.J., Parrish, D.M., and Brown, E.J. (1982). "Foodweb and nutrient dynamics in nearshore Alaskan Beaufort Sea waters." Cumulative Summary Rept. to OCSEAP.
Searing, G.F., Kuyt, E., Richardson, W.J., and Barry, T.W. (1975). "Seabirds of the Southeastern Beaufort Sea: Aircraft and Ground Observations in 1972 and 1974." Beaufort Sea Tech. Rept. 3b. Dept. of Environ., Victoria, B.C.
Solomon, H., and Ahlnas, K. (1980). Arctic 33, 184.
Spindler, M.A. (1981). "Bird populations and distribution in the coastal lagoons and nearshore waters of the Arctic National Wildlife Refuge, Alaska." Draft Arctic Natl. Wildlife Refuge Rept., Wildlife Ser. 81-4. Fairbanks.
Springer, A.M., Murphy, E.C., Roseneau, D.G., and Springer, M.I. (1982). "Population status, reproductive ecology and trophic relationships of seabirds in northwestern Alaska." Final Rept. to OCSEAP.
Timson, R.S. (1976). In "Environmental Assessment of the Alaskan Continental Shelf, Quarterly Report of Principal Investigators, April-June 1976," Vol. 1, p. 354. OCSEAP, Boulder.
Truett, J.C. (1981). In "Environmental Assessment of the Alaskan Continental Shelf, Final Report of Principal Investigators," Vol. 8, p. 259. OCSEAP, Boulder.
Woodby, D.A., and Divoky, G.J. (1981). Arctic 35, 403.

Man's Interaction

INTERACTION OF OIL AND ARCTIC SEA ICE

Donald R. Thomas

Research and Technology Division
Flow Industries, Inc.
Kent, Washington

I. INTRODUCTION

About 1 in 3000 offshore oil wells experiences some kind of blowout. Many of these are relatively harmless in terms of environmental damage. It has been estimated that the chance of a "serious" blowout incident is less than 1 in 100,000 wells drilled. Although this is a very low probability, blowouts happen often enough (for example, the Santa Barbara and the IXTOC blowouts) that the consequences must be considered.

During the next few years, many exploratory and possibly production oil wells will be drilled on the continental shelf in the Beaufort Sea off Alaska. Drilling will initially be from natural or artificial islands in relatively shallow water. While this procedure will reduce the probability of blowouts and provide a stable base for control efforts and spill containment, it is possible for a blowout to occur away from the drill hole. The 1969 blowout in the Santa Barbara Channel, for example, occurred through faults and cracks in the rock as far as 0.25 km from the drill site.

Previous regulations required that any offshore drilling in the Beaufort Sea be done during the period from November through March. Present regulations allow exploratory drilling year-round in some areas while prohibiting it during September and October in other areas. Due to logistic considerations and site-specific environmental concerns, much of the exploratory drilling will still be done during the ice season. The entire sea surface is covered by a floating ice sheet during that time, except for occasional short-lived leads of open water. Thus, sea ice will have an important bearing on the fate of oil spilled by a blowout.

There has been little practical experience with oil spills in ice-covered water. Accidental surface spills that have occurred in ice-covered water in subarctic regions, as in Buzzards Bay, Massachusetts, in 1977 (Deslauriers, 1979), have not been in arctic-type ice, which generally continues to build throughout the winter and is thicker and more continuous than subarctic ice.

THE ALASKAN BEAUFORT SEA:
ECOSYSTEMS AND ENVIRONMENTS

Recently, several Canadian oil spill experiments involving arctic sea ice have been performed.   The Canadian government sponsored one at Balaena Bay, N.W.T., during the winter of 1974-75 as part of the Beaufort Sea Project. It studied the initial spreading of the oil and its incorporation into the ice sheet, the effects of oil on the thermal regime of the ice, weathering of the oil, and cleanup techniques. An experiment in 1978 by Environment Canada in Griper Bay, N.W.T., was done to study the fate of oil spilled beneath multi-year ice. During the winter of 1979-80, Dome Petroleum staged an oil spill in McKinley Bay, in the Canadian Beaufort Sea.   Plume dynamics under the ice, the effects of gas on under-ice spreading, the formation of emulsions, and the surfacing of oil in the spring were some of the topics studied.

This paper summarizes relevant knowledge about the interactions between arctic sea ice and oil. Previous works by Lewis (1976), NORCOR (1977), and Stringer and Weller (1980) have also addressed this topic, but more recent work makes an updating desirable. An attempt is made to identify the major factors in the interaction between oil and arctic sea ice and to present them in a way that defines the scope of the problem. This report restricts itself to factors that can be expected to play a major role given a large under-ice blowout in the Beaufort Sea during winter. Blowouts that occur during the summer, in subarctic water, or beyond the continental shelf are not considered here.

II. THE INTERACTION OF OIL AND SEA ICE

An underwater blowout releasing large quantities of crude oil and gas into the water beneath the ice cover could be expected to follow a different chain of events than an open-water blowout.   No under-ice blowout has occurred yet, but from experimental work (NORCOR, 1975; Martin, 1977; Topham, 1975; Topham and Bishnoi, 1980; Cox et al., 1980; Buist et al., 1981) and observations made at accidental surface spills in icy waters (Ruby et al., 1977; Deslauriers, 1979), one can predict the sequence of events with some confidence. In general, an under-ice blowout in the winter will follow the course outlined below:

(1)   Initial phase—the release of oil and gas and their rise to the surface.

(2)   Spreading phase—the spreading of oil due to water currents and buoyancy.

(3)   Incorporation phase—the incorporation of oil into the ice cover.

(4)   Transportation phase—the motion of the oiled ice.

(5)   Release phase—the release of the oil from the ice.

Three areas and types of ice cover must be accounted for when considering blowouts in the Beaufort Sea: the fast ice zone, the pack ice zone, and the area of interaction between them.

The fast ice zone includes ice that forms near shore each year, although occasional multiyear floes (ice that has survived one or more melt seasons) or remnants of grounded ridges may be incorporated. The ice begins to form in early October, and for a month or two it is susceptible to movement and deformation by the winds.   Eventually, it becomes immobilized, protected by the shore on one side and barrier islands or grounded ice ridges on the other. Since motions and deformations occur for only a short time, the fast ice tends to be relatively flat and undeformed. It begins to melt in place in late May or June and is mostly gone by the end of July.

Further offshore is the pack ice zone. The ice in this zone is a mixture of multiyear and seasonal ice. Winds and currents cause the ice to be in almost constant motion. Cracks open to form leads that quickly freeze over with a layer of thin ice. Some leads are closed by moving ice, which breaks and piles up the thin ice to form ridges and rubble piles.

Where the moving pack ice interacts with the stationary fast ice, a great deal of ice deformation takes place. The winds tend to move the pack ice westward and toward shore, causing much shearing deformation. Many large ice ridges form in this area. Water depths here are from 10 to 30 m. Since many ridge keels are deeper than that, a band of grounded ridges often forms. Following Reimnitz *et al.* (1978), this is called the stamukhi zone.

In the rest of this section, the five phases of interaction between crude oil and sea ice are discussed sequentially.

A. Initial Phase

The blowout is assumed to consist of the continuous release, over a minimum of several days, of large quantities of crude oil and many times that amount of gas. The blowout occurs under an ice cover in the period from November through March. The blowout is also assumed to occur on the Beaufort Sea continental shelf in relatively shallow water (less than about 200 m deep).

1. Effects of Gas.

Topham (1975) simulated small well blowouts in open-water conditions using oil and compressed air. His tests showed that as gas is released in shallow water, it breaks up into small bubbles and rises to the surface, carrying oil and part of the surrounding water along to form an underwater plume. This plume is initially conical in shape, but becomes nearly cylindrical as it rises above the release point. The centerline velocities of plumes do not vary significantly with the depths or air-flow rates of the experiments (flow rates of 3.6 to 40 m^3 min^{-1} at depths of 33 to 60 m).

As the plume reaches the water surface, the vertical transport is deflected radially outward. During tests in open water (Topham, 1975), a ring of concentric waves was produced at some distance from the plume, marking the location of a reversal in radial surface currents. A downward current found there extended to a depth of about 10 m. During Dome's simulated blowout beneath sea ice (Buist *et al.*, 1981), no wave ring was observed, but downward flow occurred about 15 to 20 m from the plume. Small droplets of oil or oil-and-water emulsions are likely carried downward, but the majority of the oil from the blowout rises to the surface in drops with a mean diameter of 1 mm. One or two percent is in fine droplets approximately 0.05 mm in diameter (Topham, 1975). Drops of this size rise naturally at about 0.5 mm s^{-1}. Subsurface currents could carry the very small droplets many kilometers downstream during their slow rise to the surface. However, Buist *et al.* (1981) observed that 90 percent of the oil that was released surfaced within a 50-m radius. Dissolution is generally not considered important in the Arctic (NORCOR, 1975). The formation of stable emulsions was not observed during Dome's simulated blowout.

The first event affecting the ice cover is the collection of gas beneath it. Assuming that the ratio of gas to oil by volume is 150 to 1 at the surface, gas is released at rates near 33 m^3 min^{-1} in the case of a blowout releasing 2000 barrels per day (0.22 m^3 min^{-1}) of oil. Within minutes, large pockets of gas accumulate beneath the ice.

Topham (1977) studied the problem of a submerged gas bubble bending and breaking an ice sheet. There is little doubt that a gas bubble a few centimeters thick and a few meters in radius will crack thin ice. During Dome's experiment (Buist *et al.*, 1981), air accumulating beneath ice 0.65 m thick raised the ice until it cracked, releasing the gas. For thicker ice (up to 2 m), the situation is not so clear. In rough ice where large, deep pockets of gas can collect, the radius needed to crack the ice is a few tens of meters. In smoother ice where the gas can collect only to a few centimeters in thickness, the critical radius can extend from a few hundred meters to several kilometers.

It is likely, nevertheless, that the gas will break the ice cover in the fast ice areas. In fast ice, natural weaknesses in the form of thermal cracks probably occur every few hundred meters (Evans and Untersteiner, 1971). Thus, the gas should spread only a few hundred meters under the ice before it either cracks the ice or comes to a natural crack. Once a crack exists near the blowout site, the ice over the blowout is likely to be further fractured and broken up by turbulence or by sinking into the low-density gas-in-water mixture near the center of the plume.

A moving ice canopy may also be broken up as it passes over the gas plume. If a gas flow of 40 m^3 min^{-1} deposits 2400 m^3 of gas under the ice per hour, then the under-ice layer (if it is 0.1 m thick) will spread to cover an area with a diameter of 175 m in 1 hour, a rate comparable to ice motion of 3 km day^{-1} (about 126 m hr^{-1}). For first-year ice, the motion during that hour is probably of no significance; the ice would be broken up much as stationary ice would. If the ice is moving at several kilometers per day over a small blowout, however, it is possible that it would not break.

Some investigators (Logan *et al.*, 1975; Milne and Herlinveaux, n.d.) believe that large multiyear floes would not be broken up as they move across an underwater gas plume. Topham's work (1977) seems to support this view. Multiyear floes might break, though, if consolidated ridge keels trap deep bubbles of gas or if thermal cracks weaken the ice. Thermal cracks themselves provide an alternate path for releasing the gas.

2. Thermal Effects.

As hot oil enters the water, it breaks up into small droplets 0.5 to 1.0 mm in diameter. Most or all of the oil's heat is transferred to the water column, which in turn is carried to the underside of the ice by the gas-driven plume. Some of the heat from the warmed water will then go into melting the ice, with the greatest part of the melting occurring directly over the blowout.

In addition to the heat content of the oil, the water column (except in very shallow areas), being above freezing, can contribute to the melting of ice. The temperature above freezing of the bottom water—that is, its temperature Celsius—is generally 2 to 4 orders of magnitude lower than the temperature of the oil, but the volume of water circulated in the plume is about 4 orders of magnitude larger than the volume of oil. The total heat transported to the bottom of the ice will thus be roughly 2 to 20 times (depending on the temperature of the water) the amount from the hot oil alone.

The specific heat of sea ice is about 2010 J $(kg\ °K)^{-1}$ and that of a typical crude oil is 1717 J $(kg\ °K)^{-1}$. The heat of fusion of water (fresh) is about 334 kJ kg^{-1}. If the ice sheet has an average temperature of -10°C, it will require 354 kJ kg^{-1} to warm and melt the ice. About 206 kg °K of crude oil would supply this heat. Since the densities of ice and crude oil are about the same, each volume of oil will melt roughly one two-hundredth that volume of ice for each degree the oil is above freezing. Oil at a temperature of 100°C, corresponding to a reservoir depth of about 4000 m, would therefore melt about 0.5 m^3 of ice for each cubic meter of oil released. At least an equivalent amount of ice would be melted by water circulation. However, some of this heat would be spread over a large area by currents and by plume-induced circulation. The result is a small area where significant ice melt would take place and a much larger area with only a slightly reduced ice thickness or a decrease in growth rates.

One would expect the ice directly over the blowout to receive a major proportion of the heat from the oil. In stationary or very slowly moving ice, melting tends to weaken the ice over the blowout, making it more probable that gas bubbles trapped beneath the ice will fracture it and escape. For very large blowouts, a significant amount of ice may be melted, leaving a pool of open water directly over the blowout. Large amounts of oil could collect there, and some could spill over onto the surrounding ice surface.

The density of seawater is about 1020 kg m^{-3} and the density of sea ice is about 910 kg m^{-3}. Densities of fresh crude oil may range from about 800 to 900 kg m^{-3}. Thus, as a hole through the ice fills with crude oil, the oil overflows the top of the hole before it is filled to the bottom. However, during much of the ice season, the air is so cold that crude oil exposed to the atmosphere behaves more like a solid than a liquid. The oil would therefore be limited to a small area on the surface until it pools deep enough to begin spreading beneath the ice. Even during the spring, when the air temperature is above the oil's pour point, the snow cover and natural roughness of the ice surface would limit the spread of oil on the surface, so that spilled oil would still tend to spread beneath the ice.

B. Spreading Phase

Once oil gets underneath an ice sheet, several factors, such as the bottom roughness of the ice, the presence of gas under the ice, the magnitude and direction of ocean currents, and movement of the ice cover, control the concentration and areal extent of the oil spread. Of secondary importance are oil properties such as density, surface tension, equilibrium thickness, and viscosity. The effects of these latter properties are fairly well understood (NORCOR, 1975; Cox et al., 1980; Rosenegger, 1975; Malcolm and Cammaert, 1981), and while important to understanding the basic mechanisms of oil-water-ice interactions, they should not be as influential on the extent of oil coverage as the grosser, more variable factors.

1. Bottom Roughness and Oil Containment.

The bottom roughness of the ice will vary significantly between the fast ice, the pack ice, and the stamukhi zones. Roughness in the fast ice zone is determined chiefly by spatial variations in snow cover that cause differences in ice growth rates (Barnes et al., 1979). The stamukhi zone is dominated by deep ridge keels. In the pack ice zone, both these types of roughness are present along with frequent refrozen leads and a high percentage of multiyear floes, which have exaggerated underside relief. In addition, all ice growing in seawater has a microscale relief due to the columnar nature of new ice growth.

If oil alone is released under sea ice, or if any accompanying gas is vented, the oil begins filling under-ice voids near the blowout. As a void fills downward with oil, the oil eventually reaches a depth where it can begin escaping past neighboring "summits" of ice or through "passes" to the next void. If the ice itself is moving over the site of the blow-out, the voids may not be completely filled, and only that ice passing directly over the blowout plume collects oil.

If new ice forms in calm conditions, the underside is flat and smooth. Oil will spread underneath this ice to some equilibrium thickness, depending upon a balance between surface tension and buoyancy. Cox et al. (1980) reported test results for oil of various densities. The equilibrium slick thickness ranged from 5.2 to 11.5 mm for oils with densities in the range of crude oils. For a constant surface tension, a good approximation of slick thickness can be made using the empirical relationship (Cox et al., 1980)

$$\delta = -8.50 \ (\rho_w - \rho_o) + 1.67,$$

where δ is the slick thickness in centimeters and $(\rho_w - \rho_o)$ is the density difference between oil and water. The minimum stable drop thickness for crude oil under ice has generally been reported to be about 8 mm (Lewis, 1976). Using this value, we see that 8000 m^3 (50,000 bbl) of oil can spread under each square kilometer of smooth ice. This is the minimum volume of oil that can be contained under 1 km^2 of ice in the absence of currents or ice motion. Generally, sea ice is not perfectly smooth, so each square kilometer can actually hold more oil.

During October and November, a snow cover accumulates in drifts parallel to the prevailing wind direction. Barnes et al. (1979) found these snow drifts to be fairly stable throughout the ice season. The drifts insulate the ice from the much colder air, causing reduced ice growth beneath. The underside of the ice takes on an undulating appearance and, as ice continues to grow throughout the winter, these undulations become more pronounced, increasing the oil containment capacity.

NORCOR (1975), reporting on the Balaena Bay experiment, found ice thicker than about 0.5 m to have a thickness variation of about 20 percent. Not all of this variation will be available for oil containment. Because of natural variations in the snow cover and drift patterns, voids under the ice tend to be interconnected by passes. These passes may be at any depth within the range of ice drafts, but presumably the most likely depth is the mean ice draft.

Kovacs (1977, 1979) and Kovacs et al. (1981) mapped the underside relief of the fast ice at various places near Prudhoe Bay in the early spring using an impulse radar system. From the contour maps of the ice bottom, they calculated the volume of the voids that lie above the mean ice draft. This volume (the oil containment potential) ranged from 10,000 to 35,000 m^3 km^{-2} for areas of undeformed fast ice with no large slush-ice accumulations. The variation seemed to be related mostly to the snow cover. For areas of slightly deformed ice, the containment potential was as high as 60,000 m^3 km^{-2}. These numbers are several times the containment potential of perfectly flat ice (8000 m^3 km^{-2}).

Deformation of the inner fast ice zone occurs in the fall when the ice is thin. Most of this deformation is minor: raised rims on edges of individual floes, rafting, and a few small ridges or rubble fields. The relief, generally only a few centimeters, increases the oil containment capacity. As the ice grows thicker and stronger, deformation ceases, and the existing deformed features below the ice tend to be leveled out by differential ice growth between thicker and thinner ice.

Kovacs and Weeks (1979) observed major deformations occurring inside the barrier islands. A severe storm in early November 1978, with winds at 55 to 65 km hr^{-1} (30 to 35 knots) gusting to 110 km hr^{-1} (60 knots), broke up the fast ice, produced ice motions greater than 1 km, and built ridges up to 4 m high. During the three previous years, such events had not been observed but, obviously, they must be considered. The increased roughness created by the deformations creates more voids for the collection of oil. If frequent enough and intersecting, the ridges would limit the directions in which oil could spread or possibly trap deeper pools of oil.

In the stamukhi zone, deformational events continue to occur throughout the winter, creating a bottomside relief of many meters. Tucker et al. (1979) observed a maximum of 12 ridges per kilometer in the 20 km just north of Cross Island. If the average sail height is 1.5 m (Tucker et al., 1979) and keels are four times as deep as sails (Kovacs and Mellor, 1974), then the potential exists for pools of oil to collect that are several meters deep and from one to a few hundred meters across, assuming that ridges frequently intersect each other. The only direct evidence we have of the interaction of oil and ridges occurred in Buzzards Bay, Massachusetts, in 1977. Deslauriers (1979) observed that the spilled oil tended to be trapped between the ice blocks making up the ridges, with some oil appearing on the surface. These observations may not be applicable to large arctic pressure and shear ridges that can be several tensof meters wide with a lower probability of interconnecting voids extending through the ridges at shallow depths. This probability is even less as the ridges age and some of the interior voids freeze.

If oil collects in deep pools surrounded by ridge keels, buoyancy could force significant amounts of oil onto the surface through any openings that exist. Large volumes of gas could be trapped by the ridges in the stamukhi zone; however, it is unlikely that enough large areas will be impermeable to hold a significant volume of gas.

In the pack ice zone, the variety of under-ice relief increases. Not only are there first-year ice floes and pressure ridges, but variable amounts of multiyear ice and refrozen leads.

Underneath multiyear ice, there is an order of magnitude increase in the quantity of oil or gas that may be contained. Kovacs (1977) profiled the bottom of a multiyear floe and estimated that 293,000 m^3 km^{-2} of space existed above the mean draft of 4.31 m. Other investigators (Ackley *et al.*, 1974) also reported greater relief under multiyear ice than under first-year ice.

Refrozen leads also could hold large amounts of oil or gas. The ice in a lead is relatively thin and smooth, while the ice of the original floe has a draft up to 3 m deeper than the ice in the lead. A large lead may be several kilometers wide and many kilometers long, limiting the direction of spreading of the oil but not the area covered. A large flaw lead often forms along the Alaskan coast at the southern boundary of the moving pack. However, most leads are quite narrow, less than 50 m wide (Wadhams and Horne, 1978). Since leads do not form as perfectly straight lines but follow meandering floe boundaries or recent thermal cracks, there are generally many points of contact along a lead. Thus, if oil or gas does come up beneath a refrozen lead, or flows into it from the surrounding ice, it will usually be collected in an elongated pool rather than spreading indefinitely along the lead. The oiled ice in a refrozen lead has a high probability of being built into a ridge.

There is also some probability of oil from a blowout coming up in open water in a newly opened lead. Throughout most of the ice season in the Beaufort Sea, this probability must be fairly low. New ice begins to form immediately and, within one day, a solid ice cover will exist in new leads. Oil beneath thin ice in leads is more likely to reach the surface than oil beneath thicker ice. The ice motion that produces leads will also make them wider or close them by rafting or ridging the thin ice. Gas collecting under a lead can also break the ice.

2. Currents.

A possible contribution to the spread of oil beneath sea ice is ocean currents. Until the oil is completely encapsulated by new ice growth, currents of sufficient magnitude can move the oil beneath the ice until either an insurmountable obstruction is reached or the currents cease.

NORCOR (1975) performed some oil spill experiments near Cape Parry in March 1975 in the presence of currents of about 0.1 m s^{-1}. In one test, the ice appeared to be perfectly flat with roughness variations of 2 to 3 mm. Oil discharged under this ice spread predominant downstream to a thickness of about 6 mm. After all the oil had been discharged, movement of the oil lens appeared to stop.

A second test was performed nearby in the same current regime, but in ice with more underside relief. Troughs 0.5 m deep were present, as well as a small ridge keel downstream from the test site. This time, the oil spread downstream until one of the depressions was reached. At that point, the oil collected in a stationary pool averaging about 0.1 m in depth.

Evidently, currents of only 0.1 m s^{-1} may influence the direction of the spread of crude oil under ice, but will not greatly affect the amount of spreading.

More recently, the relation between current speed, bottom roughness, and the movement of oil under ice has been quantified (Cox et al., 1980). From flume experiments, it was determined that, for smooth ice or ice with roughness less than the equilibrium slick thickness, there is a threshold water velocity below which the oil does not move. For smooth ice, the threshold velocity was about 0.035 m s^{-1}; for ice with roughness scales of 1 mm, the threshold was 0.10 to 0.16 m s^{-1} (depending upon oil density); and for roughness scales of 10 mm, the threshold velocity was 0.20 to 0.24 m s^{-1}. For currents above the threshold velocity, the oil moved at some fraction of the current speed.

For bottom roughness elements with depths several times the slick equilibrium thickness, a boom-type containment/failure behavior was observed. The oil collected upstream of the obstruction to some equilibrium volume, after which additional oil flowed past the obstruction. The size and shape of the obstruction had little effect on oil containment. Thus, even mild slopes act as barriers to oil movement.

As the water velocity increases beneath the ice, a Kelvin-Helmholtz instability eventually occurs, in which case the entire slick is flushed from behind the obstruction. For the range of oil densities tested, the failure velocities ranged from about 0.14 to 0.22 m s^{-1}.

When roughness elements are spaced closer than the slick length for a given current speed and oil density, cavity trapping rather than boom containment occurs (Cox and Schultz, 1981). Cavities have the potential for containing more oil in the presence of currents than do simple barriers, and they retain oil at higher current speeds. Some oil was observed to remain in cavities at current speeds of 40 cm s^{-1}.

Measurements of nearshore under-ice currents reported in the literature (Kovacs and Morey, 1978; Weeks and Gow, 1980; Matthews, 1981; and Aagaard, this volume) indicate that the current speed is generally small, less than about 0.1 m s^{-1}, and will not cause significant oil spreading.

3. Ice Motion.

The motion of the ice cover over a blowout is another mechanism by which oil can be spread beneath the ice. As ice motion increases, the containment potential of the ice decreases, leading to potentially larger contamination areas. High ice velocities also increase the possibility that gas concentrations under the ice will not be sufficient to crack thick ice and will increase oil spread.

In the fast ice zone, ice motion is largely confined to the fall just after freezeup or after breakup in the spring. Kovacs and Weeks (1979) observed that motions of several kilometers can occur in the fast ice soon after freezeup, while the ice is thin and weak, during severe storms. During the majority of the ice season, motions of the fast ice amount to a few meters (Tucker et al., 1980).

A blowout in the pack ice zone is most likely to occur under a moving ice cover. The area of ice under which oil spreads depends upon many factors: the velocity of the ice; the discharge rate of oil and gas from the blowout; the amount of gas that can escape; the diameter of the blowout plume; the roughness of and amounts of different ice types and thicknesses; and the duration of the blowout. It is possible, however, to estimate the area of moving ice that would collect oil in a typical blowout situation.

Assume that a blowout releases 5000 bbl of oil during one day. If the ratio of gas to oil is 150 to 1, then a total of 120,000 m^3 of oil and gas is released during one day. If the containment potential of the ice passing over the blowout is 30,000 m^3 km^{-2}, then 4 km^2 would be contaminated if all the gas remains beneath the ice. The length and width of the swath of oiled ice depends upon the speed at which the ice is moving. The minimum width is roughly the diameter of the radial currents above the plume. If this diameter is 100 m, then the ice would have to be moving faster than 40 km day^{-1} for more than 4 km^2 day^{-1} to be contaminated. Therefore, 4 km^2 day^{-1} can be considered a maximum for this example. The actual area would probably be much smaller, since much of the gas would be released through thermal cracks or broken ice.

4. Ice Growth.

During the fall and winter, the first-year ice over the inner continental shelf is thickening as fast as 10 mm day^{-1}. For a blowout lasting several days under a stationary ice cover and in the absence of currents, this ice growth may be significant in limiting the spread of the oil. When an area of ice overlies gas and oil, the ice does not immediately begin growing beneath the oil. In the region near the blowout site, the heat from the warm oil or from bottom water circulated by the blowout plume reduces ice growth or actually melts ice. Meanwhile, unoiled ice growing outside this region tends to contain the oil.

C. Incorporation Phase

How oil is incorporated into the ice cover varies with the ice morphology and the season. The oil may be incorporated into the new ice forming in leads, may seep to the ice surface through cracks or unconsolidated ridges to be soaked up by any snow cover, and may be frozen into existing ice by new ice growth. As a secondary form of incorporation, oiled ice may be built into ridges.

1. *Oil on the Ice Surface or in Open Water.*

In the fall, new ice forms as a highly porous layer of ice crystals. Oil would rise to the surface through this porous ice and, within a few days, be trapped there as the ice solidifies beneath it. Snow would cover most of the oil through the remainder of the ice season.

There are two differences between oil trapped above and below thin ice. The first is the presence of suspended sediment in the water during the fall freezeup period. Barnes *et al.* (1982) documented the presence of sediment-laden ice within the fast ice zone. Sediment concentrations ranged from 0.003 to 2 kg m^{-3} of ice with considerable variations in regional distribution and yearly amount. Oil in the water beneath the ice cover might adhere to this suspended matter. Second, the oil on the ice surface, even when covered by snow, is subject to evaporation. The evaporation rate varies considerably, depending upon the constituents of the crude oil, the temperature, and exposure to the atmosphere. NORCOR (1975) measured evaporation rates as high as 25 percent within one month. This was for a Norman Wells crude on the surface during the winter with a few centimeters of snow cover. Rates decreased sharply after the first month, but 30 percent or more of the oil could have evaporated by spring.

Oil that surfaces in newly opened leads or in the broken ice directly over a blowout would also be trapped as new ice grows beneath it and thus be subject to weathering throughout the remainder of the winter. Oil, being less dense than sea ice, tends to overflow the tops of cracks. Cold temperatures and an absorbent snow cover limit the spread of the oil to a distance of approximately 1 m (NORCOR, 1975).

2. *Oil Under Undeformed Ice.*

Most of the oil from a winter blowout ends up beneath the ice; the gas trapped with it escapes within a day (Reimnitz and Dunton, 1979). In the spring, trapped air has been observed to escape through open brine channels within minutes (Barnes *et al.*, 1979). In the absence of strong ocean currents, the oil becomes encapsulated by new ice growth. NORCOR (1975) found that the time needed to form an ice sheet below an oil lens is a function of the thermal gradient in the ice and the thickness of the oil. In the fall, a layer of new ice surrounds the oil within 5 days. During the winter, that time increases to 7 days, and in the spring, 10 days.

Martin (1977) observed no traces of oil in the ice that forms beneath an oil lens. The skeletal layer in the ice above an oil lens does appear to become heavily oiled 0.04 to 0.06 m into the ice, but has been found to contain less than 4 percent (volume) of oil (Martin, 1977; NORCOR, 1975). This is equivalent to an oil film about 2 mm thick, or about 25 percent of the equilibrium thickness of oil under thin, smooth ice.

A layer of oil beneath sea ice tends to raise the salinity of the ice above it and lower the salinity of the new ice below it (NORCOR, 1975). The oil layer may trap rejected brine in the ice above, or, by insulating the ice from the seawater, lower the ice temperature above the oil lens. This insulating effect also causes slow initial ice growth below the oil, which results in lower salinity ice. The high–salinity ice directly above the oil likely accelerates the migration of oil into brine channels when the ice begins to warm, but the effect on ice growth appears to be minimal beyond the first few days (NORCOR, 1975).

The incorporation of oil into multiyear ice presumably occurs much as it does in first-year ice. Growth rates are lower under thick multiyear ice than under thinner first-year ice, but it has been postulated that a thick oil lens, as would collect under multiyear ice, would actually enhance ice growth due to convective heat transfer through the oil.

3. Oil Incorporated in Deformed Ice.

Oil spilled in the fall under thin ice, or in newly refrozen leads, may be incorporated into pressure or shear ridges. Some of the oil may remain in these ridges in an unweathered state through several melt seasons. The oiled ridges can travel great distances, releasing the oil along their paths, which may be advantageous since the oil would be released slowly over a greater area. This would remove the oil from the sensitive coastal regions and release it in lower concentrations elsewhere, which is desirable.

Large ridges do not generally form in the fast ice zone because the barrier islands along this part of the coast, along with grounded ridge systems, serve to protect the fast ice zone from effects of the pack ice. Exceptions certainly occur, especially in the Harrison Bay or Camden Bay regions during early freezeup before protective ridges become grounded.

The most common deformation in newly formed ice is rafting. Rafting will halve the area of oiled ice and double the average oil concentration under the ice, but its effects would be hardly noticeable at breakup. Due to ice growth through the winter, rafted and undeformed ice reach approximately the same thickness, thus all the ice breaks up and releases oil at about the same time.

Outside the fast ice zone and the barrier islands lies the stamukhi zone. This zone comprises the past, present, and future position of the active shear zone between the moving pack ice and the stationary fast ice. All observations indicate that this zone is the most heavily ridged area in the southern Beaufort Sea with ridge densities as great as 12 ridges per kilometer (Tucker et al., 1979). If, during the fall, the ice in the stamukhi zone becomes contaminated with oil, then there is a good chance of the oiled ice becoming incorporated into a ridge. Using some typical values (an average sail height of 1.5 m; an average keel depth of 4 times the sail height; average sail and keel slopes of 24° and 33° respectively; and a 10-percent void volume in ridges), the area of ice in a typical ridge profile is computed to be about 54 m^2. If the ice blocks in a ridge are 0.5 m thick, then to get 12 ridges in 1 km, a 2.3-km lateral extent of ice must have been deformed to a 1 km width. As a first approximation, then, more than one-half the area of ice in the stamukhi zone becomes ridged. Of course, the problem is more complicated than this: many ridges are built from new ice grown in leads; many ridges are much larger than the typical ridge described and are built from thinner ice. Nevertheless, the possibility of oiled ice becoming incorporated into a ridge is significant in the stamukhi zone.

We can make a similar estimate for oil in the pack ice zone. First, if the oil comes up under a large multiyear floe, there is only a small chance of it later becoming part of a ridge. Most of the ice involved in ridging has been observed to be young ice, thinner than 0.5 m (R. M. Koerner, personal communication, in Weeks *et al.,* 1971). It is possible that, when a lead opens across a multiyear floe, oil trapped in the ice nearby could drain into the open lead and later be incorporated into a ridge if the lead closes up. Kovacs and Mellor (1974) stated that there is 1 to 5 percent open water in the seasonal pack ice zone. Wadhams and Horne (1978) reported from 0.1 to 3.5 percent thin (ridging-prone) ice (less than 0.5 m), with a mean value of 0.9 percent. These percentages certainly vary with the time of year, especially in the fall, and also with the distance from shore. If we use the value of 1 to 5 percent open water and thin ice as the measure of ice available for ridging, then that is the probability of oiled ice being built into a ridge at any one time. The cumulative probability over the entire ice season is higher, but the increasing thickness of the oiled ice eventually reduces the possibility of its being ridged.

In the fall, a much larger percentage of the seasonal pack ice zone is covered by thin ice. While not all of this thin ice becomes involved in ridging, the probability is certainly greater than later in the winter.

D. Transportation Phase

Estimating possible motions of oiled ice in the southern Beaufort Sea is difficult due to the lack of data. Only a few buoy observations made by AIDJEX during the winter and spring of 1976 (Thorndike and Cheung, 1977), and some radar ranging by Tucker *et al.* (1980) during 1976 and 1977 in and near the fast ice, are in the public domain. These data are insufficient for making reliable predictions of ice motions. Statistics might be formulated using historic winds and ice motion models, but ice motions are strongly dependent on the strength of the ice sheet, and data on ice strength are very limited. However, the range of possible motions can be computed.

The fast ice lies motionless throughout most of the ice season (Tucker *et al.,* 1980). In October and November, strong winds are able to move nearshore ice. Large motions are probably not common, but one case has been reported in the literature (Kovacs and Weeks, 1979) where motions of a few kilometers were observed near shore in early November as the result of high winds. By December, the fast ice is thick enough to resist typical storms and it remains so until breakup in June or July.

Rivers begin flooding the nearshore fast ice in late May or early June. Shore polynyas form and spread from mid-June through early July. The ice sheet becomes thinner and rotten. Sometime during July, the ice becomes weak enough that winds can cause it to move. At first, the most likely direction of motion is toward the shore polynyas, as the ice is weakest in that direction. Soon, enough open water exists that the ice can move in any direction. The winds during the summer are predominantly from the east or northeast, so typically the ice is driven westward and alongshore. Maximum motions are probably comparable to pack ice motions.

Grounded ridges along the outer boundary of the fast ice, in the stamukhi zone, sometimes remain stranded throughout the summer (Reimnitz and Kempema, this volume). If not securely grounded, they are driven by the winds and currents. Ridged ice driven out to sea into the pack may last for several years and travel great distances.

The pack ice motion has a long-term westward trend. During the winter, there may be periods of days or weeks when no significant pack ice motion occurs. This happens when the pack is very consolidated and light winds have blown from the north or west for long periods. When the pack is unconsolidated, the ice has little or no internal resistance to wind and water forces, and it moves about freely. This condition, known as free drift, represents the maximum extreme of possible ice motion. Between the extremes of no motion and free drift, the motion depends upon the atmospheric and oceanic driving forces, the sea surface tilt, the Coriolis effect, and internal stresses transmitted through the ice. This last term is difficult to model for long periods of time, since small errors in velocity affect the distribution of ice and, thus, the ice strength, which in turn affects future velocities.

Thomas (1983) computed typical pack ice motions and standard deviations of daily motions using historical wind data, a range of ice conditions and ocean currents, and an ice model. The model was "tuned" so that average motions corresponded to the limited observations of ice motions. The computed trajectories showed an average westward motion of about 3.7 km day^{-1} during the fall, 1.3 km day^{-1} during the winter, 2.1 km day^{-1} during the spring, and 3.6 km day^{-1} during the summer. The standard deviation of daily motions was more than 5 km. The motions near the shore tend to be smaller than those further offshore. Along the Alaskan coast, the ice has a shoreward component of motion, but west of Point Barrow the motion turns toward the north. While the westward trend persists from month to month over many years, daily motions exhibit a great deal of meandering and back-and-forth motion in all directions.

E. Release Phase

Oil spilled in the winter beneath the sea ice is not an immediate threat to the environment as the ice contains the spill in a relatively small area away from land and insulates it from the ocean and the atmosphere. It is when the oil is released from the ice that it begins to interact with and become a danger to its environs.

For first-year sea ice, this release is well understood and has been documented by NORCOR (1975), Martin (1977), and Buist et al. (1981). The oil trapped in first-year ice may be released by two major routes: by rising to the ice surface through brine drainage channels or by escaping when the ice melts. Some oil is released from newly opened cracks or leads. The release of oil from beneath multiyear ice probably occurs more slowly. Comfort and Purves (1980) reported that, of the oil placed beneath multiyear ice in Griper Bay (Melville Island, N.W.T.), over 90 percent had surfaced at the end of two melt seasons.

1. Brine Drainage Channels.

In late February or early March, the mean temperature begins to rise in the southern Beaufort Sea. As the ice begins to warm, brine trapped between the columnar ice crystals begins to drain. Oil trapped beneath the ice probably accelerates this brine drainage by raising the ice salinity directly over the oil. Martin (1977) observed that oil released beneath the ice during the winter migrated 0.16 m upward through brine channels by 22 February. By the time the air and ice temperature approach the freezing point, in late April or May, the brine channels extend through the ice. Once the channels reach the surface and are of sufficient diameter, oil begins appearing on the ice surface. Oil released under ice with top-to-bottom brine channels also begins to appear on the surface within an hour (NORCOR, 1975).

Flow rates must be fairly low, since it has been observed that not all the oil is released until the ice has melted down to the initial level of the oil lens (NORCOR, 1975; Buist et al., 1981). An upper bound can easily be set. Oil has been observed to take about 1 hour to migrate up through about 2 m of ice with open brine channels. The brine channels were about 4 mm in diameter, so each brine channel had a maximum volume flow of 8 x 10^{-6} m^3 hr^{-1}. The brine channels were spaced from 0.2 to 0.3 m apart, so each square meter of ice contained about 16 brine channels, and the flow rate per square meter was about 0.0004 m^3 hr^{-1}. This is equivalent to an oil film 0.4 mm thick being released each hour. The actual flow rate probably is smaller.

Oil that surfaces through brine channels should primarily be found floating on the surfaces of melt pools; if melt pools do not exist when the oil surfaces, they soon form as the oil darkens the surface. Snow forms an effective barrier to the spread of the oil, but wind and waves splashing oil onto the nearby snow, enlarge the pools. Oil-in-water emulsions were observed to form in the melt pools when winds were over 25 km hr^{-1}. As much as 50 percent of the oil in a melt pool could be in the form of emulsions, but generally emulsions break down within a day after winds subside (NORCOR, 1975; Buist et al., 1981).

The rates at which the oil evaporates, emulsifies, or dissolves are considerably lower in arctic regions than at lower latitudes. Not only does the ice serve to protect the oil during the winter but, as it melts in the spring, it releases the oil slowly over periods of weeks. The ice also moderates wind effects, so smaller waves and less mixing occur in melt pools and open leads. The lower temperatures also increase the stability of the oil. In general, the process that has the most significant effect on oil quantity during spring release is evaporation.

NORCOR (1975) estimated that by early June, at some test sites of the Balaena Bay experiment, 20 percent of the oil had evaporated. By 16 June, it was reported that "the flow of oil from the ice had almost completely stopped," since the ice had melted down to the trapped oil lens in most cases. More than 50 percent of the oil had evaporated by late June.

2. Surface Melting.

Most of the undeformed first-year ice near shore melts down to the oil layer during the summer months. Any oil that does not reach the ice surface through open brine drainage channels is then released. Typically, the nearshore area first begins to open and break up around the end of June and is mostly ice-free by the end of July (Barry, 1979). Oil on the ice surface accelerates ice melting and breakup by darkening it. NORCOR (1975) estimated that ice contaminated by oil would break up about two weeks earlier than clean ice.

III. FATE OF OIL

We have seen that the vast majority of oil spilled in the Beaufort Sea during the ice season stays in the ice until spring, when it begins to appear on the ice surface. The surfaced oil begins to weather and accelerate melting. Typically, all the ice in the contaminated area has melted by mid-July, at which time about 50 percent of any remaining oil has evaporated. Emulsification, dispersion, and dissolution of the oil on the open-water surface also occurs, and silt from rivers may absorb oil and sink.

Until all the ice has melted, natural processes degrade and disperse the weathered oil slowly. The gradual release of the oil, its reduced surface area because of confinement by the ice, and short fetches for wind energy input are all contributing factors. The amounts of oil removed by natural processes are insignificant but may have a critical effect on the ecology of the area.

Once the area becomes free of ice, conditions parallel an open-water spill. The major difference is the evaporative losses of the oil by this time. Because of prevailing winds in the southern Beaufort Sea, an open-water slick is likely to be driven onshore to the west or southwest. Since the oil is partially weathered, the slick tends to be more concentrated than a recent spill from a blowout. Southerly or easterly winds drive the slick offshore, breaking it up and spreading oil over larger areas. When the winds eventually reverse, an even larger stretch of coast is in danger of contamination.

Oil deposited upon beaches is probably the second largest sink for hydrocarbons after evaporation. Sedimentation to the sea bottom is also important. Over much longer time periods, oxidation and biodegradation dispose of small percentages of the oil.

IV. THE EFFECTS OF ICE ON CLEANUP

Although this report does not address methods of oil spill cleanup in arctic waters, it is worthwhile to review the characteristics of the ice cover and oil-ice interactions that will affect cleanup activities after our model blowout.

A. Pack Ice Zone

Oil from a blowout under stationary ice might be partially collected from the blowout site if the blowout occurs near a facility that could pump and store the oil. Burning the oil and gas during the blowout could also be partially useful if recovery is impractical. If the pack ice is moving more than a few hundred meters per day, recovery would probably be impossible and burning would be difficult. In that case, ensuring gas release over or near the blowout would reduce oil spread under the ice. To help locate the oiled ice in the spring when the oil begins to surface, markers and beacons could be placed near the blowout site. Then, the oiled melt pools could be ignited, probably by air-dropped incendiary devices, to dispose of some of the oil. Most of the oil and the residue from burning would remain on the ice surface or in newly opened leads. Dispersants could be used as soon as oil appears in open-water leads and polynyas. As summer proceeds and the lighter, more toxic components evaporate, seeding with microbes and fertilization could enhance biodegradation. Since the long ice season halts or slows the natural processes acting to degrade and disperse the spill, summertime activities would be important for reducing the chances of harm in future years. The environmental contamination would probably persist for several years in any case, especially since oiled ridges may retain some oil through the summer.

B. Fast Ice Zone

A blowout and oil spill in the fast ice zone could potentially be the most harmful because that is where open water first appears in the spring, but effective cleanup may be possible. The ice does not move between November or December and the following June; currents are low, so the oil is not moved about under the ice; and the ice provides a stable work platform. Nevertheless, a large spill could cover several square kilometers. The spill area could be reduced considerably by preventive measures early in the ice season. These measures would be as simple as cleaning the snow from narrow strips surrounding possible blowout sites to promote faster ice growth and at least partially dike the spill underwater. Other methods of increasing oil containment under the ice can be postulated (skirts frozen into the ice, air-bubble systems to reduce ice growth), but none of these methods are feasible until after the ice has become thick and strong enough to resist movement by winds and be safe for surface travel.

The blowout would likely create an area of open water in the fast ice directly above it, letting gases escape and trapping a great deal of oil in the surrounding ice. This area of open water could be enlarged by blasting. If storage facilities are available, oil could be pumped directly from the pool during the blowout.

Oil from a large blowout that has spread beneath the ice, especially early in the ice season when bottomside relief is small, would be more difficult to collect during the winter. The oil can cover a large area and collects in many small pools beneath the ice. No proven technology exists for locating these pools other than trial and error drilling. The negative correlation between depth of snow cover and ice thickness (Barnes et al., 1979) would aid in the search. Other possibilities are being developed, but they will probably also be very labor intensive. Even when pools are found, it would be virtually impossible to remove all of the oil from the ice. After new ice growth has completely encapsulated the oil lens, removing it all will become impossible.

When oil rises to the ice surface in the spring, concentrations will still be so low that removing the oil would be difficult. Burning the oil at this time would be much simpler, and for small spills or remnants of large spills, a significant proportion might be disposed of in this fashion. Since the oil is released from the ice over a period of weeks, burning would have to be done several times. It is unlikely that all the oil could surface to be burned before breakup.

C. Stamukhi Zone

Oil spill cleanup in the stamukhi zone depends on many factors. Much ridge building takes place, but a grounded ridge can extend the fast ice boundary seaward. The greater bottomside relief tends to concentrate the oil in the region inside grounded ridge systems. Oil within ridges may be impossible to clean up, since ridges may be able to hold oil for several years. This could be advantageous, since the oil would be released slowly over several seasons and over a large area as the ridges drift with the pack, reducing contamination at any one place and time. Oil trapped in deep pools behind ridge keels should be recoverable, but would require considerable effort. The distance from shore and the difficulty of surface travel would be obstacles to cleanup.

The most successful cleanup would involve concentrating oil in a small area. Since it would be difficult to enhance the bottomside relief in the stamukhi zone, one could only rely on ridges to provide this concentration. If the oil is not contained by natural features, or cannot be collected directly from the blowout site, springtime burning of surfaced oil must be considered. For small amounts of oil this can be effective, but for very large spills the majority of the oil will remain. Even after cleanup and evaporation, as much as 40 to 50 percent of a large spill remains in ridges, on unmelted ice floes, or on the water surface. If the pack retreats northward, conventional open-water cleanup methods and dispersants might remove more of the oil. If the pack remains near shore through the summer, cleanup would concentrate on the beaches and open-water lagoons behind the barrier islands. Release of oil from the ice is likely to occur in subsequent summers, making cleanup a long-term, wide-area project.

REFERENCES

Ackley, S.F., Hibler, W.D. III, Kugzruk, F.K., Kovacs, A., and Weeks, W.F. (1974). In "AIDJEX Bulletin," No. 25, p. 75. University of Washington, Seattle.
Barnes, P., Reimnitz, E., Toimil, L., and Hill, H. (1979). Open-File Rept. 79-539, U.S. Geol. Survey.
Barnes, P.W., Reimnitz, E., and Fox, D. (1982). J. sed. Pet. 52, 493.

Barry, R.G. (1979). *In* "Environmental Assessment of the Alaskan Continental Shelf," Final Rept. Vol. 2, p. 272. NOAA, Boulder.

Buist, I.A., Pistruzak, W.M., and Dickins, D.F. (1981). *In* "Proceedings of the Fourth Arctic Marine Oilspill Program Technical Seminar," p. 647. Environmental Protection Service, Ottawa.

Comfort, G., and Purves, W. (1980). *In* "Proceedings of the Third Annual Marine Oilspill Program Technical Seminar," p. 62. Environmental Protection Service, Ottawa.

Cox, J.C., and Schultz, L.A. (1981). *In* "Proceedings of the Fourth Arctic Marine Oilspill Program Technical Seminar," p. 3. Environmental Protection Service, Ottawa.

Cox, J.C., Schultz, L.A., Johnson, R.P., and Shelsby, R. A. (1980). *In* "Environmental Assessment of the Alaskan Continental Shelf," Final Rept. Vol. 3, p. 427. NOAA, Boulder.

Deslauriers, P.C. (1979). *J. Pet. Tech. 31*, 1092.

Evans, R.J., and Untersteiner, N. (1971). *J. geophys. Res. 76*, 694.

Kovacs, A. (1977). *In* "Proceedings 1977 Offshore Technology Conference," Vol. 3, p. 547. Houston.

Kovacs, A. (1979). *In* "Environmental Assessment of the Alaskan Continental Shelf," Ann. Rept. Vol. 8, p. 310. NOAA, Boulder.

Kovacs, A., and Mellor, M. (1974). *In* "The Coast and Shelf of the Beaufort Sea" (J.C. Reed and J.E. Slater, eds.), p.113. Arctic Institute of North America, Arlington, VA.

Kovacs, A., and Morey, R. M. (1978). *J. geophys. Res. 83*, 6037.

Kovacs, A., Morey, R., Cundy, D., and Dicoff, G. (1981). *In* "Proceedings POAC-81," Vol. 3, p. 912. Univ. Laval, Quebec.

Kovacs, A., and Weeks, W.F. (1979). *In* "Environmental Assessment of the Alaskan Continental Shelf," Ann. Rept. Vol. 3, p. 181. NOAA, Boulder.

Lewis, E.L. (1976). *In* "Pacific Marine Science Report 76-12," Institute of Ocean Sciences, Victoria.

Logan, W.J., Thornton, D.E., and Ross, S.L. (1975). *In* "Beaufort Sea Technical Report 31B," Dept. of the Environment, Victoria.

Malcolm, J.D., and Cammaert, A.B. (1981). *In* "Proceedings POAC-81," Vol. 2, p. 923. Univ. Laval, Quebec.

Martin, S. (1977). "Department of Oceanography, Special Report 71." University of Washington, Seattle.

Matthews, J.B. (1981). *J. geophys. Res. 86*, 6653.

Milne, A.R., and Herlinveaux, R.H. (n.d.). *In* "Crude Oil in Cold Water," (R.J. Childerhose, ed.). Beaufort Sea Project, Dept. of Environment, Sidney, B.C.

NORCOR Engineering and Research, Ltd. (1975). "Beaufort Sea Technical Report 27." Dept. of the Environment, Victoria.

NORCOR Engineering and Research, Ltd. (1977). "Technology Development Report, EPS-4-EC-77-5." Environmental Impact Control Directorate, Ottawa.

Reimnitz, E., and Dunton, K. (1979). *In* "Environmental Assessment of the Alaskan Continental Shelf," Ann. Rept. Vol. 9, p. 210. NOAA, Boulder.

Reimnitz, E., Toimil, L.J., and Barnes, P.W. (1978). *Mar. Geol. 28*, 179.

Rosenegger, L.W. (1975). "Beaufort Sea Technical Report 28." Dept. of the Environment, Victoria.

Ruby, C.H., Ward, L.G., Fischer, I.A., and Brown, P.J. (1977). *In* "Proceedings POAC-77," p. 844. Memorial Univ. Newfoundland, St. John's.

Stringer, W., and Weller, G. (1980). *In* "Proceedings of the Third Annual Marine Oilspill Program Technical Seminar," p.31. Edmonton.

Thomas, D.R. (1983). "Flow Research and Technology Report 252," Flow Industries, Kent, WA.

Thorndike, A.S., and Cheung, J. Y. (1977). "AIDJEX Bulletin No. 35." Univ. of Washington, Seattle.

Topham, D.R. (1975). "Beaufort Sea Technical Report 33," Dept. of the Environment, Victoria.

Topham, D.R. (1977). *J. appl. Mech. 44*, 279.

Topham, D.R., and Bishnoi, P.R. (1980). *In* "Proceedings of the Third Annual Marine Oilspill Program Technical Seminar," p. 87. Environmental Protection Service, Ottawa.

Tucker, W.B., Weeks, W.F., and Frank, M. (1979). *CRREL Report 79-8*, U.S. Army Cold Regions Res. and Eng. Lab. Hanover, New Hampshire.

Tucker, W.B., Weeks, W.F., Kovacs, A., and Gow, A.J. (1980). *In* "Sea Ice Processes and Models" (R.S. Pritchard, ed.), p.261. Univ. Washington Press, Seattle.

Wadhams, P., and Horne, R. (1978). "Scott Polar Research Institute Technical Report 78-1." Cambridge.

Weeks, W.F., and Gow, A.J. (1980). *J. geophys. Res. 85*, 1137.

Weeks, W.F., Kovacs, A., and Hibler, W.D. III (1971). *In* "Proceedings POAC-71," Vol. 1, p. 152. Norwegian Institute of Technology, Trondheim.

Index

461